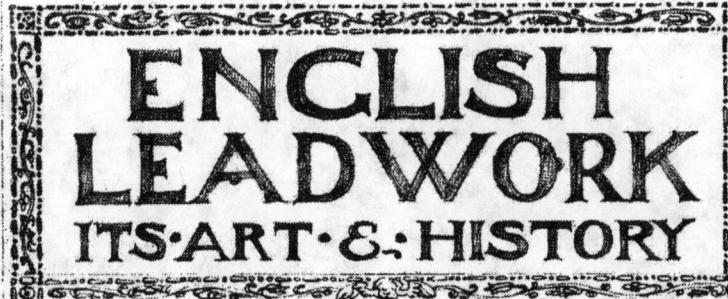

ENGLISH LEADWORK
ITS·ART·&·HISTORY

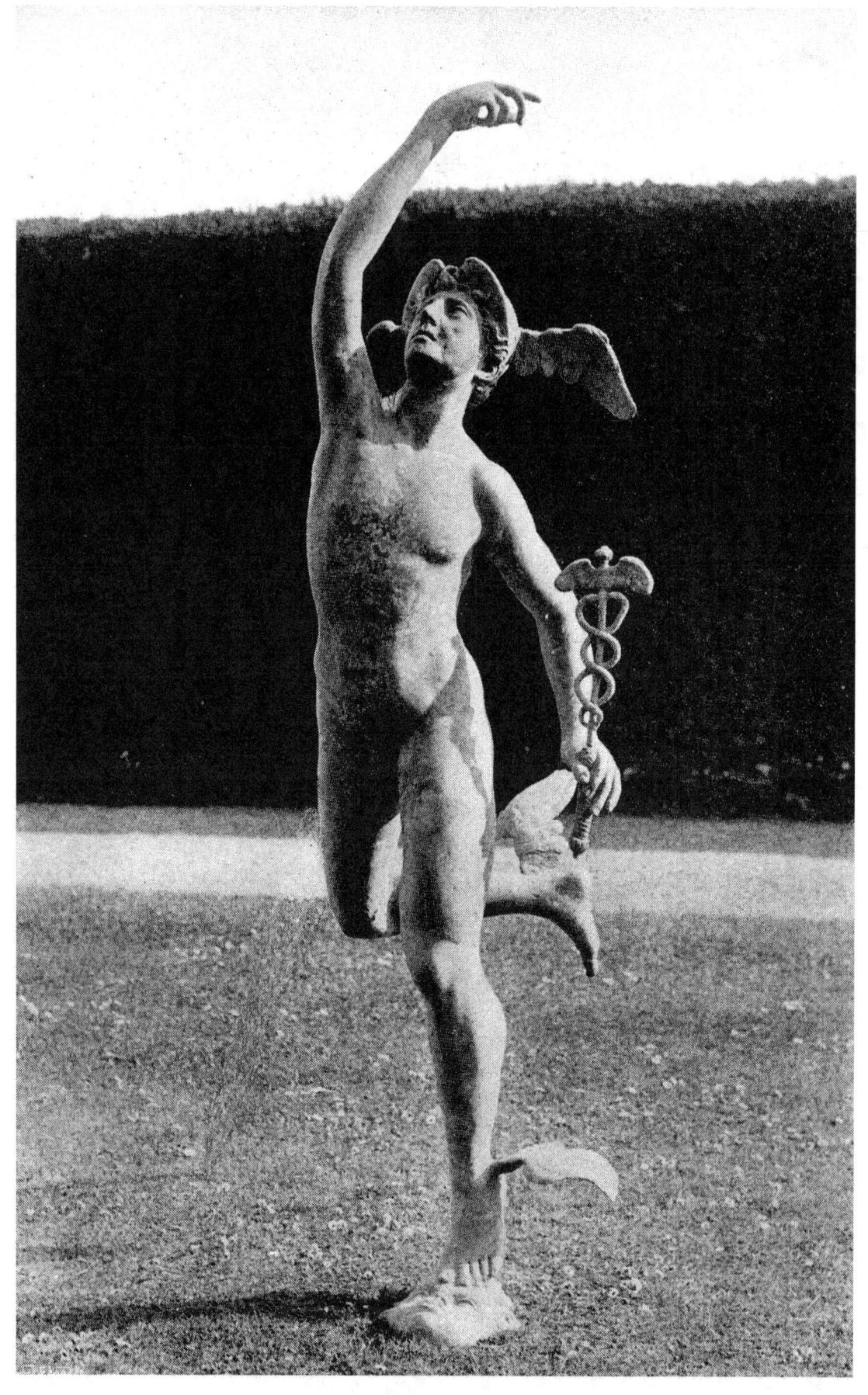

FLYING MERCURY AT HOLME LACY, HEREFORDSHIRE.

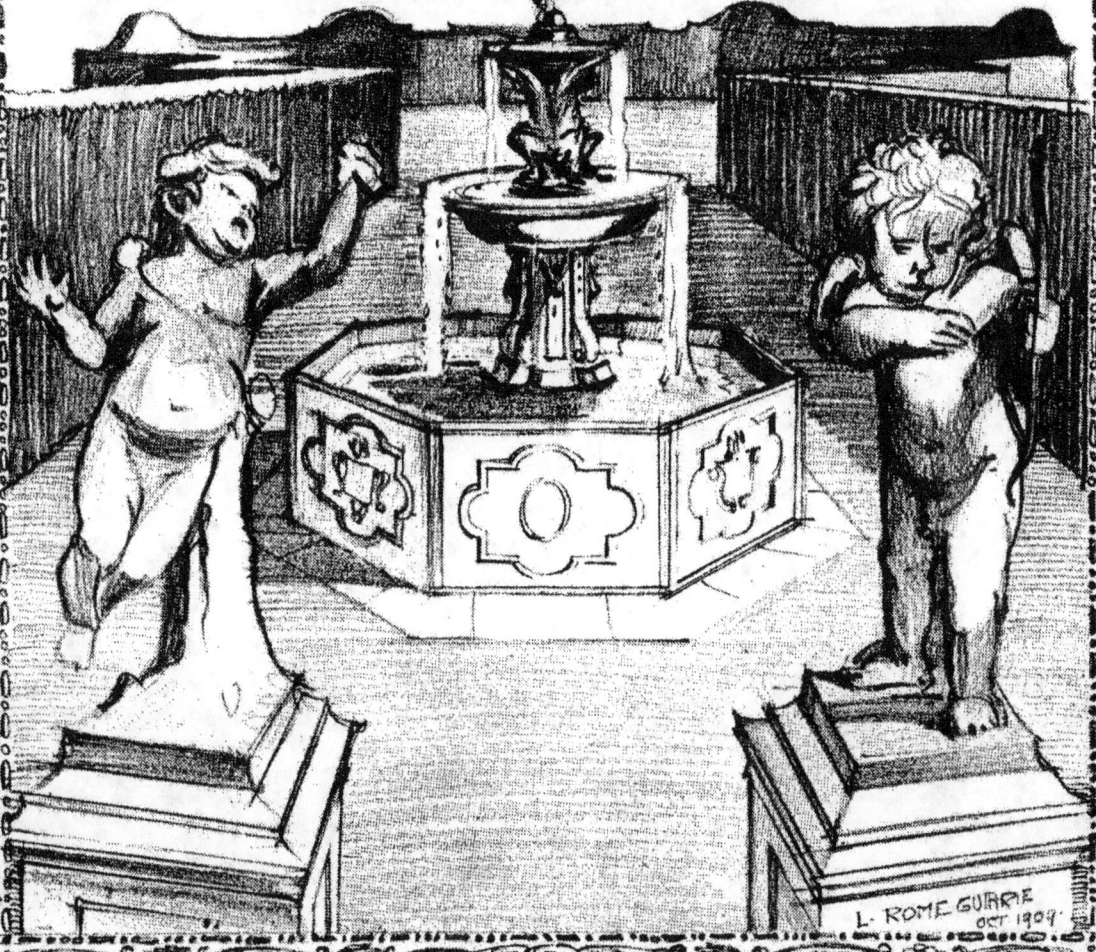

ENGLISH LEADWORK
ITS·ART·&·HISTORY
LAWRENCE·WEAVER·F.S.A.

L. ROME GUTHRIE
OCT 1909

CLASSIC EDITIONS

This edition digitally re-mastered and
published by JM Classic Editions © 2007
Original text © Lawrence Weaver 1909

ISBN 978-1-905217-75-5

"But thou, thou meagre lead,
Which rather threatenest than dost promise aught,
Thy paleness moves me more than eloquence ;
And here choose I."

THE MERCHANT OF VENICE.

PREFACE.

THE growing sense of the decorative value of lead in architecture and the garden has created a demand for a larger history of the leadworker's art, which shall show, with some fulness, what has been done in the past.

Of books on the technical side of leadwork there has been no lack; the sanitary plumber has a library ready to his hand. The art and history of leadwork have found but one protagonist, my friend Professor Lethaby, but he is a host in himself. His little book, published in 1893, and long out of print, reminded us of the forgotten spirit of old leadwork with so just a perception and so stimulating a sympathy, that I can do and would do no more than write myself down his disciple.

Professor Lethaby relied on sketches, chiefly from his own charming pencil, for his 76 pictures. The 441 illustrations of this volume are almost exclusively from photographs or measured drawings. If haply this book be found to have merit, it will be, I think, in its presentment for the first time of a full series of the chief uses of lead which demand the judgment of the artist as well as the capacity of the craftsman.

The scheme of the book has been to put into the hands of the architect, the sculptor, the garden designer, and the worker in lead, a book of some practical use. I have endeavoured to lay just so much stress on the historical side of my subject, as will show the development of design and treatment, while connecting the work with the workers and the days in which they worked. Details of a purely archæological character I have tried to exclude from the text, and Roman coffins and the like have been slightly dealt with. For the antiquary a Bibliography has been added, and the notes there given will perhaps be of use in clearing the ground for the student. For the owners and lovers of gardens I have attempted to identify some of the work of the sculptors of the eighteenth century who did so much for the architectural side of gardencraft.

The material which is available for illustration is so great in amount (particularly in pipe-heads, cisterns, and statues) and so scattered, that there are doubtless omitted both from illustration and reference many admirable examples, but a book has its limits. My collection of photographs contains many examples which I should have included but for the fear of overloading.

Those who are familiar with a cistern here and a statue there may look for them in vain: I can only hope that every important class of subject is represented. I have made but small reference to traditional methods of working lead as belonging rather to the technical that the artistic history of the metal's uses.

For such matters I refer the student to my friend Mr F. W. Troup's admirable lectures, and notably that published in "The Arts connected with Building." Had I dealt with such details, I could but have borrowed from him. One side of the history of leadwork, viz., the story of the Worshipful Company of Plumbers, with the place of the craft among the City Guilds, I have omitted altogether. Some day this fascinating branch of the subject will doubtless secure such an historian as the allied craft of the Pewterers found in Mr Charles Welch, F.S.A. It was, however, too big to include, and too important to trifle with, so I have left it.

Mine has been largely the function of the compiler, and for such work the help of many is needful. It has been given so widely and with such freedom and kindness that I make personal acknowledgments in a following note.

My thanks are due to scores of people who have suffered me gladly when I pestered them for information, and wandered with my camera about their churches, houses, and gardens.

The formal dedication is out of fashion, but the spirit which prompted it is always fresh. I lay down my pen with a lively sense of the sympathy and forbearance of those who have allowed me to dedicate to leadwork the leisure hours of many years—my mother and my wife.

LAWRENCE WEAVER.

14 NORTHWICK TERRACE,
　　ST JOHN'S WOOD, N.W.,
　　　　November 1909.

NOTE OF ACKNOWLEDGMENT.

THE majority of the photographs that illustrate "English Leadwork" are either from the large collection which I acquired from Mr W. Galsworthy Davie, or were taken by myself with the help of my life-long friend, Mr Benjamin H. Bedell. For other photographs, drawings, and information, I am indebted to various helpers, some of whose names appear in the text. Amongst others I now acknowledge the kindness of the following :—Captain Charles Lindsay (for the fine series of pipe-heads at Haddon Hall); Viscount Dillon, V.P.S.A., Lord Bolton, F.S.A.; the Rev. W. Woodlock, S.J.; the Rev. E. Hermitage-Day, the Rev. T. S. Cunningham, the Rev. Athelstane Corbet, Miss E. Morton, Miss H. M. Knox, the Editor of the "A. A. Sketch-Book," Lieut.-Col. C. Field, Lieut.-Col. G. B. Croft Lyons, F.S.A., and Messrs G. Harry Wallis, F.S.A.; J. Starkie Gardner, F.S.A.; W. Niven, F.S.A.; Charles Angell Bradford, F.S.A.; Albert Hartshorne, F.S.A.; Philip M. Johnston, F.S.A.; Leonard Stokes, F.R.I.B.A.; Alfred Harris, J. H. Allchin, S. G. Hewlett, George Clinch, F.G.S.; R. Eden Dickson, Alexander A. Inglis, W. D. Haydon, William Kelly, F. W. Troup, F.R.I.B.A.; A. R. Goddard, Arthur T. Bolton, F.R.I.B.A.; Stanley H. Page, H. T. Austin, W. S. Curr, Ambrose P. Boyson, J. C. Brand, Geo. P. Bankart, C. King.

As some of the illustrations have appeared in magazine articles, I have to thank the proprietors of the *Architectural Review*, the *Burlington Magazine*, *Country Life*, *The Journal of the Royal Institute of British Architects*, and others, for facilitating arrangements for reproduction here.

To the *Architectural Review* I am indebted for the use of the initial letters of the chapters. It is probably inevitable that some who have aided me with illustrations have not been mentioned in the list above. To such I can only temper my thanks with full apology.

Many hard things, mostly unjust, have been said about publishers, both before and since Byron's savage witticism. That author, however, is happy whose work materialises in the hands of Messrs Batsford, of whom I can only say, in Ferdinand's words, that they "make my labours pleasures."

L. W.

CONTENTS.

INTRODUCTION.

THE uses of lead in the earliest times were so various, that a stout volume might be made which would lead us to Egypt and Assyria, show the pigs of lead stacked on the quays of Tarshish, make us see the Spartan of the sixth century B.C. casting his little votive figures, and surprise the prehistoric man plugging his earthen pots with lead. English leadwork, however, is large enough both as subject and title; my text and illustrations rarely stray abroad, and then only for a passing comparison.

The art of leadwork is as living as it is individual. Its chief applications are in architecture, where they are many and necessary. They begin with the severely practical, as in roofing and water supply pipes. They range through the objects which blend the useful and the decorative, such as fonts and pipe-heads, and reach the purely decorative in garden ornaments. The illustrations that follow are designed to show that with few exceptions their subjects present two marked characteristics. The material is fit for its uses, and its varied treatments befit the material.

It has been objected to lead that it is a metal little individual. It has been suggested that everything made in lead would be better in some other medium; that, in fact, lead's function is to take, for economy's sake, the place of some richer material. This attitude is founded on an imperfect study of the products of the leadworker's art, as a rapid survey will show.

The fonts illustrated in the first chapter, when seriously considered from the aspect of their possibility in other materials, give answer enough. The general character of the arcaded bowls with large figures is admittedly like that of the stone fonts of the same period. There is, however, a delicacy of modelling in the floral decoration and in the detail of the robes, combined with a general softness of effect, which would be impossible in stone. The fineness of detail might be obtained in marble, but it would be joined with a certain harshness unavoidable in delicately wrought stone. There remains the alternative of bronze, but bronze calls for treatment more defined and less homely than suits the character of lead. Bronze is the metal of the grand manner, a fitting substance for the effigies of kings. Lead has a lower place, but can take on a gentle dignity and simplicity incapable of transference to another period. How, if not in lead, could the motifs of the Pyecombe and Warborough fonts have been expressed?

If the history of pipe-heads set out in Chapters II. and III. be rightly considered, they are seen to have given what is the most attractive field for the right use of lead in the minor building arts. Chapter XIII. shows many good modern examples which have caught the spirit of the old work without slavish imitation. Despite, however, much precept from those who seek to raise the level of the crafts, very small is the number of people who make pipe-heads of merit, and this complaint is true of all leadwork which has artistic possibilities. The fault lies rather with the average plumber than with the average architect. There is a clear enough call for good design and for a return to sound and traditional methods, but nearly all the " ornamental " leadwork done at technical schools is unspeakably bad. In more than one of the books on plumbing which have won a deservedly high place, hints on "ornamental" work are given by instructors, who are past masters in technical mysteries. Most of the examples used to mould the decorative sense of the student are wholly bad. Until the authorities of technical schools realise that the craft of leadwork must be taught by one who is an artist, as well as a technical expert, these grievous productions will be thought by the rising generation of plumbers to be "artistic." There are, of course, honourable exceptions. Professor Lethaby, Mr F. W. Troup, and others have struggled manfully to fill London County Council students with a wise spirit, and individual architects have sought to instil into the mature plumber some right feeling for his material. In practice, however, if good leadwork is wanted, the few firms who specialise are almost the only sources of supply. The Worshipful Company of Plumbers has done as much as, if not more than, any City Company to support and improve the craft it represents. If the Company would devote to some instruction in artistic righteousness a tithe of the energy which it gives to improving technical conditions, a good and greatly needed work would be done.

In the field of roofing, and as a covering for spires, lanterns, and domes, the long range of illustrations shows the yeoman service of lead to the larger needs of architecture. In this connection it is well to remember what Sir Christopher Wren wrote in 1708: " Lead is certainly the best and lightest covering, and being of our own growth and manufacture, and lasting, if properly laid, for many hundred years, is without question the most preferable." He was then seventy-six, and the dictum is quoted from a letter to a friend, which set out the gist of his vast experience in building. It is fair to say that for many buildings lead is still the "most preferable" to-day. As to its possibilities in the future, the subject of Fig. 406, and Mr Starkie Gardner's bridge (Fig. 405) are full of encouragement.

In garden leadwork the decorative idea is supreme, and expresses itself in fountains, cisterns, vases, and statues. It may be true that for some of the portrait statues lead was employed because it was cheaper than bronze. So much may be conceded, but as to garden statues it is fair to affirm that it is a more suitable material. It has a gentle unobtrusive quality which harmonises with the domestic air of gardens. Bronze would be, under English skies, an absurd material for the engaging triviality of *The Kneeling*

Slave, or the rather stodgy ladies who represent the arts at Hardwick. If Bassanio was a little uncivil to "thou meagre lead," at least its paleness moved him more than eloquence. This paleness is manifest in garden ornaments as a silvery grey patina, and forms one of the most delightful features of lead, which in England at least must be regarded as the characteristic garden metal. Even for portrait statues in lead there seems no reason for undue apology. One may admit the coarser treatment that lead demands, and the absence of such finely modelled sinew and vein as bronze makes possible, but no one will affirm that good lead is less good than bad bronze. If, sometimes, where money is strictly limited, a better artist and a cheaper material were employed, instead of a feeble artist and a costly material, our public places would not be the losers. Where the pedestal of a portrait statue is to be decorated by less important figures of an emblematic sort, why cling to a uniformity of metal? With the portrait figure in bronze, the lesser figures in lead would not only yield a pleasant diversity of effect, but also by contrast heighten the dominance of the greater statue?

Before closing this introduction, I would plead for lead as offering to the designer and craftsman a field of opportunity too much neglected. Since for six centuries it held a place, small but distinguished, in the history of the building arts, it is not unreasonable to hope that it will win it back, and renew a sleeping but imperishable tradition. In matters artistic and architectural, the pursuit of novelty is apt to make for trouble. The sense of material that ought to be the basic sense in craftsmanship has been debauched by the fatal facilities of modern manufacture. In urging the claims of lead, the need of soft and simple modelling must be emphasised. In view of the Norman fonts it sounds like attenuated paradox to speak of lead as a novel material. As, however, lead was almost forgotten during the nineteenth century, it offers problems which are virtually new, and demands fresh thought which will be stimulated by study of the old work.

ENGLISH LEADWORK:
ITS ART AND HISTORY.

———•◄►•———

CHAPTER I.

FONTS.

Destroyed and Incorrectly Described Examples—Geographical Distribution—Classification by Design—Detailed Description of the Thirty Existing Ancient Fonts—Various Font-like Vessels.

FONTS never fail of interest. They necessarily take a high place in Christian art, for they are the place of the first sacrament of the Church, and they afford singular decorative possibilities. Their ecclesiastical significance is comparable only with that of the altar, yet unlike the altar the font fortunately has not been the battle-ground of iconoclastic zeal to any marked extent.

In so far as fonts sometimes bear figures, they have been open to puritanical disapproval, and have suffered from the "axes and hammers" of the righteous. Their material, however, has never been the shibboleth of theology, which has made the English stone altar an affair of ancient history, and a lost vehicle of religious art and symbolism.

Among English fonts the thirty of lead which remain have an important if a small place.

The greatest enemy of lead fonts, as of all lead objects, has been the intrinsic value of the material. The discarded stone font makes a convenient trough for watering animals, or will pleasantly decorate the parsonage garden when used as a flower-pot, but the lead font has higher uses. It can be turned into many bullets. There may be no present occupant of the bench of Bishops who, in his youth, converted a lead font into slugs for the shooting of rooks, but there is a stain on one episcopal conscience to-day in the matter of the fingers of the lead statue of a heathen god. Doubtless, therefore, in less enlightened days lead fonts have gone piecemeal on the same charming errand.

Lead was much beloved of Henry VIII.'s Commissioners, as is obvious from the grim tale of fodders from conventual roofs, which added so markedly to the value of the monastic spoils. Monasteries would not have had fonts except where their naves or chapels were put to parochial use. Edward VI.'s Visitors, however, who purged the parish churches at the abolition of the chauntries, were probably not innocent in this matter. They would scarcely have omitted (from their inventories of superstitious objects removed) a storied font which so obviously meant money, if it could be done away without too violent a local outcry. In those spacious days the Severn Valley was rich in spoils of leadwork from the roofless churches, for the river was the highway to the Continent. Perhaps it is because it was a drug on the market that there is spared to Gloucestershire the largest number of lead fonts, nine in all out of the total of thirty, and six of Norman date. Unforgettable also are the economic ecstasies of the churchwarden era, and the iconoclasm of the Commonwealth, responsible for the destruction of many. In 1878 when St Nicholas-at-Wade in Thanet was "restored," the lead font was also restored to its original condition of pig lead. The lead fonts once at Chilham, Kent, and at Hassingham, Norfolk, have gone the same ruinous road. Clifton Hampden, Oxfordshire, knows its lead font no more; about 1840 it was decreed "unshapely" (lead will get unshapely sometimes, but does not resist being put into shape again) and was hurried to its doom. In 1828 there existed in the church at Leigh, Surrey, a lead font, but it has since disappeared.

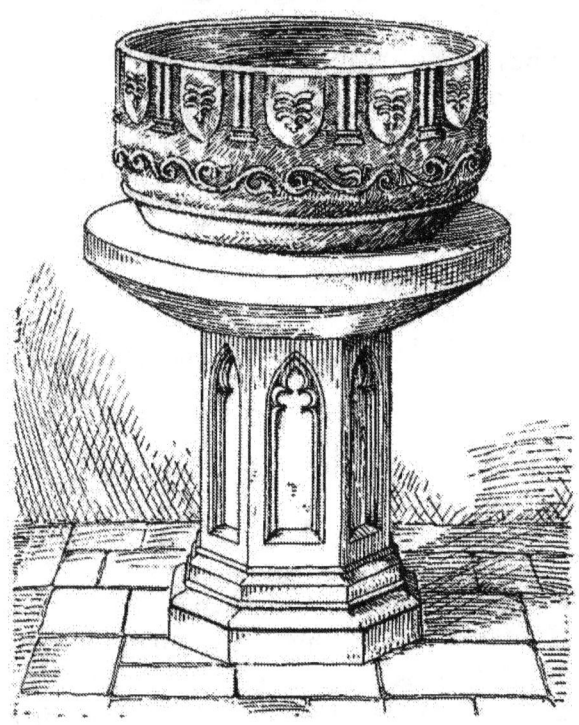

FIG. 1.—Font (destroyed in 1891), St Mary's, Great Plumstead.

Woolhampton Church is included in some lists as possessing a font "in which the lead is placed over stone and pierced, leaving an arcade and figures showing against the stone background." We may trust that this is the case, and that some day we may see so delightful a treatment. It is, however, doubtful. About sixty years ago the present church was built, encasing a Norman building. The opportunity was seized to bury the font under the floor of the north transept, *as they could not sell it.* The "oldest inhabitant" is responsible for this information, and the advisability of digging for his hidden treasure has been suggested to the vicar. Pending a little spadework there is no more information than is here given.

As recently as 1891 another has disappeared, but this by mischance, for when St Mary's Church, Great Plumstead, was burnt, the font was melted.

As far as can be ascertained this is the only destroyed lead font of which any record remains. Amongst Cotman's drawings of Norfolk antiquities there is a sketch, and another engraving exists in a fine collection of pictures of fonts in the library of the Society of Antiquaries. A drawing from the latter is here reproduced (Fig. 1). Apparently the fire which encompassed its final destruction was not the first malevolent act in its history. It was when drawn (and Cotman's drawing agrees) much mutilated. The top of the font had been neatly sliced off. The upright objects round the bowl appear to be columns, which originally carried arches.

The other ornaments are unusual, consisting of shields under the (theoretical) arches, and a band of fat scrollwork encircling the bowl.

If the elements really needed to consume a lead font, it is fortunate that an example already so much damaged was chosen for their sport.

Among reputed lead fonts which have been noted in various lists those at Clewer, Cherrington, Swymbridge, Chirton, Wansford, Pitcombe, Marton, and Avebury are not of lead. Clunbridge, Gloucestershire, which is sometimes described as possessing a font dated 1640, is probably a misprint for Slimbridge. The latter is, however, of date 1644, and there is no place named Clunbridge in Gloucestershire.

Altogether fire and the devices of the wicked have left us but thirty. Of these, ten are made from three patterns (with some small variations), leaving twenty-three separate designs. We may classify the thirty in two ways :—

    I. By their geographical distribution, and

    II. By the general character of their design.

Arranged by counties they are as follows :—

Berkshire.—Childrey, Long-Wittenham (thirteenth century), Woolstone (Norman).

Buckinghamshire.—Penn (date uncertain).

Derbyshire.—Ashover (Norman).

Dorset.—Wareham (Norman).

Gloucestershire.—Frampton-on-Severn, Llancaut (preserved at Sedbury Park, Llancaut Church being in ruins), Siston, Oxenhall, Tidenham, Sandhurst (these six are Norman, and all cast from the same patterns), Haresfield (fourteenth century), Down Hatherley, Slimbridge (Renaissance).

Hampshire.—Tangley (Renaissance).

Herefordshire.—Burghill (probably Norman), Aston Ingham (Renaissance).

Kent.—Brookland (Norman), Wychling (probably Early English), Eythorne (Renaissance).

Lincolnshire.—Barnetby-le-Wold (Norman).

Norfolk.—Brundal (probably Early English).

Oxfordshire.—Dorchester (Norman), Warborough (thirteenth century).

Surrey.—Walton-on-the Hill (Norman).

Sussex.—Edburton, Pyecombe (Early English), Parham (Decorated), Greatham House, Pulborough (date uncertain).

It is worthy of note that there is no lead font north of Lincolnshire.

Classification by design gives us the following arrangement of the thirty :—

a. Eleven, the chief feature of which is a large arcade, generally with prominent figures under the arches.—Frampton-on-Severn, Siston, Oxenhall, Tidenham, Llancaut, Sandhurst (Gloucestershire), Dorchester (Oxfordshire), Burghill (of Burghill all is restoration save the top of the arcade), Walton-on-the-Hill (Surrey), Wareham, Ashover.

b. Six, arcaded, but with other important decoration.—Brookland, Warborough, Long Wittenham (the last two from the same patterns with variations), Edburton, Pyecombe (these two from the same patterns with variations), Haresfield.

c. Three, not arcaded, with figure decoration.—Childrey, Brundal, Eythorne.

d. Nine, without figures or arcading, but with various decorations.—Wychling, Woolstone, Barnetby-le-Wold, Parham, Tangley, Slimbridge, Down Hatherley, Aston Ingham, Greatham House (Pulborough).

e. One, without any decoration.—Penn.

Class A.—Fonts with Large Arcades and Prominent Figures.

The six Gloucestershire Norman fonts are tub-shaped and cast from the same patterns.

Only those at Oxenhall (Fig. 2) and Sandhurst (Fig. 4) are illustrated, as the

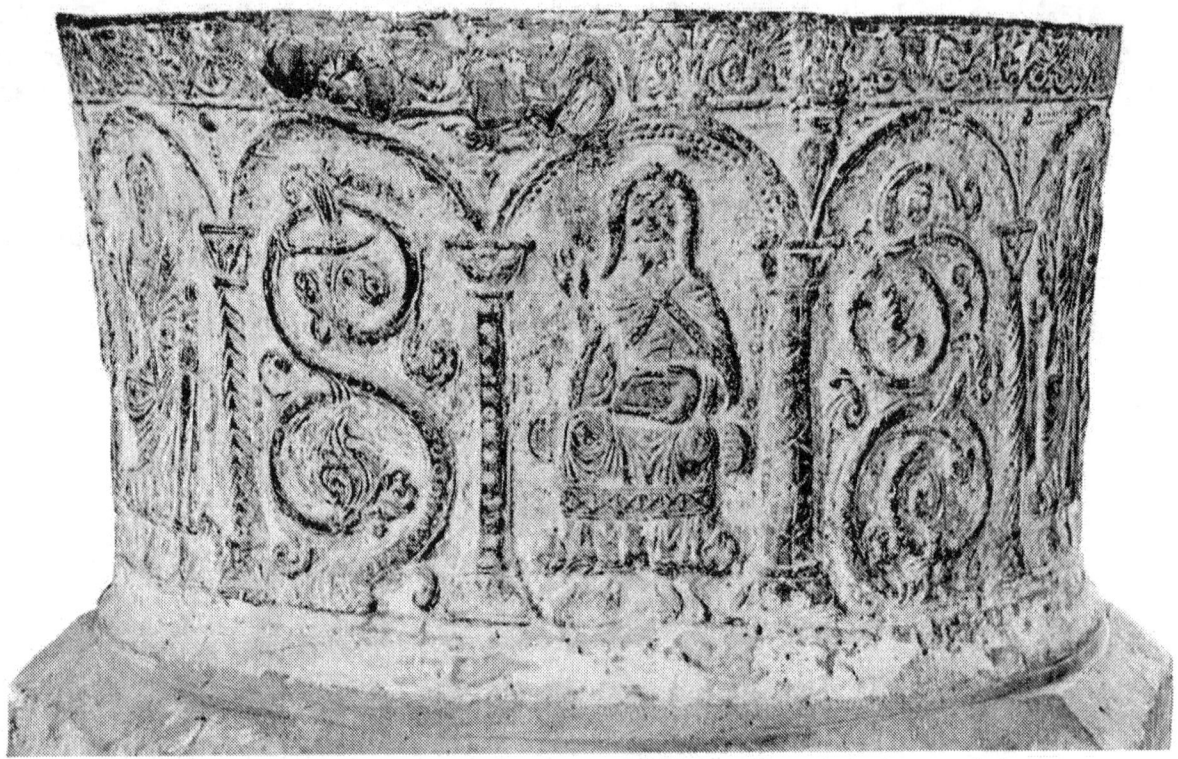

Fig. 2.—Oxenhall, Gloucestershire.

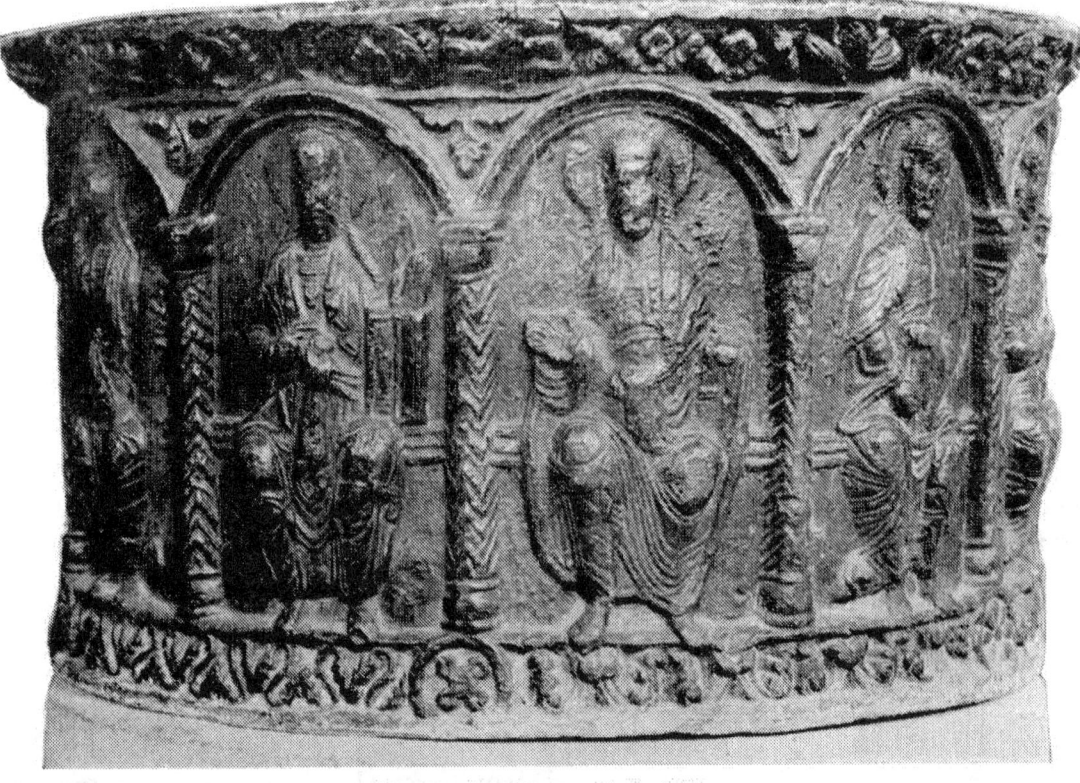

Fig. 3.—Dorchester, Oxfordshire.

others are the same. With the exception of these four, which it would be superfluous to illustrate, this chapter includes one or more photographs of every existing ancient lead font so far recorded.

Four of the Gloucestershire fonts have an arcade of twelve, six arches being filled with scrollwork of a vigorous snake-like pattern, and six with seated figures. The latter are of great interest. Two figure patterns only have been employed. In both, the right hand is lifted in benediction, while the left hand holds a book, sealed in one figure, unsealed in the other—an Apocalyptic suggestion. The robes are richly ornamented, and Dr George Ormerod suggested that the figure represents the Trinitas, but a more likely interpretation is Christ enthroned.

The Llancaut example has ten arcades only, and the Sandhurst font eleven (six with scrolls and five with figures). The friezes are all decorated with a delicate floral pattern.

The existence of these six fonts all cast from the same mould is a pleasant example of the stock pattern in the twelfth century. They suggest that the stock pattern is not in itself (if we accept the teaching of history) an evil thing. The odious character of most of the stock patterns of the last century, particularly of those which took their inspiration from the dreary atmosphere of the fifties and the Great Exhibition, has caused a not unnatural feeling that no architectural detail is tolerable unless it is designed *ad hoc*. Where it is a matter of hand-wrought objects this nervousness of repetition is likely to stimulate fancy and make for variety. Where, however, casting in metal is concerned, it seems a more reasonable method to encourage repetition, as it enables a greater amount of thought and effort to be expended on the original pattern than is economically possible ordinarily if only one object is made. The Norman craftsman evidently did not fear

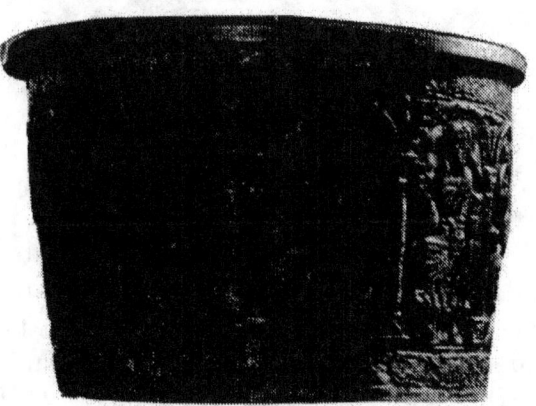

Fig. 4.—Sandhurst, Gloucestershire.

to scatter replicas of his lead font once he was satisfied, as he might well be, with the original pattern. If six examples have persisted for about eight hundred years, it is reasonable to suppose that there were originally two or three times six made from the pattern. One cannot help wondering what shrieks about stock patterns would rend an outraged architectural heaven, if twelve or more modern churches were made to-day the artistic dumping ground of one pattern of font.

Among the many treasures of the Abbey Church at Dorchester, Oxfordshire, is an arcaded Norman font similar in general character to the Gloucestershire type. Fig. 3 shows the complete font, and Fig. 6 a part of it, the latter to emphasise the peculiar beauty of the fall of the robes.

The arcade is in eleven bays with a different figure seated under each arch. The number suggests the faithful apostles, but as each figure is nimbed, and as the hair falls on both sides of the face in all, it seems more likely that the modeller intended to represent our Lord in different attitudes.

Here we have the same *motifs* of books and benediction. Two of the figures,

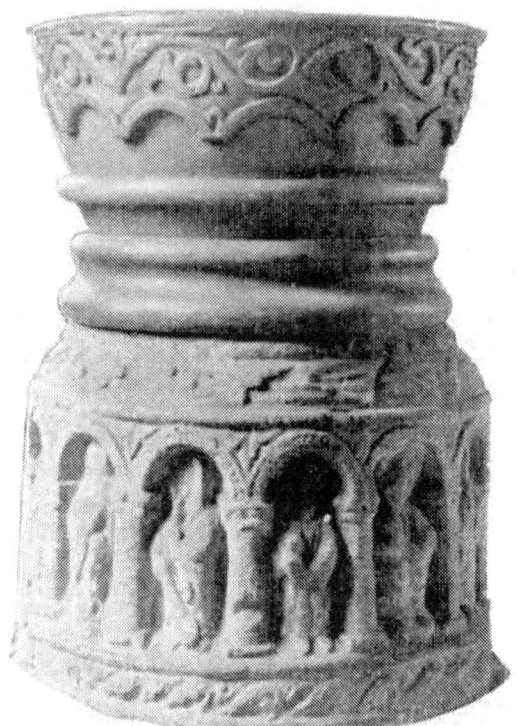

FIG. 5.—Burghill.

however, hold keys. Had this been so in only one case, Saint Peter would reasonably have been indicated. As there are two, they probably symbolise the keys of Hell and of Death in the hand of Christ.

The general treatment of the figures on these two fonts is that of Anglo-Saxon times, and this date was claimed by the late Dr George Ormerod for the Gloucestershire fonts (he wrote actually of the Tidenham example, but Oxenhall is identical), and by the late Professor Freeman for the Dorchester font.

The architectural treatment of the arcading suggests Norman work, however.

In the history of art there must be few examples of conservatism so marked as in the case of the leadworker, and it is likely that we have here a Norman plumber using Anglo-Saxon casting patterns.

Patterns persist, and there is a natural tendency to use old ones rather than to make new ones in a rising style. To take a modern instance, present-day ironfounders of the unwiser sort discovered *L'Art Nouveau* some eight years ago. Designers of the "glue and string" school rushed to the rescue. New patterns were made at great cost. The result is that, though *L'Art Nouveau* is "dead and damned," its stringy tulips will sprout for many years on the fire-places of Suburbia. For this we have to thank the permanence of casting patterns. Fortunate, however, the same permanence which has preserved for us Anglo-Saxon modelling to give interest and beauty to a Norman font. It is probable, moreover, that the Gloucestershire and other fonts now described as Norman belong to the end of the twelfth century, if not to the beginning of the thirteenth.

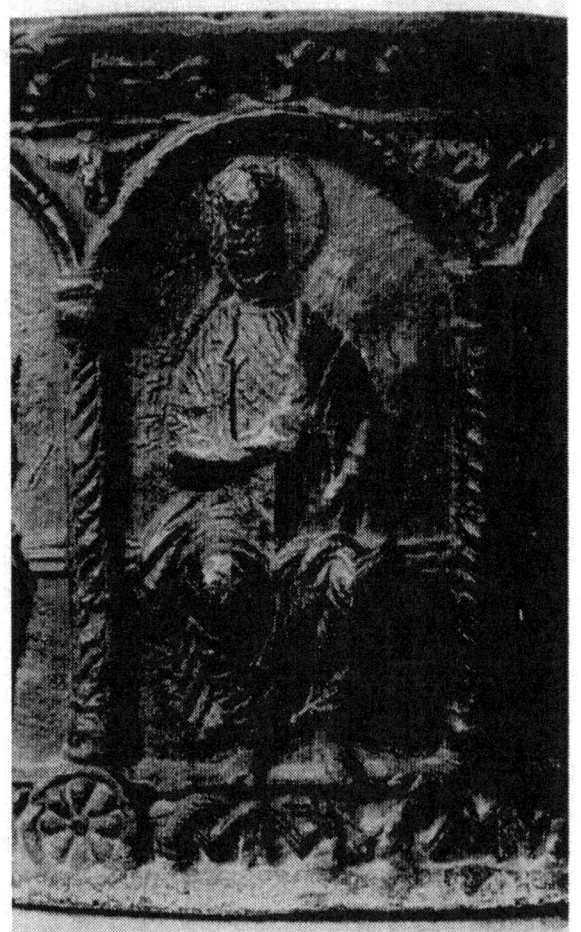

FIG. 6.—Dorchester, Oxfordshire.

The font at Burghill, Herefordshire (Fig. 5), is interesting rather for what it was, and for what its stone base suggests, than for any present beauty. Early in the nineteenth century the tower of the church fell and seriously damaged the font, which was placed in the vestry for safety. In 1880 it was restored, but in the effort to straighten the lead the lower part, which was very thin, perished. The upper part was then attached to the aggressively moulded bowl which was made for the purpose. The curves on the lower edge of the border appear to be the tops of lost arches. There were thirteen of them, and the contemporary stone base also has thirteen arcades;

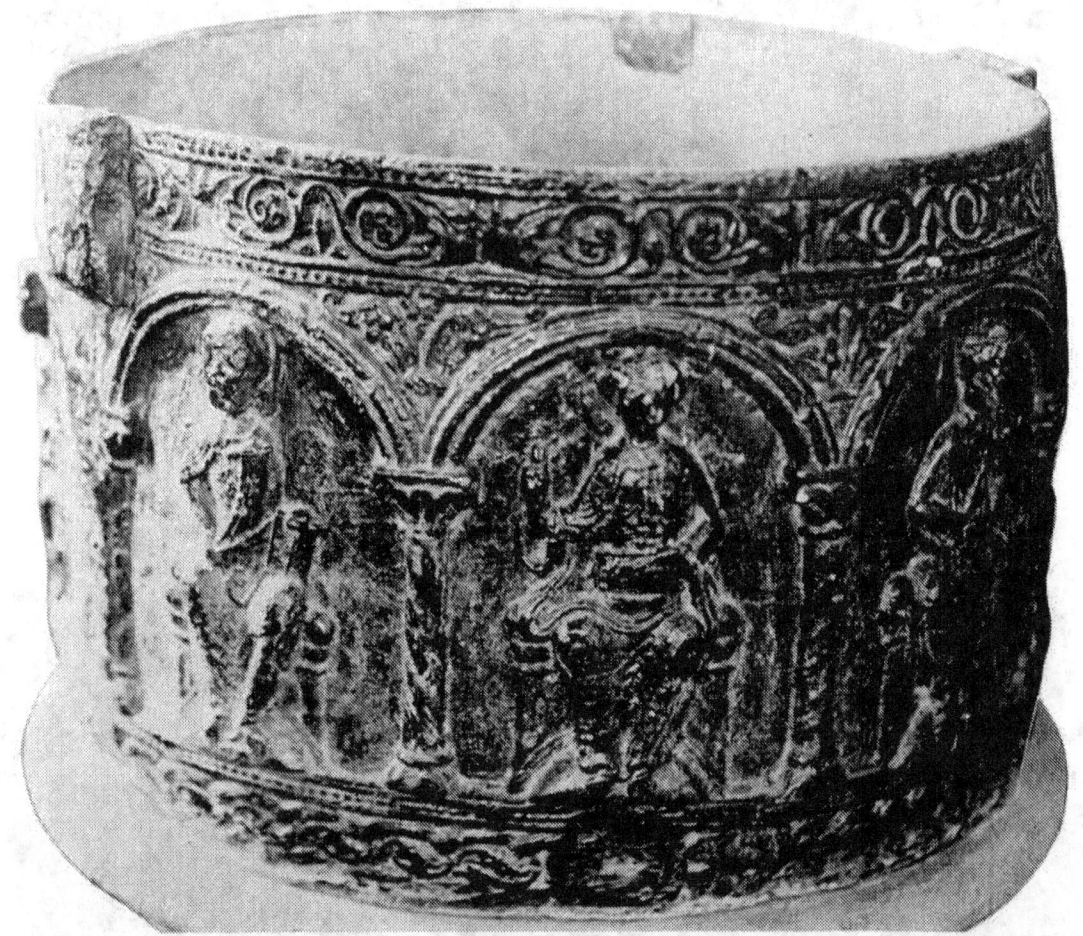

FIG. 7.—Walton-on-the-Hill, Surrey.

they were probably designed together. The figures on the base, though much mutilated, appear to be those of our Lord and the apostles, and the lead arcades possibly repeated these figures or contained scrollwork similar to the alternate panels of the Gloucestershire Norman fonts. The carving of this base affords an excellent comparison between stone treatment and the treatment of like designs in lead (compare Figs. 3 and 5).

Walton-on-the-Hill, Surrey, has a magnificent example. Only three patterns are employed for the twelve seated figures, which have no nimbus. All three hold books, and two have the right hand uplifted in benediction. The top band of ornament, enclosed by lines of beads, is rich, and the spandrels have delicate ornament. It

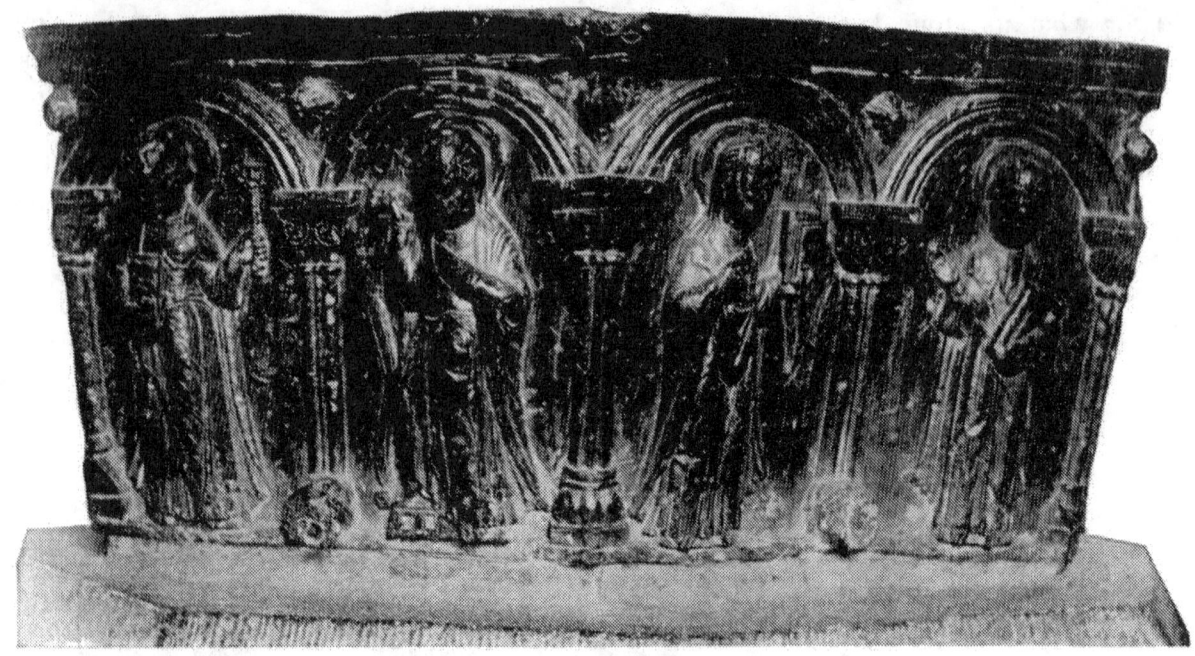

FIG. 8.—Wareham, Dorset.

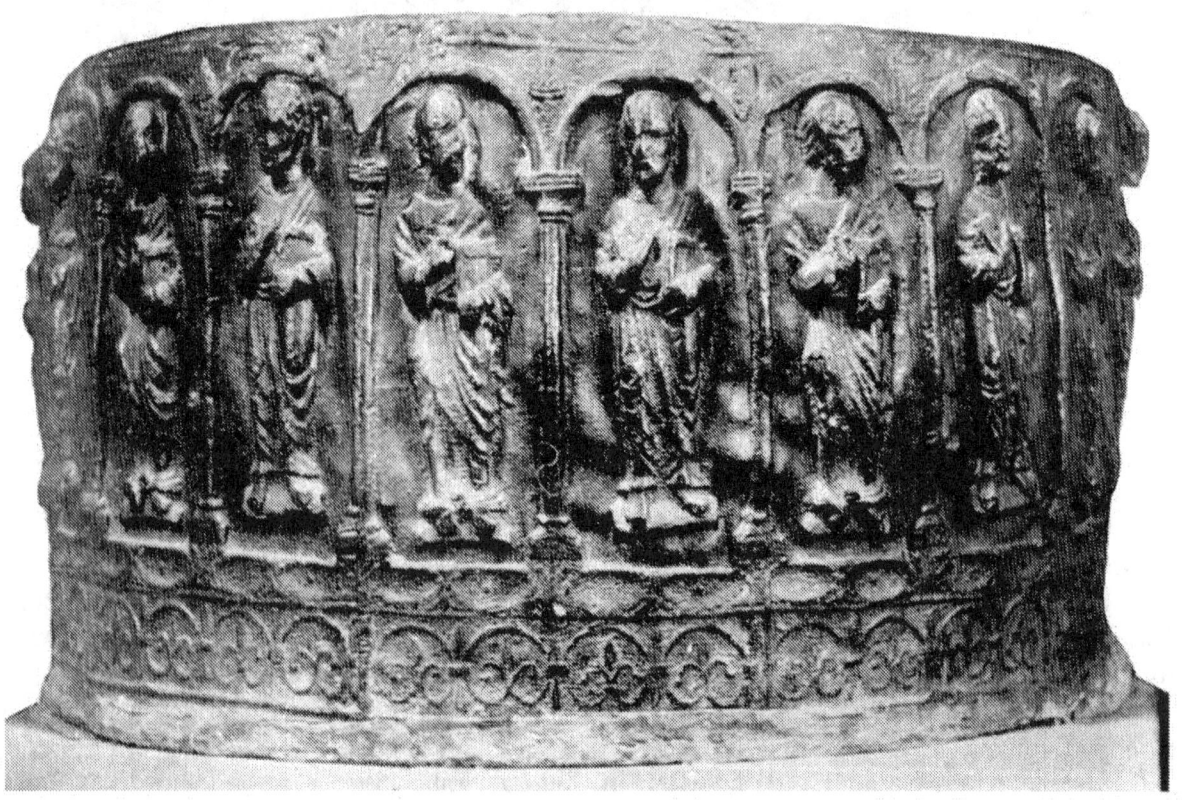

FIG. 9.—Ashover, Derbyshire.

is curious that, of the thirty, only two lead fonts should be other than round. The bowl at St Mary's Church, Wareham, Dorset, is hexagonal, and twelve boldly modelled figures stand under the round-headed arcading. None has the nimbus, but as one holds a square-headed key, the figures are doubtless St Peter and the eleven apostles. There are no other marked evangelistic symbols; either scrolls or books or both are in the hands of the eleven. It is to be noted, though, that the figures are cast from separate patterns, and do not repeat, as for instance at Walton-on-the-Hill, Surrey, where three patterns are repeated four times.

It is worthy remark that no lead font is octagonal. The Wareham font stands on an octagonal base, which suggests that either the bowl or the base came from another church, the bowl probably, as being conveniently portable. The number eight was symbolically the number of regeneration (why so is not clear), but this symbolism did not attack fonts generally until the Perpendicular period Symbolically lead fonts are weak. There is none either with the seven or the two sacraments, and the symbolism of the Brookland font is cosmic rather than Christian.

The font at All Saints' Church, Ashover (Fig. 9), has been described as a stone font with leaden statues. This is perhaps a little misleading. The figures are not attached direct to the stone, but the stone bowl is covered by the lead casing which the figures decorate. For the twenty figures under the arches two patterns only were used. They are simply draped, and have neither mitre nor nimbus. Each carries a book, but the right hand is against the body and not lifted in benediction. The modelling is remarkable for its bold relief, which is about $\frac{3}{4}$ inch in the figures. The top band of ornament has been damaged greatly, but the lower border is unhurt and beautiful. It is probably late twelfth-century work.

Class B.—Fonts with Arcades, but with other Important Decoration.

The example at Brookland, Romney Marsh, may fairly claim to be the most interesting of lead fonts, if not, indeed, of all English fonts. It is 6 feet in girth, and its double arcading bears the signs of the zodiac in the upper tier, and delightful busy figures, illustrative of the labours of the months, below. The heads of the arches bear the names of the signs in Latin and of the months in French, and as there are twenty arcades, eight appear twice, the duplicates being from March to October. This perhaps suggests that the patterns were not made for the purposes of this font. If they were, and an arcading of twelve only had been used, the bowl would have been about 14 inches in diameter. This is smaller than any of the others, which vary from $18\frac{1}{2}$ inches at Down Hatherley to 32 inches at Barnetby-le-Wold. The mouldings running round the upper part of the bowl are thrice broken by added panels, which are much rubbed but appear to represent the Resurrection. They are evidently an afterthought. The plumber's priestly client perhaps thought the decoration secular rather than spiritual, and called for these additions, unwillingly done may be, for one is crookedly fixed.

The creatures of the zodiac and the scenes are freshly and gaily modelled. Dealing with them in order, beginning at the middle of the large illustration (Fig. 12), to the right of the seam and reading to the right, we have—

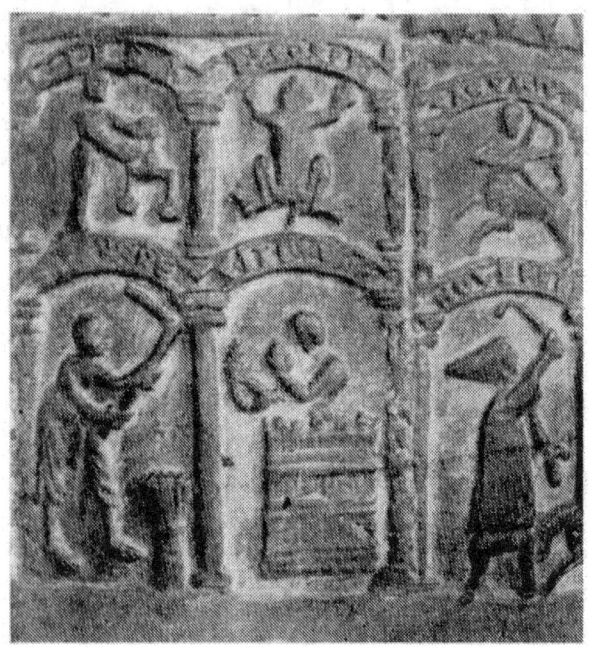

FIG. 10.—September to November

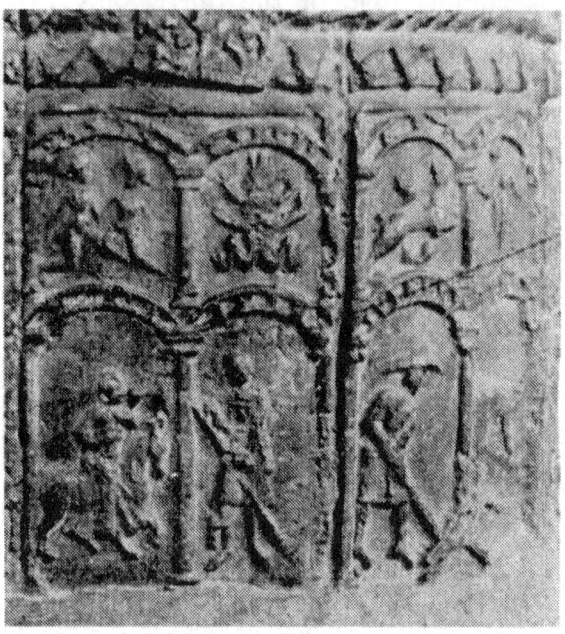

FIG. 11.—May to August.

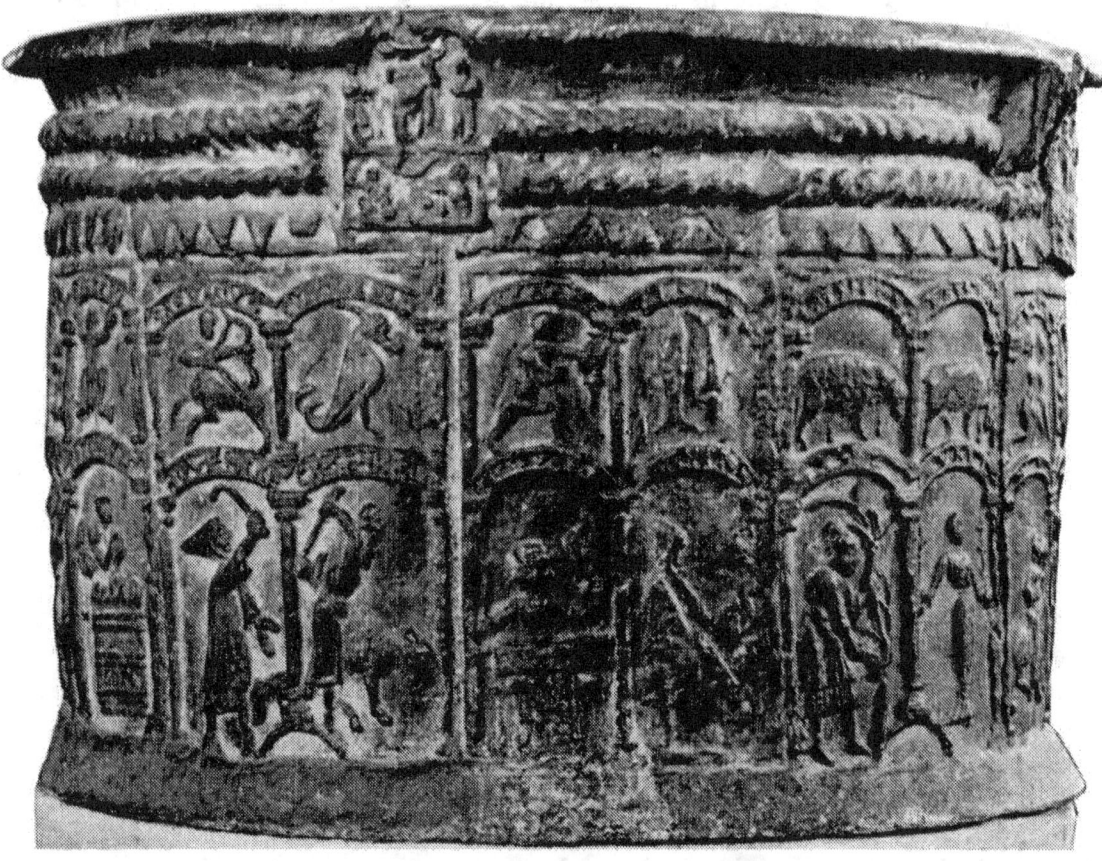

FIG. 12.—Brookland: October to December, and January to May.

Aquarius—January.—Above, Aquarius upturns his waterpot vigorously; below, two-headed Janus drinks farewell to the old year, and welcome to the new.

Pisces—February.—Above, the usual two fishes reversed; below, a seated hooded figure warms his feet at the chimney.

Aries—March.—Above, a patient-looking ram; below, a delightful hooded figure pruning a vine. (The lettering above the arch is incorrectly given as Capricornus.)

Taurus—April.—Above, the bull, almost as lean as Capricorn; below, a girl of slender graceful figure stands with tall lilies in her hand. She doubtless is a symbol of Rogation-tide. The "gang-days" fall generally in May, but sometimes in April.

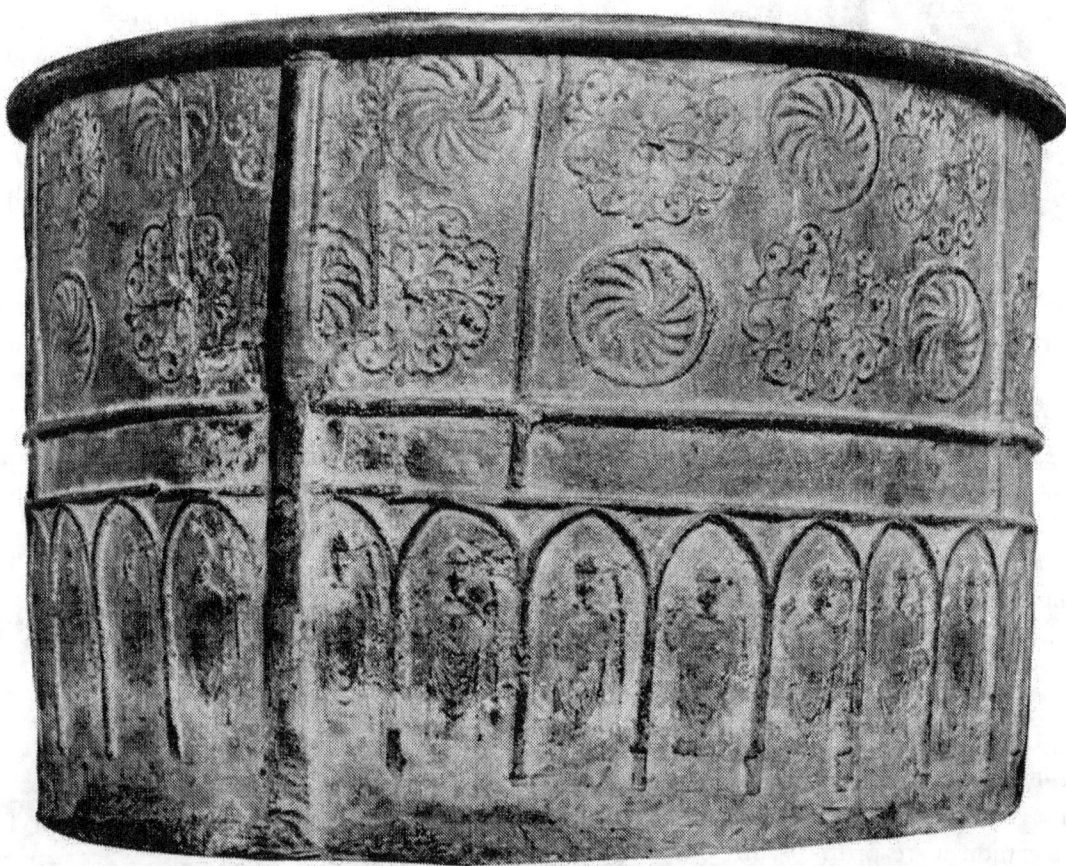

Fig. 13.—Long Wittenham.

Passing now to Fig. 11 we find, reading from the left—

Gemini—May.—Above, the twins, naked children; below, a knight on a rather small palfrey, with a hawk on each wrist.

Cancer—June.—Above, the crab is fortunately labelled, for it would not have been suspected; below, a man mows with a scythe, whetstone at side.

Leo—July.—Above, a leopard-like lion; below, a man in a wide-brimmed hat is raking hay.

Virgo—August.—Above, Virgo has a slim girlish figure, with a spike of corn in one hand and a vindemiatrix in the other; below, a man bends down reaping.

Passing now to Fig. 10, and reading from the left, we get—

Libra—September.—Above, Justice with bandaged eyes holding even scales; below, a thresher with flail uplifted over the sheaf.

Scorpio—October.—Above, the scorpion is a harmless creature, a frog save for his tail, which doubtless does the necessary stinging; below, a figure treads the wine-press, or perhaps a cider vat.

Sagittarius—November.—Above, a centaur fires his shaft behind him; below, a swineherd in a delightful conical hat is apparently beating down acorns for pannage.

Capricornus—December.—Above, Capricorn is an amazing creature (see to the left of the seam in the large illustration) and might have come out of the Bad Child's Book of Beasts; below, a man is killing a wolf with an axe, a winter sport now happily fallen into disuse.

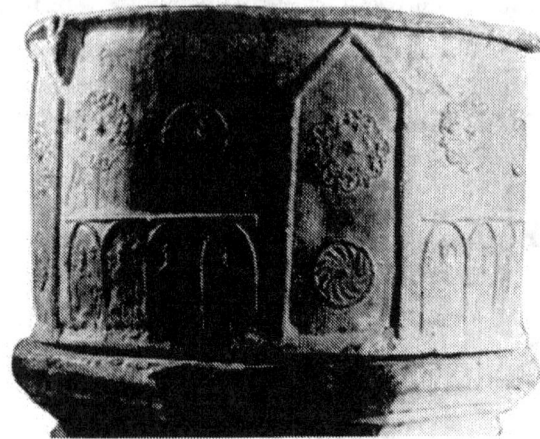

FIG. 14.—Warborough.

The stone font at Burnham Deepdale has similar subjects for the labours of the months, with some differences of treatment.

An odd feature of the architectural treatment of the Brookland font is, that every third pillar of the arcading stands on a loop.

The secular character of this font having impressed a clerical correspondent, he asked whether it expressed the following idea:—That the sequence of the months represents man's temporal existence, and that baptism creates the spiritual life which should inform our external life. The idea that the temporal life is shown as a microcosm of the eternal is delightful, but quite unlikely to have been in the plumber's mind. The twelfth-century men were probably little conscious of such subtleties, and just modelled the things they felt best and knew best and loved best, to the glory of God and with the artist's pleasure in doing a job well.

The Warborough font is most decorative and came from the same plumber as the font

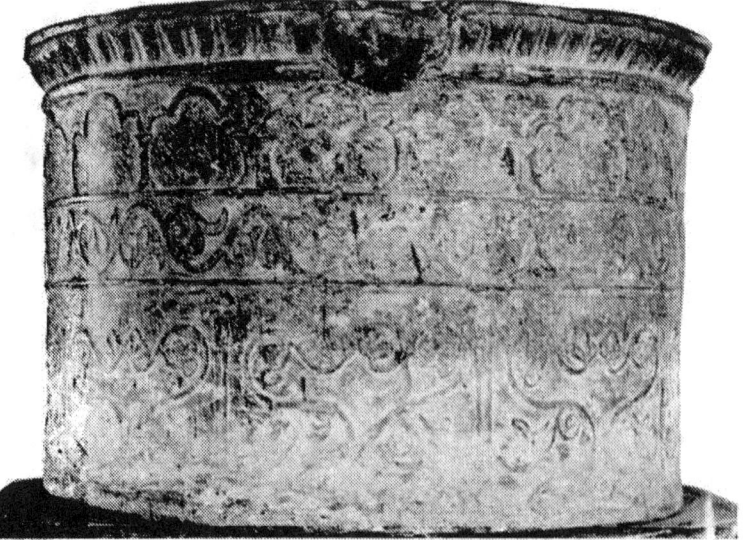

FIG. 15.—Edburton.

at Long Wittenham, to be described next. Several of the ornaments are the same, though their arrangement varies. Both, too, have the pointed arcade at the bottom, and bishops apparelled as in the Childrey example, with the right hand in the act of blessing. The big middle feature of the Warborough bowl is a somewhat angular arch. Of the two circular ornaments, which appear under it and elsewhere on the bowl, one is a wheel with curved spokes, and one a beautiful geometrical design which suggests lacework. Mr Lethaby describes this font as Norman, but the decoration seems more appropriate to the late thirteenth century. This bowl is of the maximum depth that is found, viz., 16 inches, and has only one seam. The circumference was cast in one piece, whereas most of the lead fonts were cast in four pieces (in addition

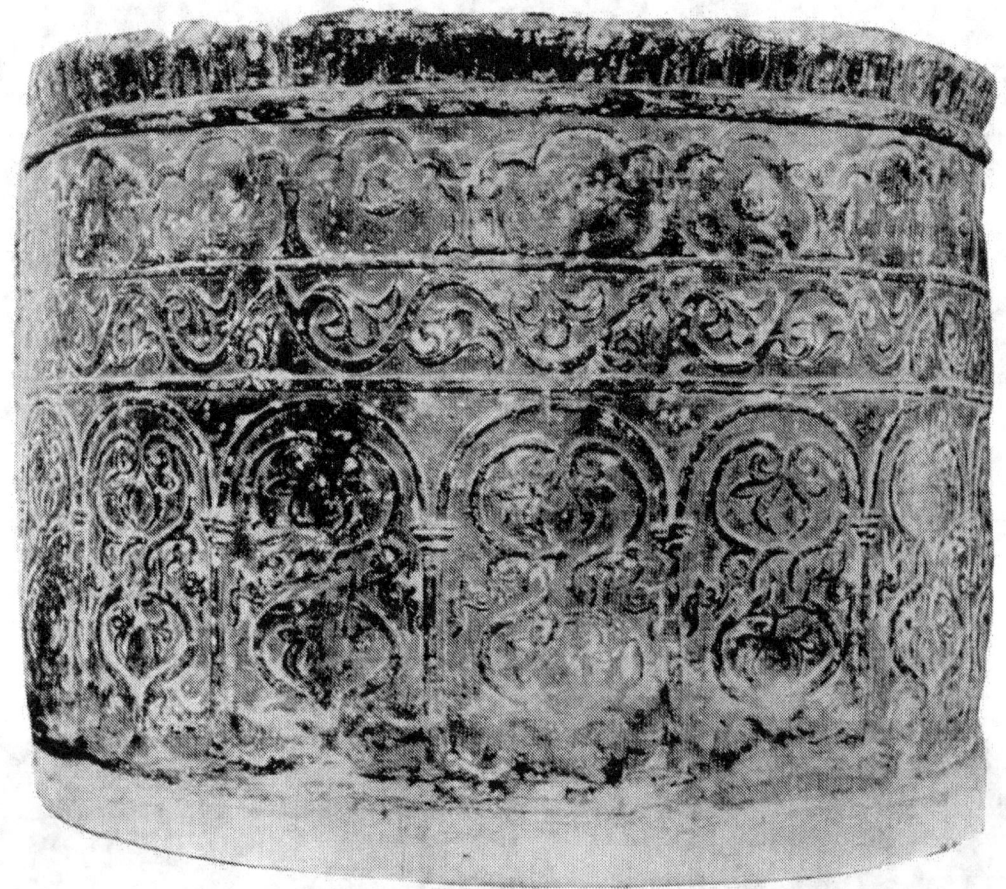

Fig. 16.—Pyecombe.

to the bottom) and joined. At Woolstone, however, there are two seams, and at Walton-on-the-Hill we find three. At Warborough, as with most of the lead fonts, there are the marks of the locks of the covers, which were made compulsory by Edmund Cantuar. in 1236.

At Long Wittenham (Fig. 13) the tall arches are omitted. The upper half is divided into compartments and more plentifully decorated with wheels.

The Edburton and Pyecombe fonts help to keep up the high archæological reputation of Sussex. They lack figures altogether, and are probably the work of a Norman plumber of about 1200 or later. Both fonts have the heavy fluted rim, the upper

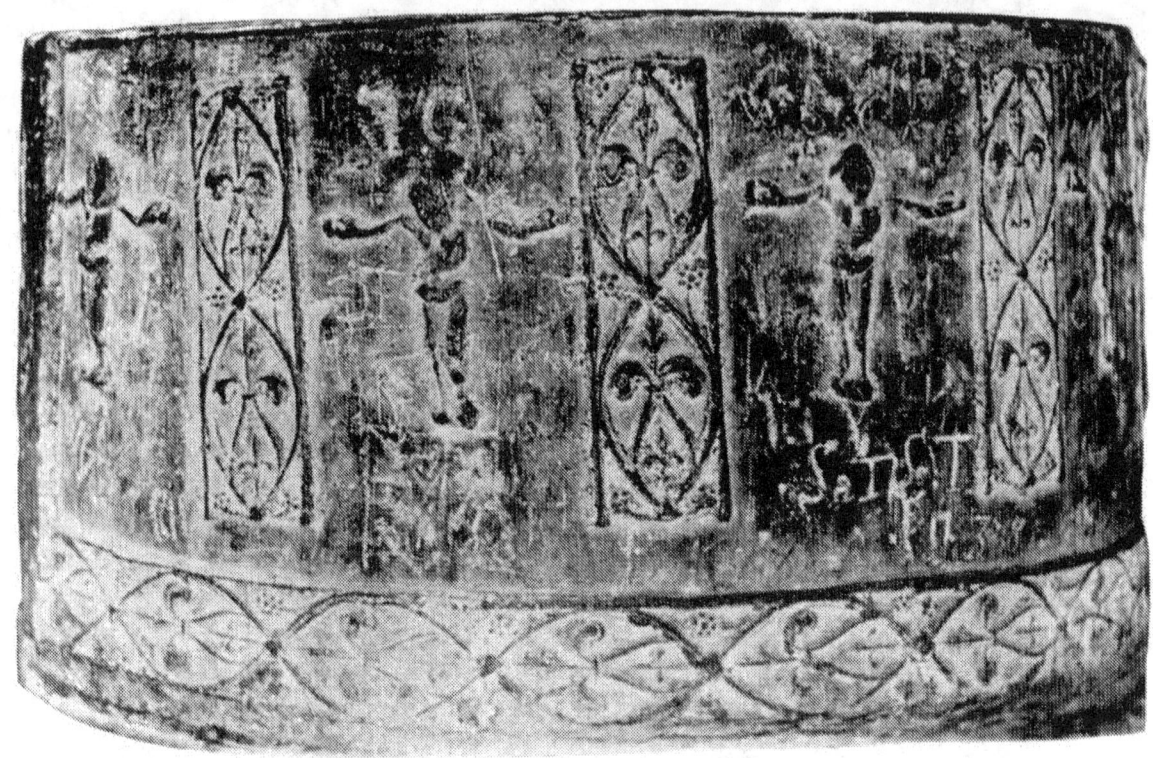

Fig. 17.—Brundall.

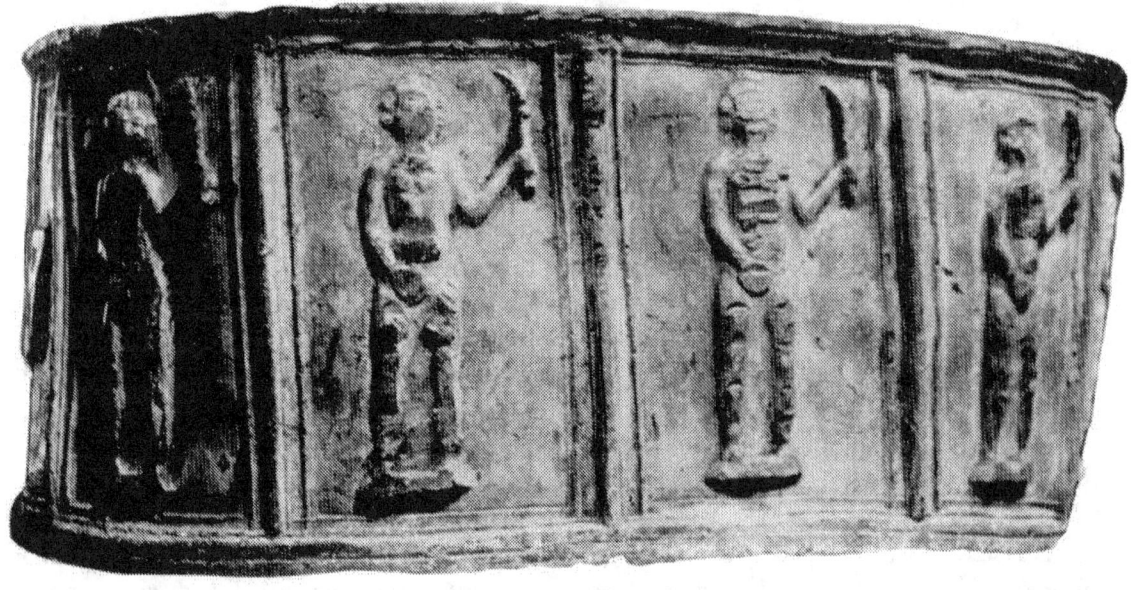

Fig. 18.—Eythorne.

arcading and the narrow middle band of scrollwork, but there is no slavish likeness in detail or size. The lowest band differs in the two, the Pyecombe font (Fig. 16) having an arcading of fifteen, with floral work within the arches; the Edburton example (Fig. 15) shows the scrolls without the arches.

The Pyecombe bowl is 6 feet in circumference and 15 inches deep, that of Edburton is 5 feet and 13½ inches respectively.

Though distinctively Norman in character, the coming of Gothic is apparent in

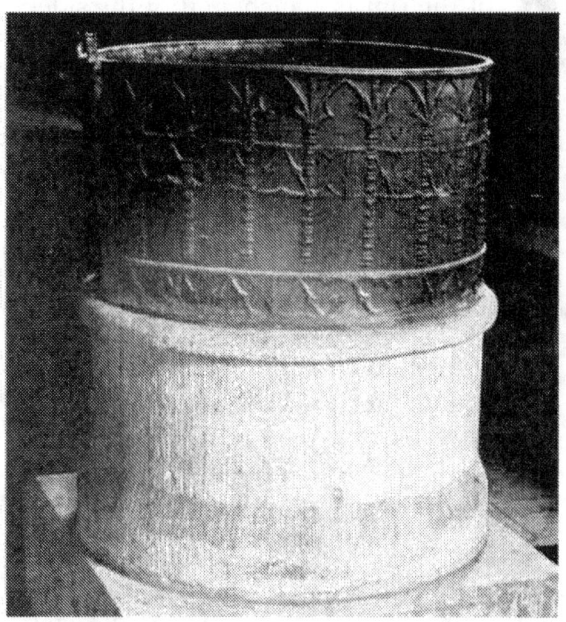

FIG. 18A.—Haresfield, Glos.

FIG. 19.—Eythorne.

the trefoil heads of the upper arcading. The general effect is perhaps a little suggestive of embroidery, but very successful.

The decoration of the Haresfield font (Fig. 18A) is paradoxical, and raises a somewhat difficult question of date. The arcading has the character of fourteenth-century work, while the buttoned vertical shafts suggest the seventeenth. Several authorities consulted vary in their attribution of date, but as the cusping can hardly be post-Gothic,

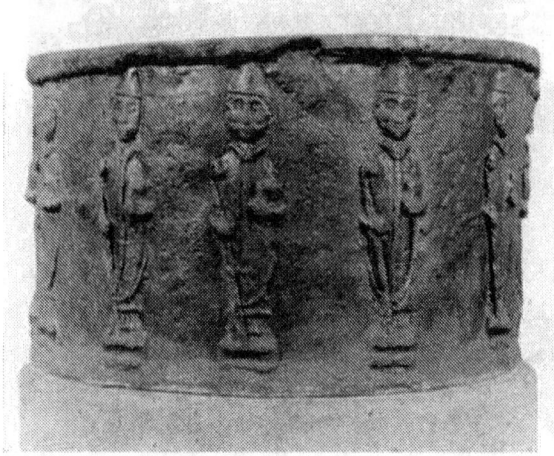

FIG. 20.—Childrey.

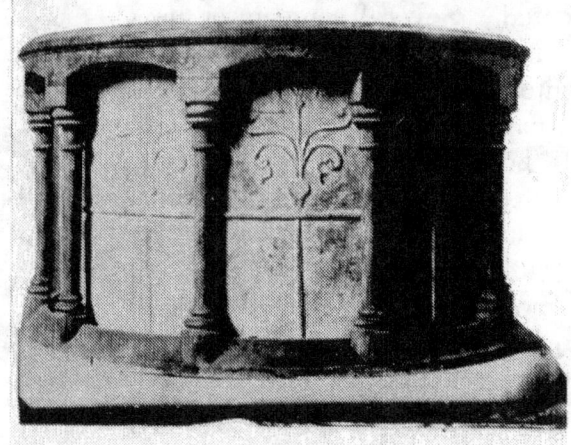

FIG. 21.—Wychling.

and as there are instances of such turned shafts being used in fourteenth-century wood-work, the earlier date is here adopted.   This font has appeared in some lists as being of bell-metal, but incorrectly.   Its diameter is 24 inches, the thickness of the rim is ½ inch, and of the sides generally a little over ¼ inch.

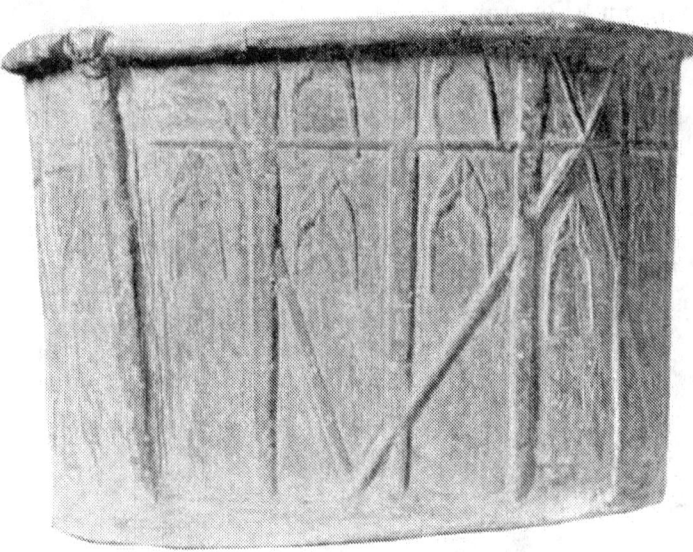

FIG. 22.—Woolstone.

Class C.—With Figure Decoration but without Arcading.

The Childrey font (Fig. 20) is very simply treated. The twelve bishops who stand on low pedestals round the bowl all wear mitre, alb, and chasuble, and all carry a crozier in the right hand and a book in the left. The modelling is of a rather elementary sort.

The Brundall bowl (Fig. 17) is the only lead example left to Norfolk, a county rich in fonts.   It is probably of late in the thirteenth century, and is the only one bearing an image of the crucifixion.   The fleur-de-lys treatment of the lower border and of the vertical panels is as delightful as it is naïve.   A notable feature of the Christ figures is that they are impressed.   The font is in two thicknesses, the outer one very thin and the inner heavier and later.

The Eythorne font has a figure of unusual type, seven times re-peated.   Several conjectures have been made as to who is repre-sented, but, as the figure is nude, perhaps Adam is the most likely. He holds a torch in his left hand. There is no difficulty in settling the date, for the artist has written it large, 1628, on four panels, a numeral to each panel.   A sugges-tion that the seven light-bearing figures are in some way symbolic may well be dismissed.   In 1628

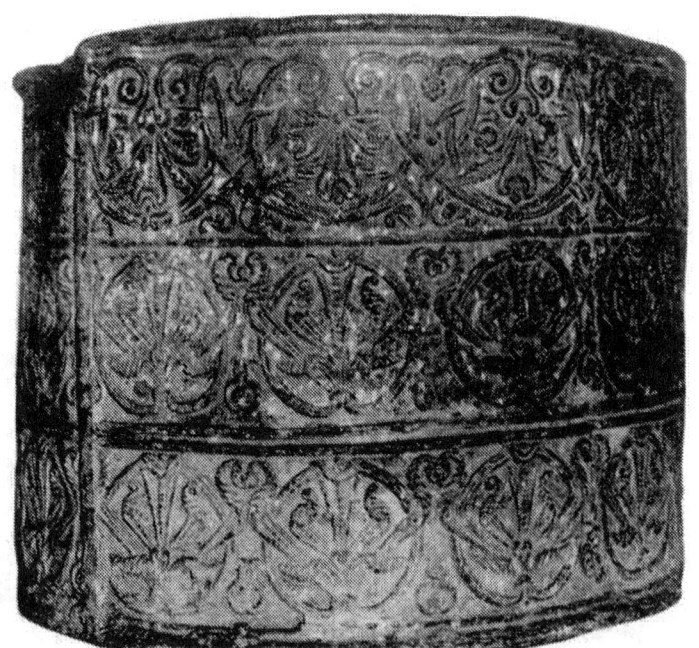

FIG. 23.—Barnetby-le-Wold.

the sense of religious symbol was not very acute.   The bowl is shallow, 10 inches only in depth, and is much battered and out of shape.   It no longer fulfils its use, a modern

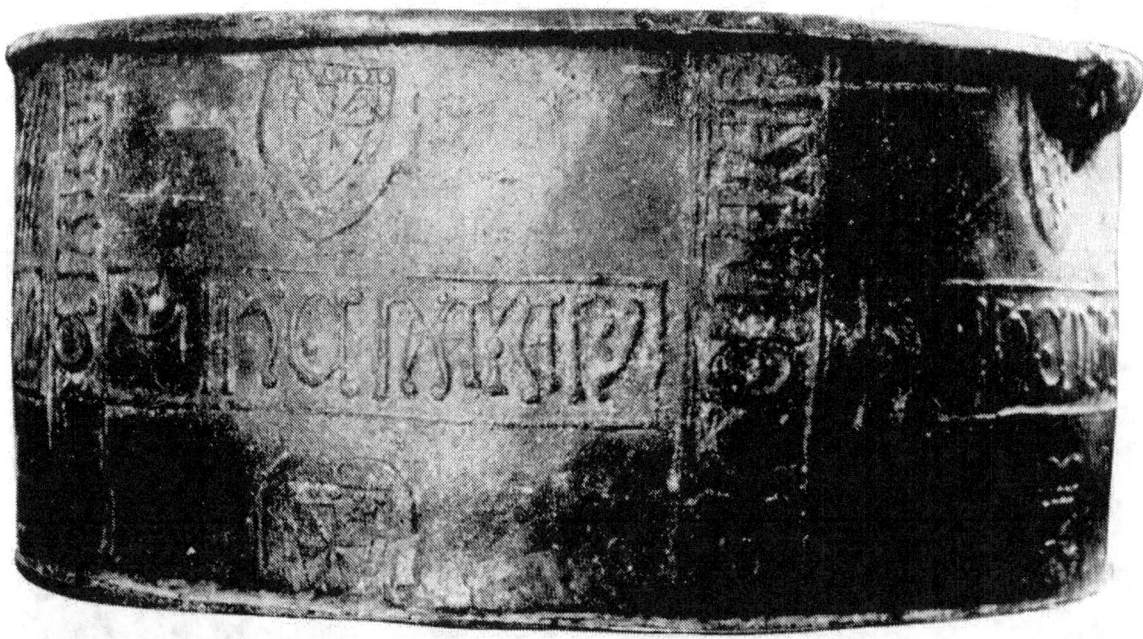

FIG. 24.—Parham, Sussex.

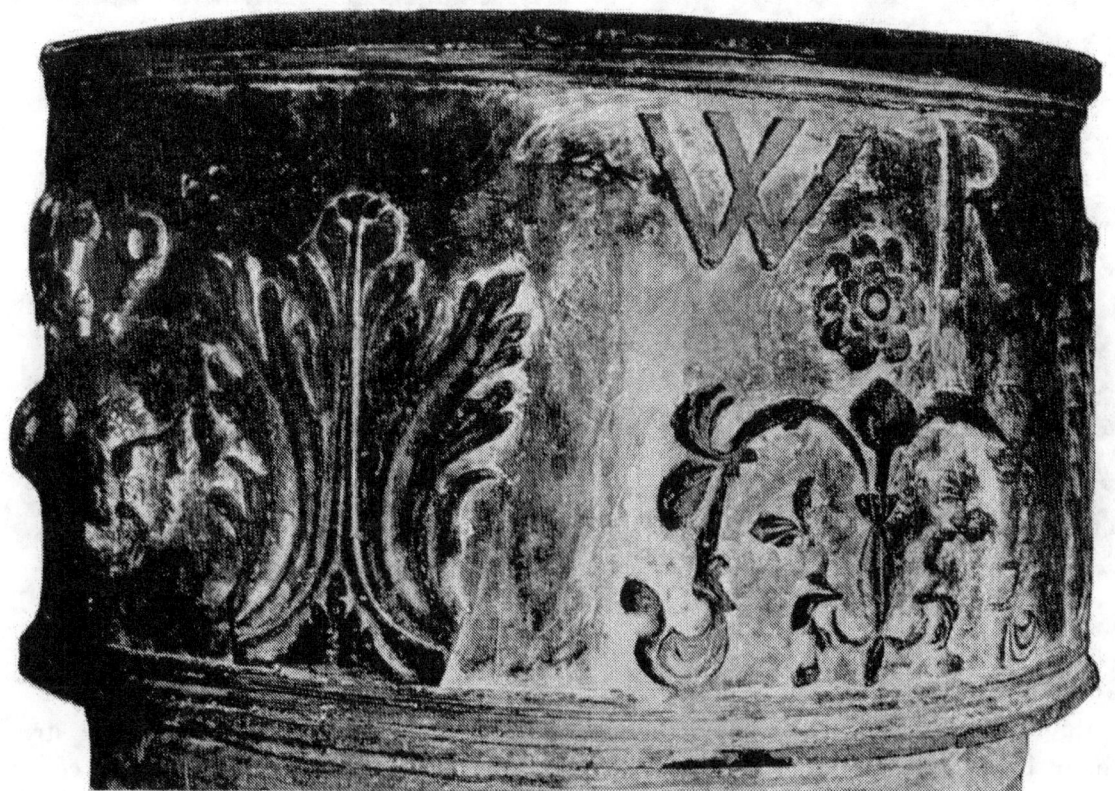

FIG. 25.—Aston Ingham, Herefordshire.

stone font has taken its place. Of the five post-Reformation lead fonts it is notable, in that it is alone in possessing figure decoration.

Class D.—Consisting of Nine Fonts without Figures or Arcading.

The Wychling bowl (Fig. 21) is a good deal disfigured by the rather aggressive modern woodwork which has been added, presumably to keep the leadwork in shape.

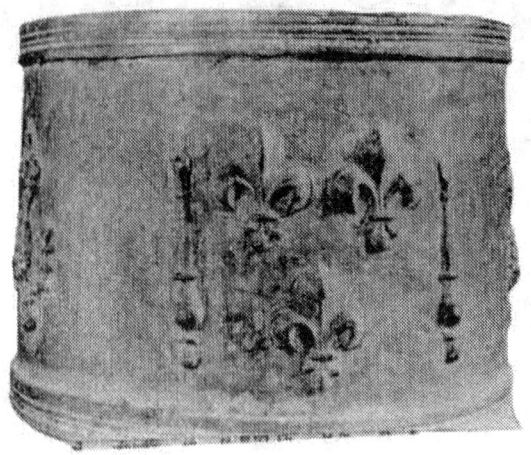

Fig. 26.—Tangley, Hants.

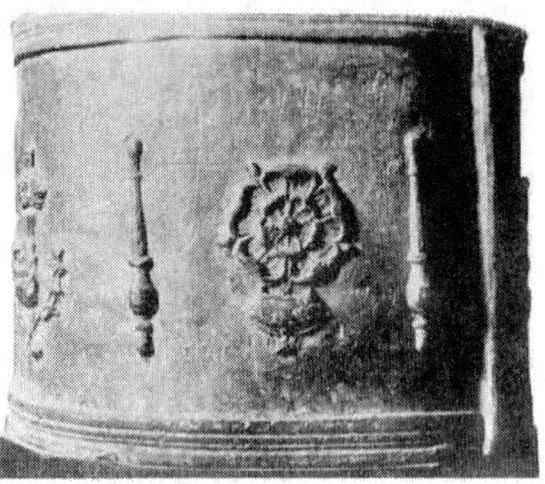

Fig. 27.—Tangley, Hants.

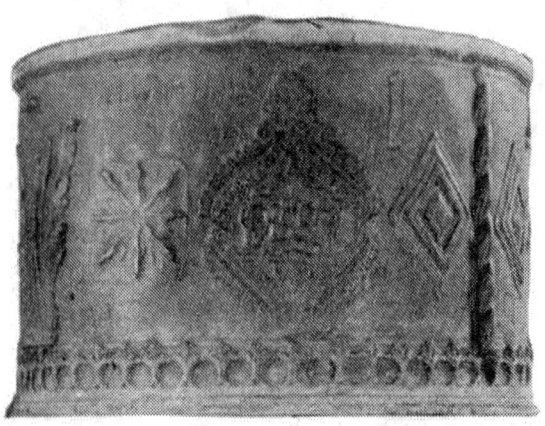

Fig. 28.—Down Hatherley.

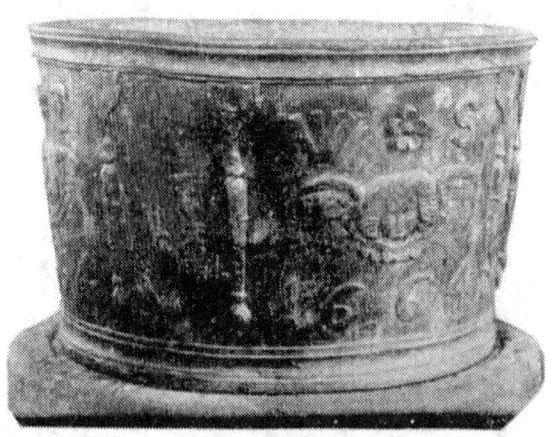

Fig. 29.—Slimbridge.

It is the simplest of the pre-Reformation fonts, and, though difficult to date (the stringy ornament has a curiously modern look), it is probably of the end of the thirteenth century. It is an example of the chequered history of metal fonts. The rector states that the font was found when he restored the church, built into a lot of brickwork and "providentially saved from the bricklayers and smashers." Restorers have so often proved the most finished of "smashers" that it is refreshing to find a church where these vocations have been kept distinct.

At Woolstone, Berkshire (Fig. 22), is the most architectural of the lead fonts. It altogether lacks figure work, and is in effect a sketch of a church. A narrow band separates the top part of the bowl, which is divided into an arcading of twelve pointed arches. These, as do the thirteen arches below the horizontal band, possibly represent windows. At the bottom of the bowl is a single arch—the door. As there are ten bold perpendicular straps and eight sloping thwarts, the church represented may be an early timber building which preceded the present church of All Saints'. One does not look in the thirteenth century (which may be conjectured to be the date of this font) for so pious a sense of archæological record as this bowl suggests. It gives one furiously to think how much greater would be our knowledge of pre-Conquest buildings if mediæval

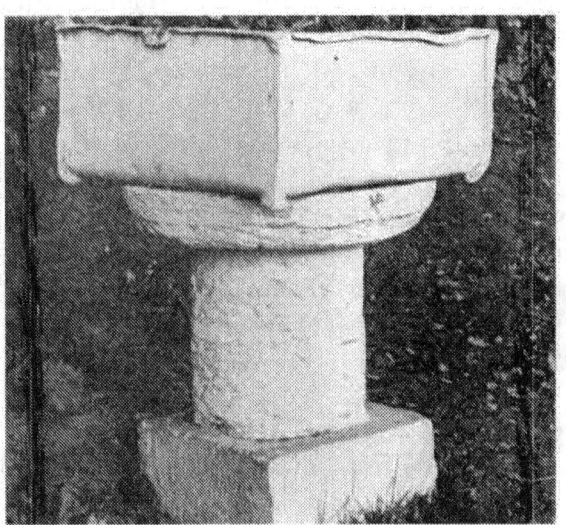

Fig. 30.—Greatham House.

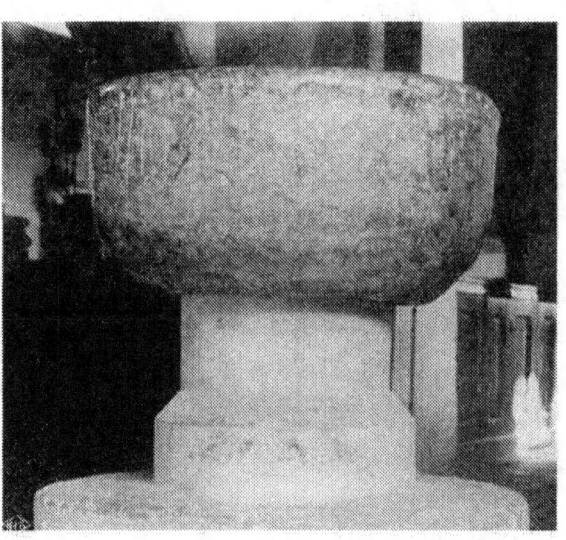

Fig. 31.—Penn, Buckinghamshire.

builders had made a practice of picturing in their new work the lineaments of the buildings they had destroyed. A modern and dreary instance of this is the tablet set up in the City showing the passer-by what manner of church was Saint Antholin's, Watling Street, before the passion for destruction took it from our ken. The Woolstone font, however, is infinitely sounder in principle, for the story of the lost church is told simply and unaffectedly, and the font is a witness of new effort and a continuing tradition of sanctity. A good deal less can be said for the St Antholin's tablet, which witnesses but to destruction and silence. Still, hideous as it is, it is better than nothing. It is proper to add that some antiquaries reject the theory that the Woolstone font illustrates an earlier church.

At Barnetby-le-Wold (Fig. 23) the decoration is very conventional but eminently suited to the material. This font was lately rescued from a coal cellar. It had been put to the base use of a whitewash tub, so has enjoyed the extremes of colour sensa-

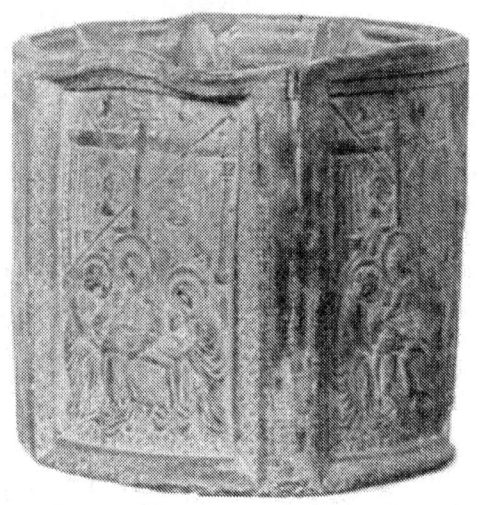

FIG. 32.   Gloucester Museum.

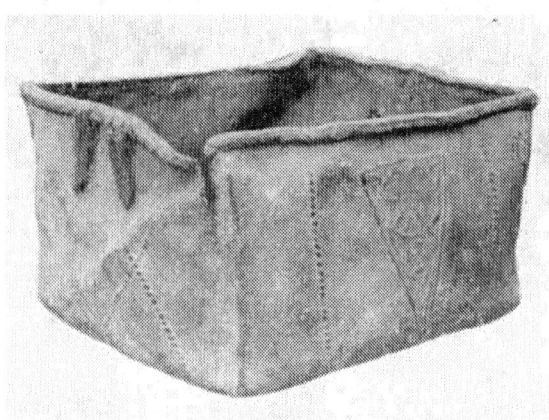

FIG. 33.—Lewes Castle.

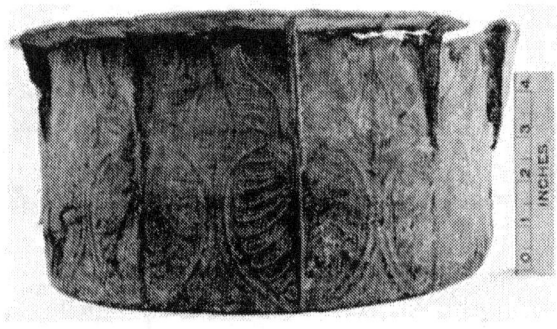

FIG. 34.—Maidstone Museum

tion.   The two lower bands are alike in pattern and differ from the top band.   It is presumably Norman.

The font at Parham (Fig. 24) is the only example unquestionably of the fourteenth century, and stands alone in treatment. There exists not only no other font, but no lead water butt even, which relies, as this does, chiefly on lettering as decoration.   The font is divided vertically and horizontally by long panels, each bearing the legend "H. C. Nazar" (Jesus Nazarenus) in beautiful Lombardic lettering.   The spaces so enclosed are filled with the shield of arms of one Andrew Peverell, who was knight of the shire in 1351 and probably gave the font.

The Tangley font is sparingly decorated in a matter-of-fact way.   Six strips of baluster shape divide the bowl, and the ornaments between are two roses (Fig. 27), three crowned thistles, and three fleurs-de-lys (Fig. 26). With such treatment it is safe to assign the work to early in the seventeenth century.

Slimbridge (Fig. 29) is quite in the cistern manner, with date, initials, and rosettes.

Down Hatherley font (Fig. 28) is very small, but the ornament is ambitious.   Round the bottom there runs a band of Tudor cresting, which might well have been used, and probably was used, to decorate rain-water heads.   The stars are of a type familiar on London cisterns, and the lozenges are of a pleasant formality.

Interesting too, among the late examples, is that of Aston Ingham (Fig. 25). The date 1689 appears on the bowl as do the initials (unpleasant habit) of the givers of the font, W. R. and W. M.   The acanthus leaves are good, which can scarcely be said of the scrappy leafwork below the initials. There are also the inevitable cherubs and rosettes.

For the font which stands on the lawn at Greatham House near Pulborough, Sussex (Fig. 30), little can be said.   It has fallen to the low estate of a flower-pot. It was disestablished some forty years ago, when Greatham Church was restored, and nothing by way of date can be hazarded, for it is a simple unassuming thing and reveals nothing. Rectangular, built up of sheet lead ¼ inch thick, and with little feet at the corners, its only ornaments are small circles on the faces. It has been suggested that this example was never anything more than the lead lining of a stone font. Its rudeness of construction makes this theory a reasonable one, but it seemed on the whole better not to exclude it.

Class E.—Without Decoration.

The font at Penn, Buckinghamshire, has only lately been added to the list of lead fonts (Fig. 31). It is unique in this respect, that it is the only one rounded at the bottom. It altogether lacks decoration, but has been scratched all over with dates and initials, and amongst them is 1625.

How much earlier than 1625 the font was made is a matter of pure conjecture.

The history of the discovery of this font is instructive and has elements of hope The bowl was coated thickly with colour, and had always been supposed to be of stone. The discerning knuckle of the vicar tapping it suggested that it was not stone, and the point of a knife confirmed his suspicion. It may very well be that other lead fonts exist which are masquerading as stone, and, provided that the clerical penknife be gently used, other surgical experiment in the same direction may increase our list.

Font-like Vessels.

There remain the vessels that have sometimes been described as fonts, the use of which, however, seems doubtful.

The lead vessel in the Gloucester Municipal Museum (Fig. 32), though given in Mr Lethaby's list as a font, must be abandoned to some other use. It was found at the old Woodchester Church in Gloucestershire. It is formed of four panels 7½ inches square attached to a circular base, which probably is a later addition. The facts militating against its being a font are :—

1. It has no markings on the edge where hinges or locks might have been attached.
2. It is much smaller than any known example, and
3. The decoration is unusual for a font.

It might, of course, have been a portable font; but if so it probably would have had handles. It weighs 20 lbs. 3½ oz. Alternative suggestions are, that it was a stoup or a reliquary or a lavabo. For its own sake it deserves illustration. The modelling is of an exquisite delicacy. The scene, framed in a border of trailing vine leaves, is the Deposition from the Cross. The dead Christ is on the knees of the Blessed Virgin, and His head and feet are supported by two kneeling figures probably representing St John and St Mary Magdalen. Above the figures and set round the cross itself are the scourge, the crown of thorns, the sponge-bearing rod, the cock of Peter's denial, and other emblems of the Passion. Notable, too, are little busts of Herod and of the High Priest, both of villainous mien.   Herod is crowned, and Caiaphas wears a mitre and a spiky beard.

With regard to the vessel at Lewes Castle (Fig. 33), it is probably Anglo-Saxon. The evidence of its use as a font is slender, in fact confined to the existence of a cross in the triangle of ornament.   There are the remains of iron handles; which seem to show that it was not an ossuary, a reliquary, or a stoup.   It may have been a salt-cellar, but its use must remain conjectural.

Another vessel at Maidstone Museum was dredged from the Medway some years ago.   It is rather damaged, and it also had iron handles.   The decoration is mystifying. It has a classical feeling, and might be Romano-British.   At such a date, however, the river was the font, as objection was taken to still water for baptism.   To the early Christians running streams were as the rivers of living water.   In any case for so early a date the font would be too small.   If it is to be saved as a font, a later date must be assigned.   Perhaps it is of early Norman date, but it is an altogether vague and dubious object.   There remains the chance of its being post-Reformation (an anti-climax after talk of Romano-British).

Some years ago Mr Roach Smith described a lead vessel found at Felixstowe which he thought belonged to the tenth century.   It had lost its rim, but seems to have retained some traces of two or three flanges.   It was 6 inches high, 31 inches in circumference, and had an iron handle.   There were four ornaments on the outside, each being a stiff-stalked plant with leaves and flowers at its base, and also two branches, each like the central stem, ending in three leaves.

The majority of stone fonts were lined with lead, and it is reasonable to assume that some such linings were decoratively treated as has been done by Mr Bankart on the inside of some modern lead fonts which are illustrated in a later chapter.   None seems, however, to have been recorded.

On the outside of a discarded stone font preserved in the church of Waldron, Sussex, there is an incision of about 8 inches in length.   In the upper part of this are small holes which may have served to secure a lead inscription, such as is found in some mediæval tombstones, and as remains of lead were found inside the basin, this theory is probably correct.

It has been stated that the font at Chobham, Surrey, is of lead with wooden panels. It can only be described as of lead in the same way that any lead-lined wood font would be.   The bowl is entirely cased in, and it is impossible to say whether the outside of the lead is decorated.   For this reason it has been excluded from the list.

In the writing of this chapter the author has to express his great debt to Dr Alfred Fryer, F.S.A.   Without his help, both in counsel and in illustration, it would have been very incompletely done.   The least that can be done is to make clear (it is common knowledge to those whose hands are grimy with the dust of archæological "Proceedings") that Dr Fryer's excursions into the history of fonts in general are typical of all that is best in the study of our national antiquities.

CHAPTER II

RAIN-WATER PIPE-HEADS

Early Uses of Down-pipes—Hampton Court—Windsor Castle—Haddon Hall—Knole Park—Dome Alley, Winchester—Hatfield—Guildford—St John's, Oxford—The Character of the Early Work.

THE design and treatment of rain-water heads may be divided roughly into two historical periods, one extending from the earliest examples of the middle of the sixteenth century until about 1650, and the other including the work of the second half of the seventeenth and the first half of the eighteenth centuries. After 1750 there is nothing of much interest except a few local schools, as, for example, those of Aberdeen and of Shropshire. In these and other scattered centres, the craft, instead of dying down into simple dulness, sometimes borrowed conventions from other sources, such as plasterwork, and produced examples which often lack a sense of material, but are not without decorative charm.

The first period (with which this chapter deals) began before the Renaissance touched the plumber's art. It continued until the new ideas were established, and may fairly be called the Augustan age of English leadwork. During the thirteenth and fourteenth centuries the English craftsman in lead had to some extent lost the pre-eminence which the lead fonts of the twelfth century had won for him. We can show nothing to compare with the delicate crockets and leafwork of French mediæval roofs, which Burges so faithfully recorded. When, however, stone gargoyles were abandoned for external lead down-pipes and heads, the English plumber came into his own again, and at a time when his ideas of design were markedly fluid.

Plumbers were conservative craftsmen, a reputation which they enjoy to-day. It is constantly found that leadwork, judged by design and treatment, is fifty years or more behind the stone carving and plasterwork contemporary with it.

The reason for this is, doubtless, that no foreign leadworkers were imported with Torrigiano, or with the German craftsmen who followed when the Italians fell into evil political odour. Even had they come, they would have brought no tradition to disturb the English treatment which had held sway since the thirteenth century. The Gothic tradition, which persisted so long in the shells of buildings, and was discarded for Renaissance treatment at first only in such details as stone carving, continued long in the details of leadwork.

The foreign leadworker's art and fancy rioted in crestings and finials, but pipes and pipe-heads seem to have left him cold. It is characteristic of the practical genius of English building that the external down-pipe is a distinctively English method of disposing of rain water. The only interesting foreign rain-water head known to the author is from a sketch of a Belgian example. It might be of the seventeenth century. Here the design is influenced by the grotesque gargoyle, which was sometimes, even in mediæval work, made entirely in lead instead of, as usually, in stone. In Italy there are no rain-water pipes except modern iron ones of the worst type. Though the Romans were often careful to conduct the rain water falling on roofs to the ground by pipes instead of shooting it off by projecting spouts, there is no evidence that these pipes were other than of stone or terra-cotta. They used lead freely for service pipes, but apparently not for rain-water pipes. Viollet-le-Duc, under "Conduite," says that in the fourteenth century lead rain-water pipes were in use in England, but nowhere else, and sketches a most unconvincing lead head and length of square pipe. He unfortunately does not suggest where the head is to be found, and there is in England nothing so early by two centuries. It has been said

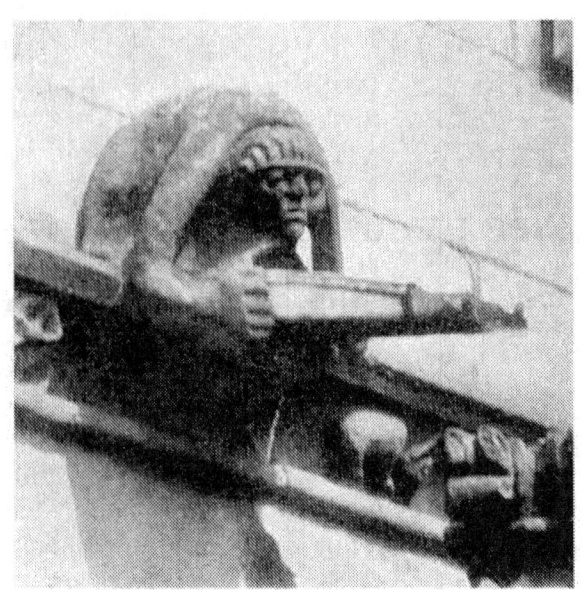

FIG. 35.—Gresford Church.

that fragments of pierced work in Gothic patterns, found at Fountains Abbey, formed parts of pipe-heads; but the fragments in question seem rather to be parts of lead-ventilating quarries. There is, however, an earlier reference than Viollet-le-Duc to English rain-water pipes. Henry III. in 1241 (see the Liberate Roll) writes to the Keeper of the Works at the Tower of London: "We command you to . . . cause all the leaden gutters of the great tower through which rain water should fall from the summit of the same tower to be carried down to the ground, so that the wall of the said tower, which has been newly whitewashed, may be in no wise injured by the dropping of rain water nor be easily weakened."

The use of lead down-pipes grew probably rather from a desire to save water for domestic use than to avoid the splashing down on the wayfarer's head of the discharge from projecting spouts. The use of porous building stone, liable to erosion through the water being

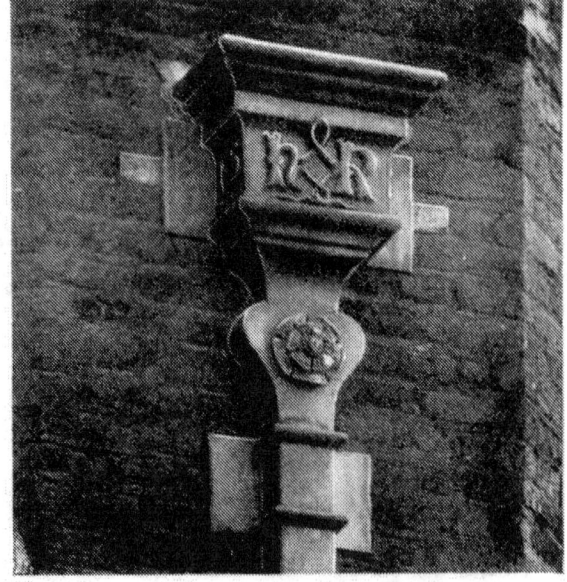

FIG. 36.—Hampton Court.

blown against the walls in its fall, would tend to the same end. Viollet-le-Duc shows a lead pipe of the thirteenth century in a vertical stone chase, sufficiently set in to allow of thin pieces of stone coming in front of the pipe in alternate courses of the masonry.

The fixing of the pipe on the face of the wall is apparently a later development, due to the greater simplicity of the method and the recognition of its decorative possibilities.

Where down-pipes were not used, the lead covering the roof gutters was often dressed through the opening in the parapet, lined the channel of the gargoyle, and extended beyond it, as on Gresford Church (Fig. 35). In other cases, as at Uffington Church, the gargoyle was a long lead channel supported on an iron stay (illustrated in Twopeny's drawings of "English Metalwork").

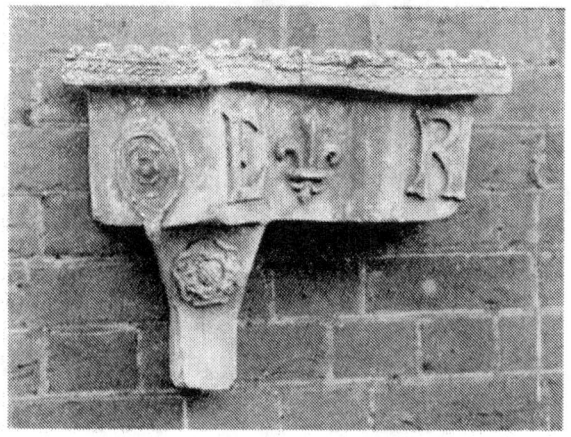

At Hardwick the lead gargoyles are bulged, slit, and twisted to the form of an Elizabethan puffed sleeve.

At Lincoln Cathedral is a great parapet gutter, illustrated in Chapter V.

FIG. 37.—Windsor Castle.

On the Mayor's Parlour, Derby, there is a curious nicked and curled lead gutter, with short round tapering spouts hanging from it at intervals. These spouts discharge

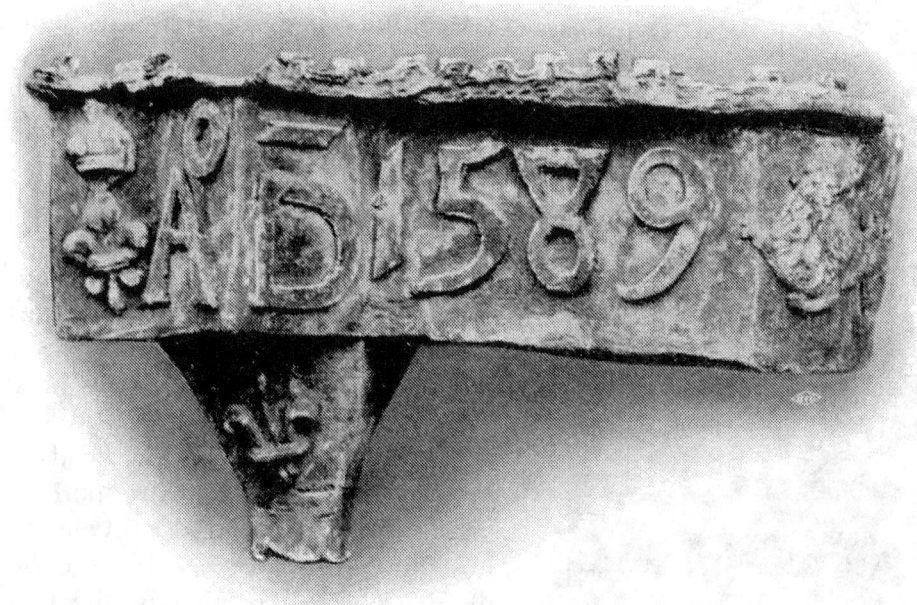

FIG. 38.—Windsor Castle.

the water clear of the face of the building. This house is probably of the last quarter of the fifteenth century, and the little spouts are interesting as being embryonic down-pipes.

Both Mr Reginald Blomfield and Mr Starkie Gardner, when writing of leadwork, refer to the head at Hampton Court Palace (Fig. 36), which bears the initials "H. R.,"

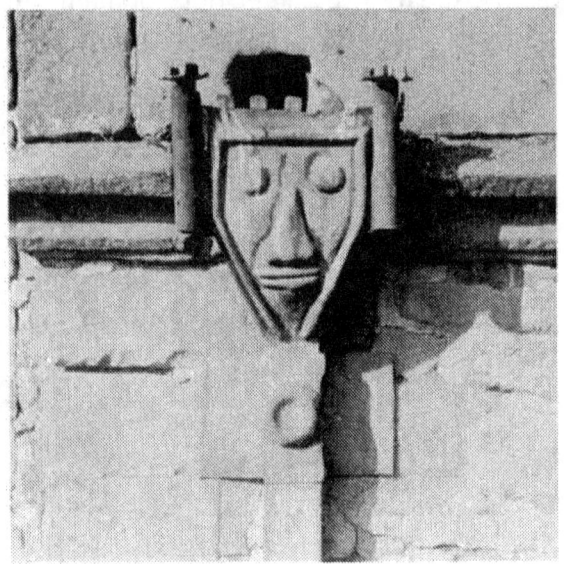

FIG. 39.—Haddon Hall.

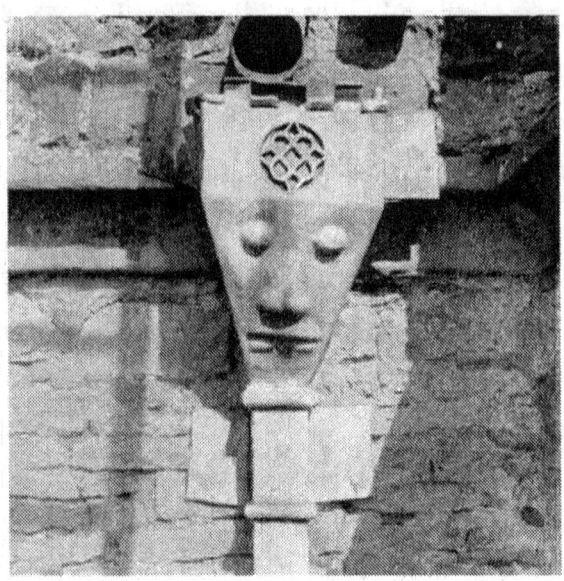

FIG. 40.—Haddon Hall.

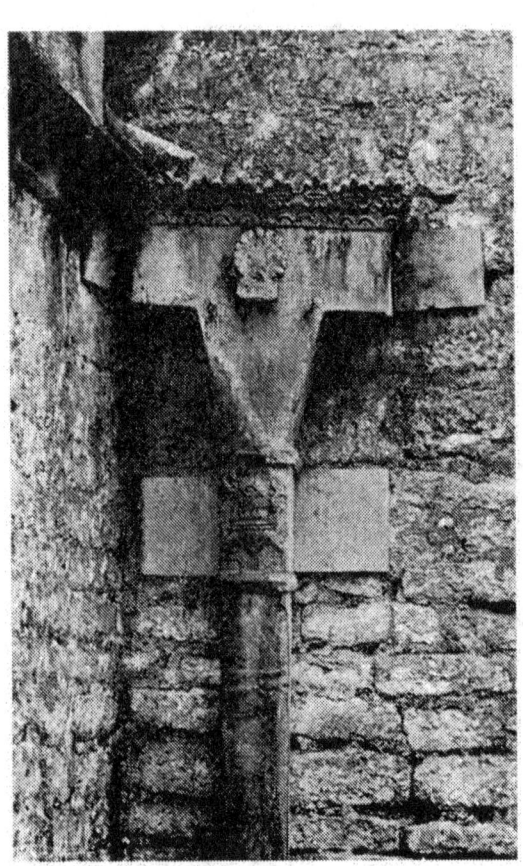

FIG. 41.—Haddon Hall.

and the date 1525, as being probably the earliest remaining, and with such authorities one does not lightly disagree. Examination, however, proves that so far from being of the sixteenth it is certainly of the nineteenth century. It is fresh looking, and the arrises are sharp. The resident surveyor, Mr Chart, to whom these suspicions were communicated, says that about forty years ago there flourished at Hampton Court a strenuous master plumber who renewed with some ferocity. Doubtless the existing heads are approximately like the originals, but the top mouldings are ugly and suggest the Victorian plumber at his coarsest. There are no authentic early heads with the same mouldings.

Amongst the earliest heads are two at Windsor Castle, which are purely in the old manner (Figs. 37 and 38). One is dated 1589 in bold figures, and both were originally on the Elizabethan portion of the Castle on the north front, now part of the Royal Library. They were taken down in February 1904, repaired, and photographed. The lion prances in vigorous mediæval style, and is a very blithe piece of modelling. All the letters, ornaments, and cresting are applied. The plan of the heads is curiously irregular and interesting.

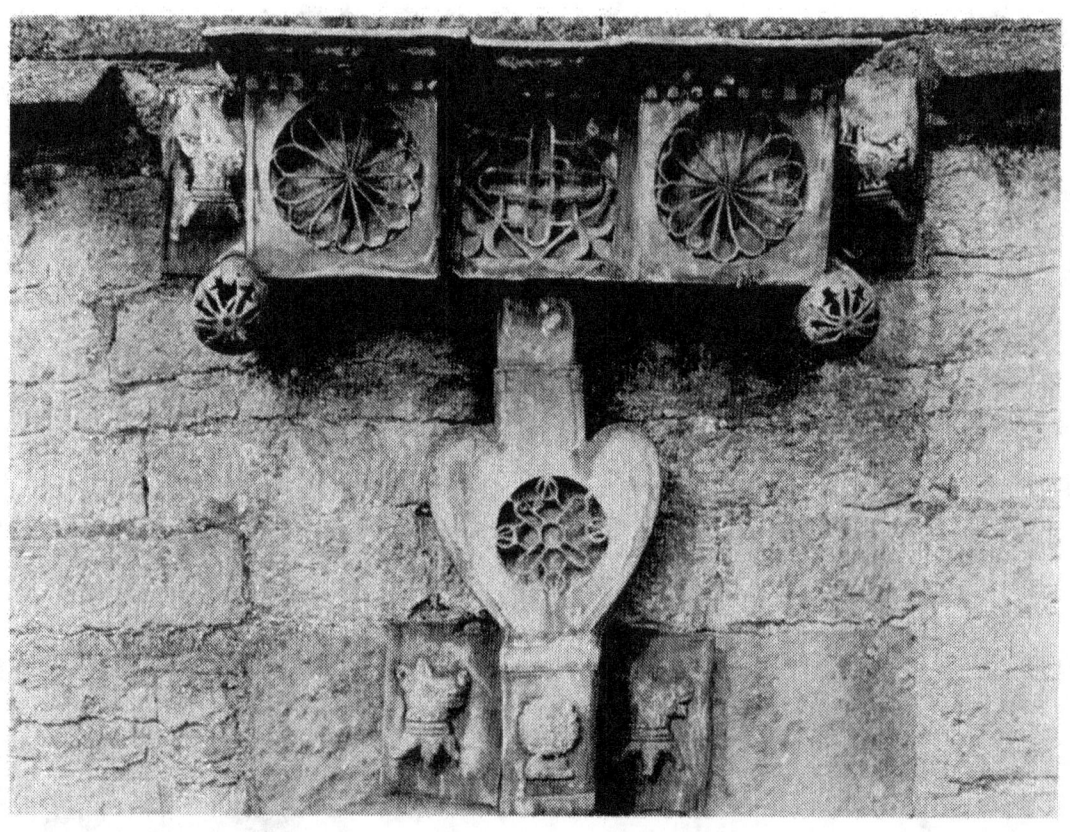

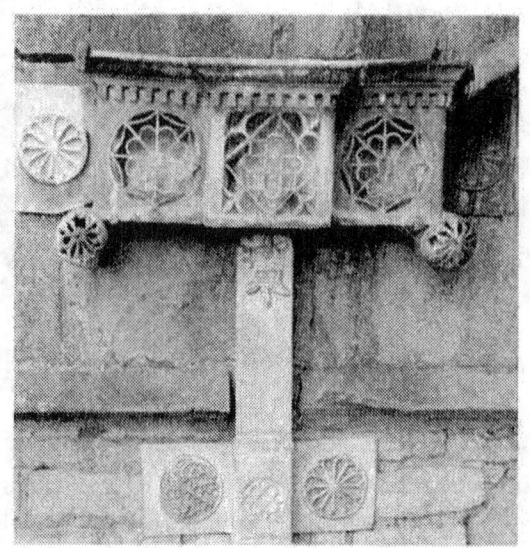

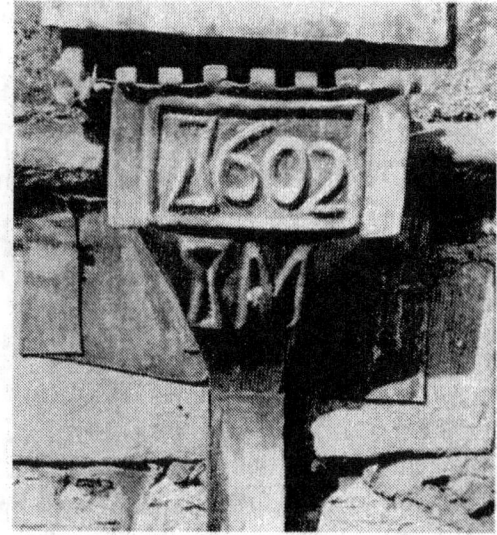

FIGS. 42-44.—PIPE-HEADS, HADDON HALL.

Amongst other early dated heads there is (or was, it may have disappeared recently) one of 1583, at Chard, with simple battlemented cresting and four pendants.   At Burton

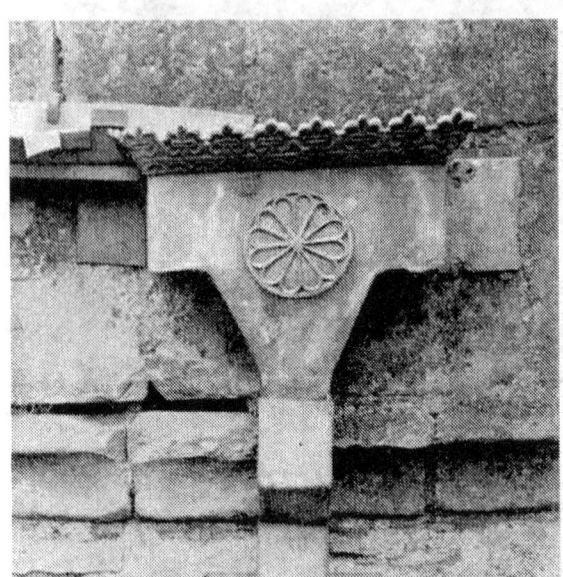

FIG. 45.—Haddon Hall.

Agnes are some fine heads bearing date 1603, and there are simple battlemented examples of 1609 on the east side of the tower at Langley Marish, Bucks, and of 1631 on a gabled house at Swindon.

At Haddon Hall the lead heads are numerous, and like most things there, a liberal education. The continuous building which enables us, as we move from one room to another, to step from one century to another, and to see the development of treatment and feeling, say of wood panelling, in its best expressions, does us the same kindness with the leadwork. The heads range from about 1580 to 1696, and beginning in work of purely Gothic feeling run on to the stiff vase-shaped heads which are the common form of the eigh-

teenth century. The later heads are illustrated in the next chapter.   Among the earlier ones some are direct descendants of the stone gargoyles. Indeed, the gargoyles have been disestablished in their favour. The lead spouts from the stone figures which originally discharged clear of the building were shortened, and now discharge into pipe-heads. In two cases the craftsman manifestly has been influenced by the gargoyle idea, and has fashioned the front of the heads as more or less human faces, one of a settled melancholy (Fig. 40), the other expressing a slightly humorous dissatisfaction (Fig. 39). They are altogether a pretty jest in lead, and save for the two laughing masks, prophetic of Dr Johnson, on an example of 1699 at Durham Castle, there are few heads which are frankly amusing.

The spirit of the mediævalist was evidently abroad when they were conceived (about 1600). We have here a grim pleasantry very different from the polite wit which suggested the arabesque masks of a few years later (see Fig. 84). In Fig. 45 is shown a head on the Great Hall, Lower Court. A long embattled gutter discharges into

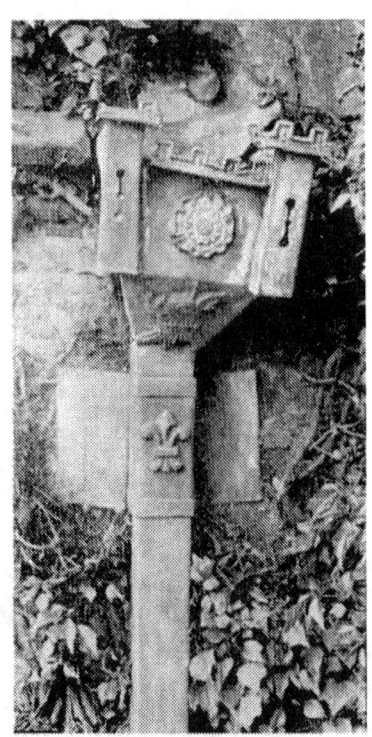

FIG. 46.—Haddon Hall.

one end. The head has a fleur-de-lys cresting and a tracery disc on the front, but no trace of Renaissance treatment. Dr Charles Cox, in a paper on Derbyshire Plumbery, has illustrated a head similar to that of Fig. 45, but without a gutter, and with a circular disc of a rather richer tracery than the simple wheel pattern of Fig. 45. He dates it as probably of the first half of the sixteenth century, possibly of the time of Sir Henry Vernon, who died in 1515. The total absence of Renaissance feeling makes this theory plausible, and if it can be maintained the head is the earliest extant. But one may be sceptical. The Eyam Hall heads have a very similar fleur-de-lys cresting, but one is dated 1676. This is cited as showing that the quite Gothic treatment does not necessarily indicate early work.

Mr Lethaby figures in his book a head the same as this example, but he shows no gutter with it. Moreover, the top pipe socket bears, in his sketch, the Vernon boar's head erased, whereas the only existing head which has the boar's head on the top socket has a peacock displayed instead of a tracery disc on the front (Fig. 41). If the Manners' peacock is indigenous to the head on which it is now fixed, it dates the heads somewhere probably not earlier than 1577, when Sir John Manners went to live at Haddon on the death of his father-in-law, certainly not earlier than 1567, when he married Dorothy Vernon, and so demolishes the idea of a head of 1515. Probably a safe date is 1580.

If the page is here somewhat overcharged with names and dates, it is by way of illustrating the slow impact of the new ideas and the permanence of the Gothic spirit.

The finest heads at Haddon Hall are unquestionably those on the north side of the Lower

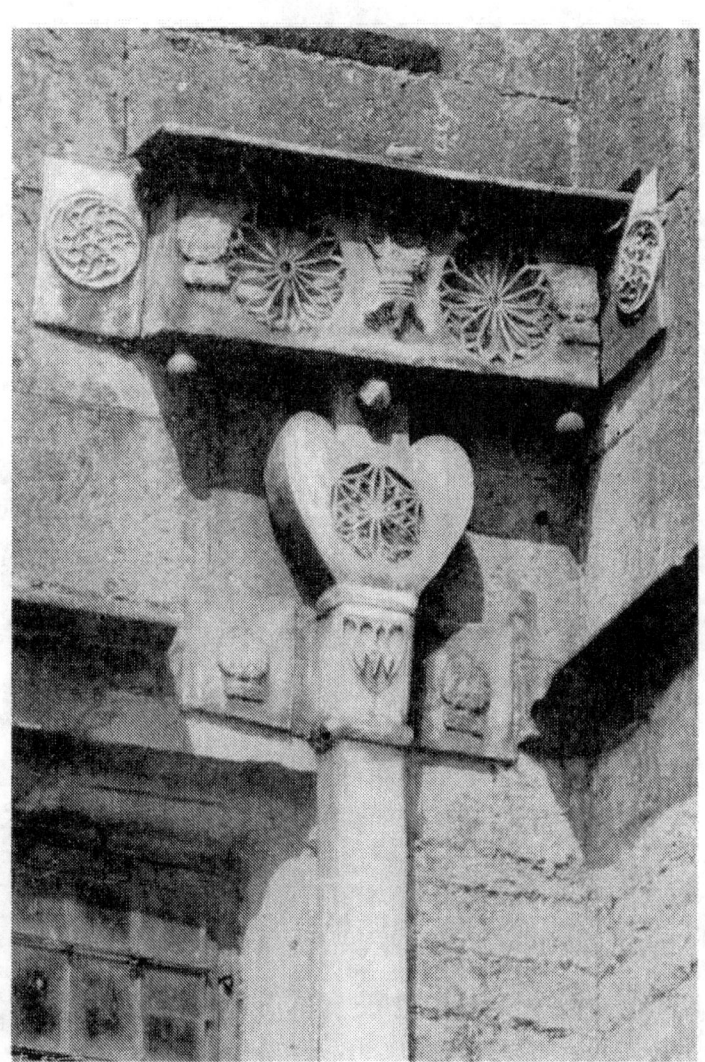

FIG. 47.—Haddon Hall.

Court (Figs. 42 and 47). A delightful feature is formed by outer fronts of pierced tracery, which produce lights and shadows of amazing grace. This tracery, and the delicate cornice with dentils, form one of the happiest possible combinations of the traditional Gothic with the new ideas. The effect is sumptuous, and we can scarcely find an example in the minor arts where the overlapping of the styles leaves a result so harmonious. The mediæval tradition was dying, but, like Nature in autumn, was beautiful even in death. The new style was finding its way somewhat uncertainly, but with all the riotous delight of the child playing a new game. If some of the new forms were curious and hybrid, all had the fascination of experiment and the vigour of youth.

Turning to Fig. 47, the three pendant knobs, the middle one polygonal while the outer ones are round, are a pleasant relief to the line of the underside of the bowl. The head of Fig. 42 is similar, save for the pierced cylinders which appear to carry it. These deserve a word. It has been suggested that they carry the heads. They are simply thin, hollow cylinders, and could only support the heads if they were the casings of oak plugs, of which there is no evidence. They

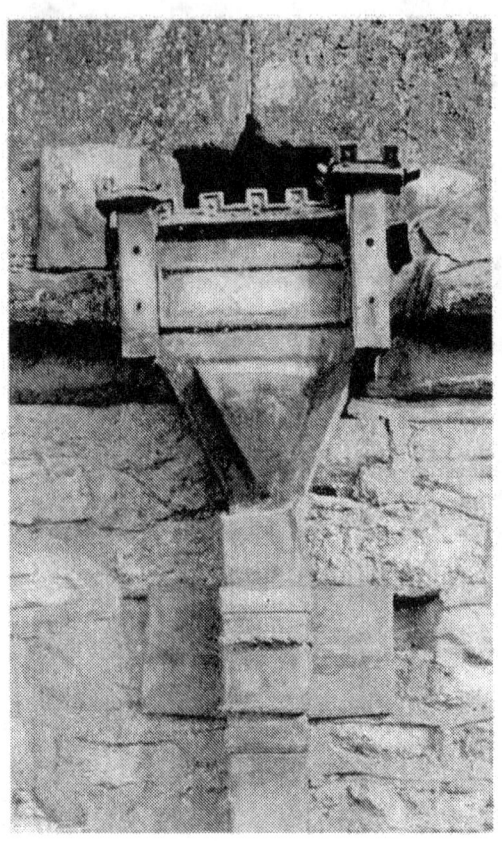

Fig. 48.—Haddon Hall.

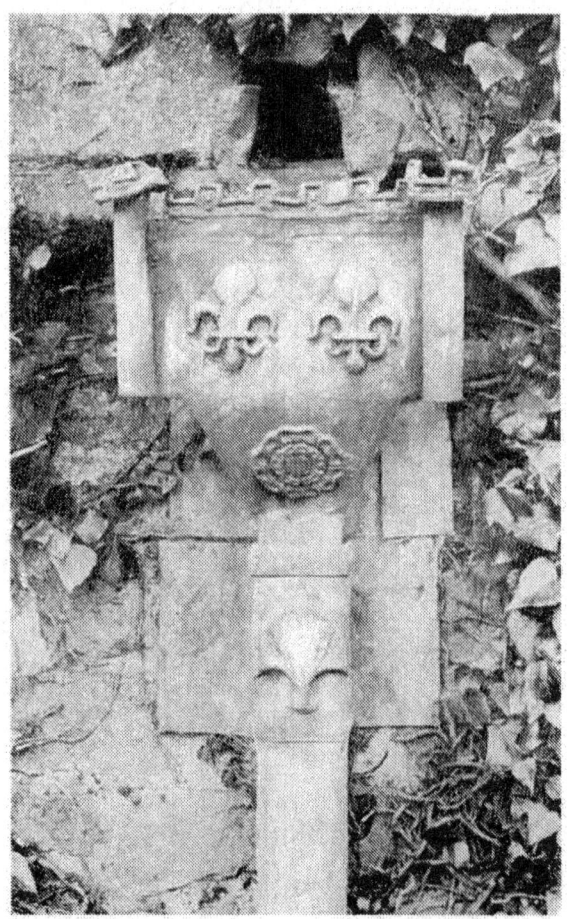

Fig. 49.—Haddon Hall.

are wiped on to the heads. The actual supports, where there are any other than nails, are plain iron staples driven under the heads. The theory of oak plugs seemed so plausible, and indeed so practical, that the heads at Bolton Hall, which have similar cylinders, when taken down at the recent rebuilding, were examined to ascertain if there was any sign of plugs, but there was none. As similar cylinders occur at Coventry, and these have no plugs, they may be taken to be purely ornamental. Moreover, if these cylinders had a constructive significance, they would scarcely have been omitted from the head of Fig. 47 if they were needful for that of Fig. 42. The example of Fig. 43 is interesting by reason of the heart-shaped funnel being omitted.

Still less touched by the rising manner, but of a graver kind, is the castellated head decorated with fleurs-de-lys of Fig. 49, which is probably of the same date as that of Fig. 44. The latter is fixed in the Upper Court, and the initials are those of Sir John Manners, whose elopement with Dorothy Vernon goes far to support our claim to be a romantic people.

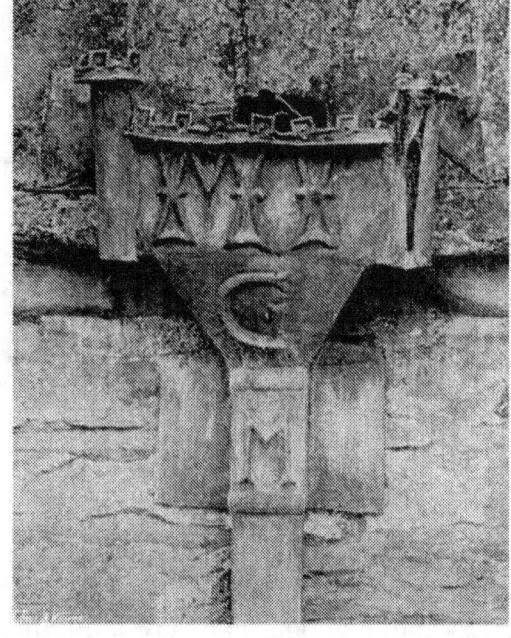

FIG. 50.—Haddon Hall.

The heads of Figs. 46 and 48, though on the same general lines of mimic castles, have each that touch of difference which gives a lively interest.

The example of Fig. 50 is a little baffling in its lettering M.I.G. M.I. probably stands for Sir John Manners, and the G. beneath for Grace or George. Grace, the eldest daughter of Sir Henry Pierpoint, married Sir John's eldest son, Sir George, on 2nd April 1594.

Not only the heads, but the pipe sockets show a wealth of care and invention. One is shown in Fig. 51, the shield bearing the arms of the Pembrugge family, *a barry of six*. Clearly the Haddon plumbers were historically minded, for it was about the middle of the fourteenth century that a Vernon married a Pembrugge.

Some are decorated with discs of tracery (Fig. 53), and the Vernon's boar's head alternates with shields of arms, interlaced diamonds, fleurs-de-lys, and even with the heart ornament of Fig. 54, which will gladden the (happily now discredited) disciples of *L'Art Nouveau*.

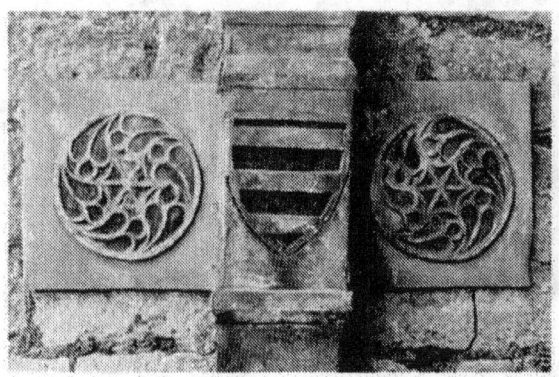 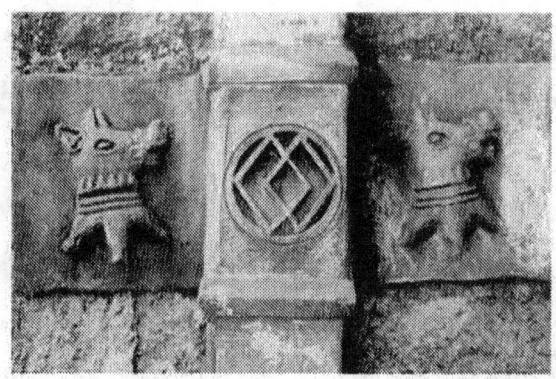

FIGS. 51 and 52.—Pipe Sockets, Haddon Hall.

In the case of some sockets the tracery disc is separate, and the nail goes both through it and the plain ear into the wall. In other cases a piece has been cut out of the plain ears and the disc soldered on from the back. In others, where new ears were necessary, the tracery discs, instead of being cast perforated, were cast with a solid back, and this heavier casting was then fastened to the new ears. The pattern for this heavier casting was probably an original disc mounted on the original plain ear, the mounting piece being trimmed round to the outline of the disc.

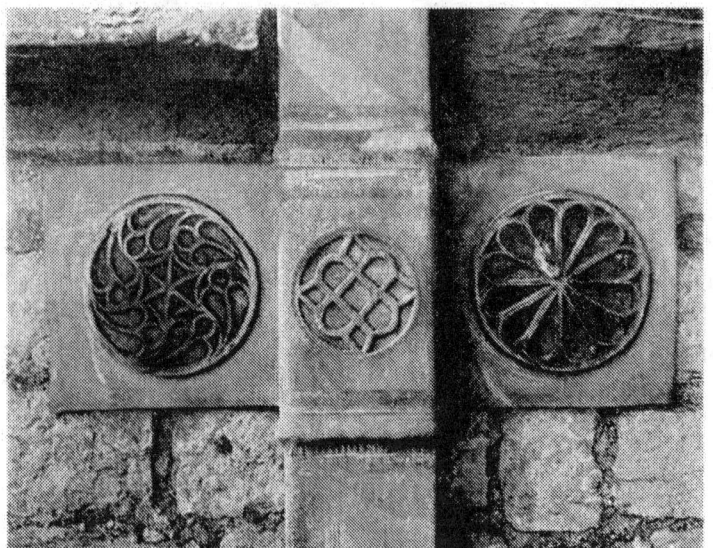

However splendid the work at Knole and Hatfield, there is a quality about the earlier heads at Haddon Hall which stirs us to positive affection. There is a wealth of pure invention, a sense of material so just, a humour so spontaneous yet gently sardonic, an historic revelling in the coats-of-arms of forgotten heiresses that must move us to amazement. Truly these seventeenth-century plumbers were Admirable Crich-tons in their craft.

Three later examples from Haddon are illustrated in the next chapter.

FIG. 53.—Haddon Hall.

While Haddon Hall provides the finest group of heads regarded as an historical series, Knole Park, Sevenoaks, certainly gives us the finest series of heads of one period. Dating from 1604-1607 there are forty-seven in all, including some thirty different types. These heads not only touch the highest point of decorative charm, but from their wealth of treatment reach the limit of dexterous craftsmanship. So excellent is the workmanship, that in spite of the delicacy of much of the detail and the great number of parts of which each head is made up, most of them are to-day in very fair condition. The examples here illustrated show the complete control of the man over his material, and his

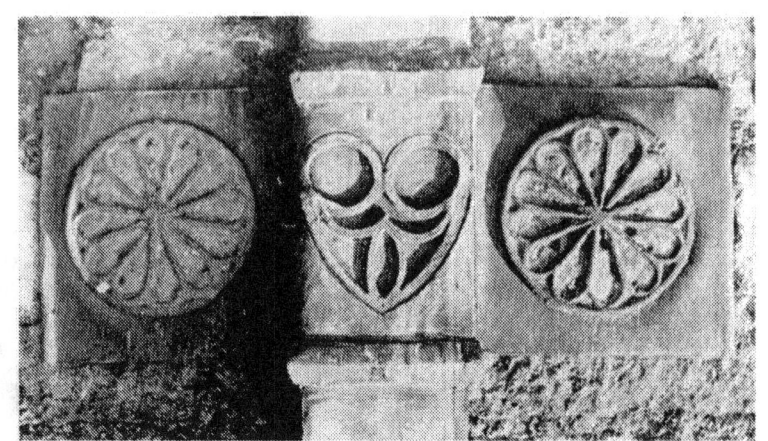

FIG. 54.—Haddon Hall.

vigorous facility when dealing either with broad and simple, or with delicate and almost feminine treatment. The lacework effect of the head in Fig. 55 is of happiest possible contrast with the masculine grip of the example in Fig. 57, with its chequers and chevrons outlined in bright tinning. In the photograph of the former there is a certain harshness due to white paper having been put into the pierced turrets, when they were photographed, but without it the delicate network would not have had full justice. It will be noted, too, how in the plainer pattern the strength of the simple lines of the design are lightened by the little embattled cresting and cable moulding, a detail much beloved in the early seventeenth century and always successful.

However richly decorated the work of this period it is always restrained, never insistent. Pierced work like lace applied flat, flat pierced panels forming false fronts and throwing sharp shadows, pierced turrets, pierced pendants finishing in polygonal balls, solid turrets innumerable, chequers, chevrons, **8**'s and strapwork in bright tinning

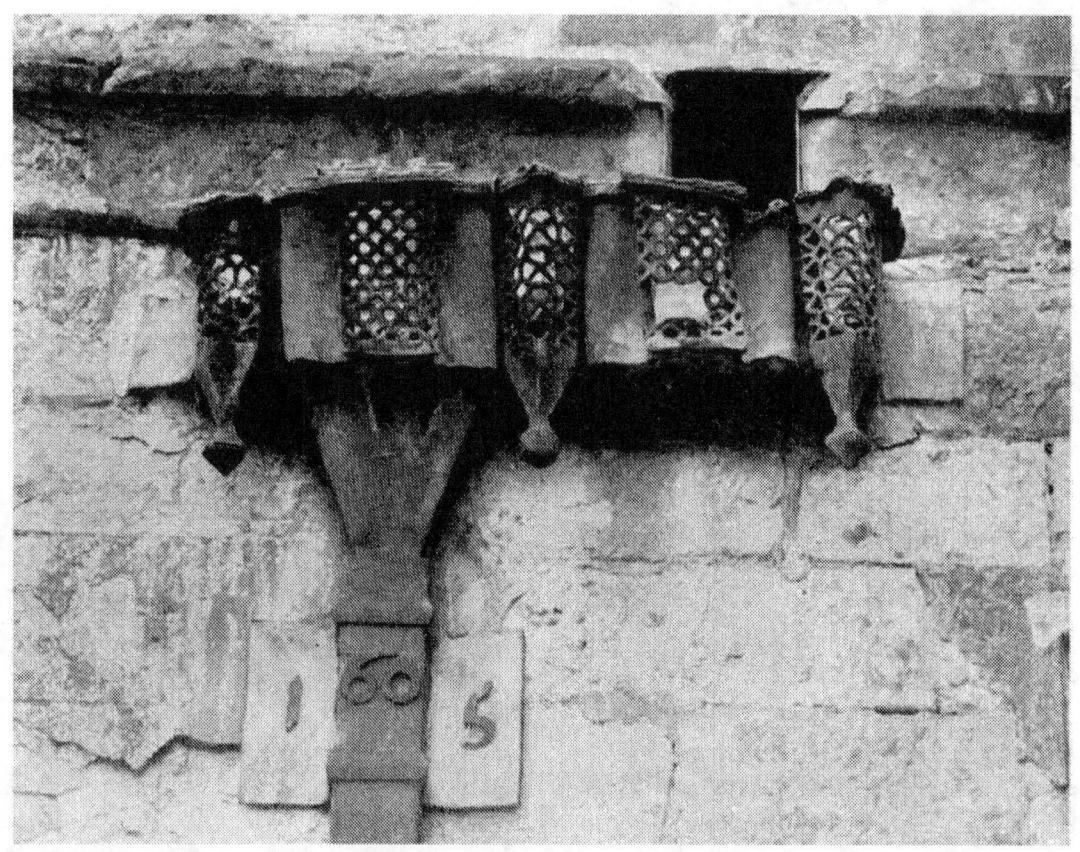

Photo. Essenhigh Corke.

FIG. 55.—Knole.

plans irregular or balanced, all go to make up a variety of treatment that indicates the apogee of the leadworker's art. The detached pierced work is perhaps the most effective by reason of the bright spots of light, which alternate with sharp shadows and touch the whitening lead to silver.

On the south front at Knole two heads have pierced and twisted terminals which match the characteristic early Jacobean stone finials (Fig. 56). They bear, as do many others, the initials, arms, and crest of Thomas Sackville, Earl of Dorset, who enlarged and beautified Knole. Another on the south front has incised bands and straps, which were probably filled originally with black or coloured mastic. The east front has eight heads, all small and of one type, but each with some difference in treatment. The Water Court has several, one particularly noticeable for its engaging plan, its great length, and the outlet at the extreme left. The Stone Court and Green Court heads are large and rich. One bears pentacles, said to be significant of Thomas Sackville's masonic interests. This is problematical; the pentacle is probably there simply as a pleasant geometrical ornament very suitable for tinning.

When we go from the series of courts to the entrance front we find no heads or down-pipes. The water is projected by plain long gargoyles to the ground, indi-

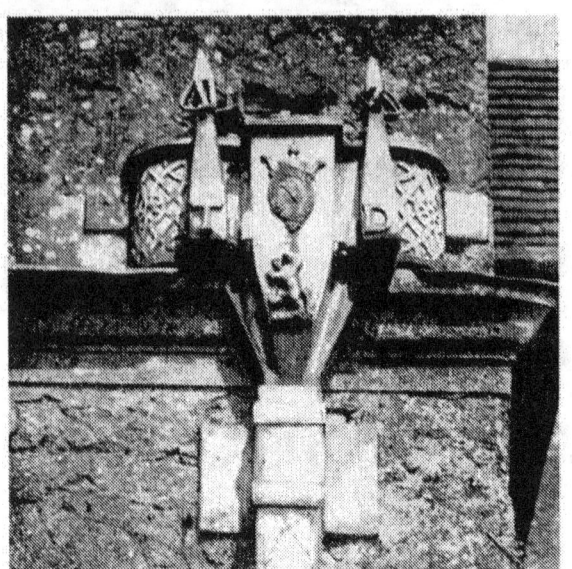

FIG. 56.—Knole.

cating that while the necessary pipes were treated as richly as could be, when pipes were not essential to convenience and habitability, the builder dispensed with them altogether.

The date of early lead heads is not always so clear as at Windsor. Mediæval feeling died hard in leadwork. Not only did the spirit of the Renaissance work in spasms, but it was so local in its incidence that the dating of sixteenth and seventeenth century work is a perilous enterprise, and "about" a word of Mesopotamian blessedness. "About" 1580, then, we may place the engaging gutters and heads at Winchester in Dome Alley. Fig. 58 shows the delightful arrangement whereby the water issues from the valley of the roof under a decorated lead apron into the long vine pattern gutter, and is discharged into the side of a frankly funnel-shaped head, and

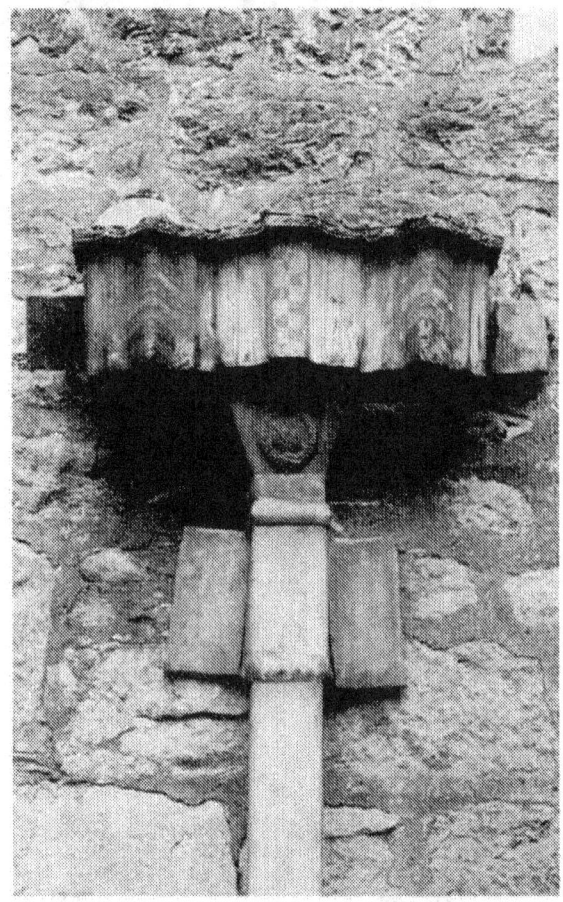

FIG. 57.—Knole.

so through a down-pipe reaches the ground. The traditional manner still holds sway here. The Tudor rose and the leaves, strewn over the surface in a pleasantly casual fashion, are richly and happily modelled. The pomegranates which decorate the pipe

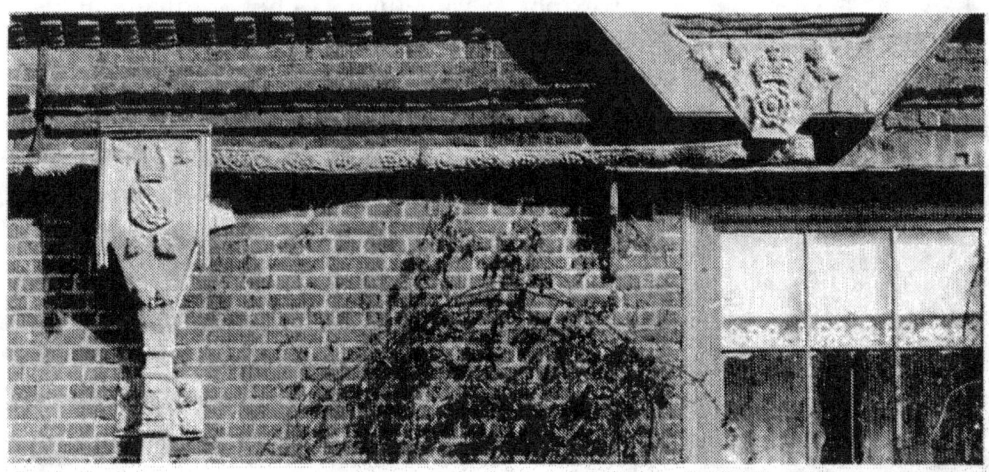

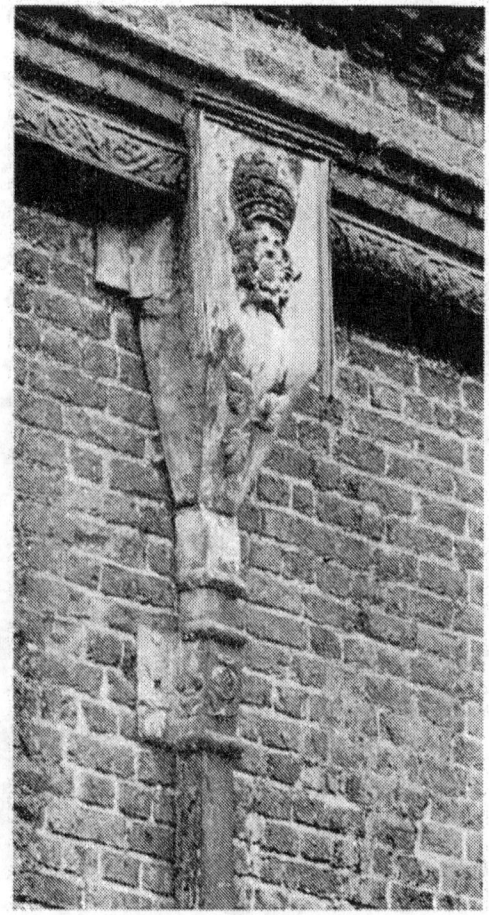
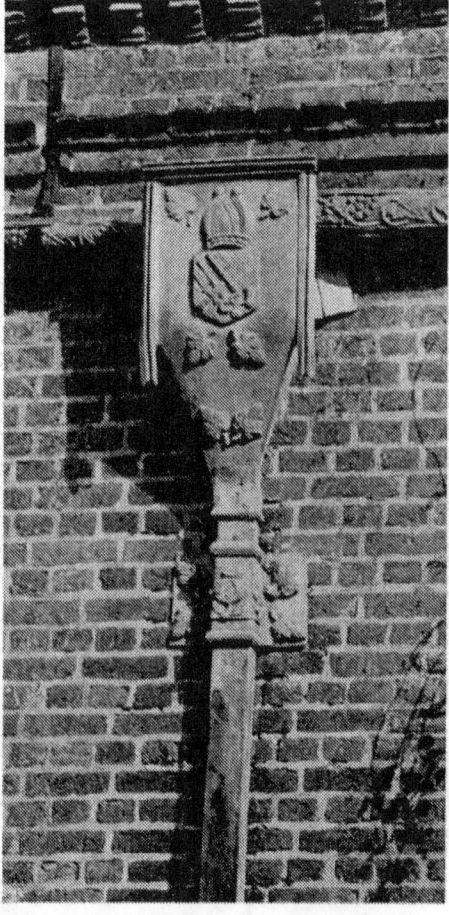

FIGS. 58-60.—PIPE-HEADS, GUTTER, AND APRON, DOME ALLEY, WINCHESTER.

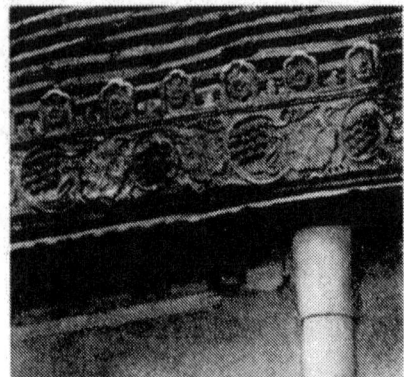

FIG. 61.—Gutter, Coventry.

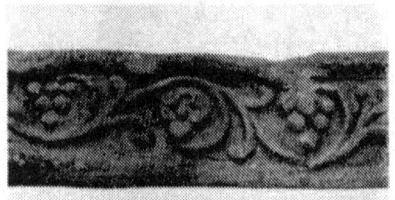

FIG. 62.—Gutter, Bramhall.

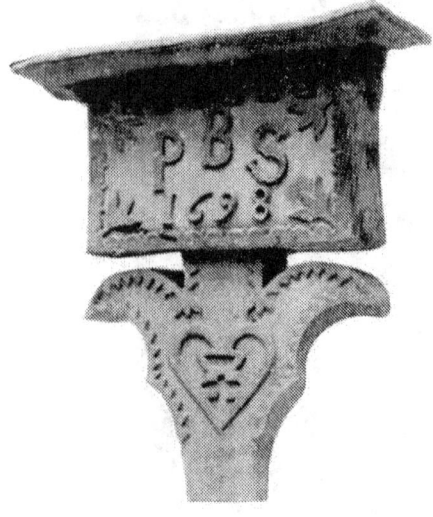

FIG. 63.—Pipe head, Bramhall.

sockets perhaps have an ecclesiastical significance, unless they are taken as representative of Catherine of Aragon or Queen Mary. The buildings of Dome Alley are probably Elizabethan. The original gables were later cut down to their present form. There is nothing in the treatment of the heraldic charges to contradict the idea that the leadwork is of Queen Mary's reign, as has been claimed by Mr Aymer Vallance, F.S.A., but it is more likely to be later. The triangular aprons are unusual, and seeing that they date probably from the alteration of the gables, it may be that the lead-work is as late as about 1620.

The heads have lost the knobs at the top and curls at the bottom, which Twopeny's drawing, made in 1833, shows. They are 3 feet high, and 16 to 17 inches wide. The gutters are in various lengths, some about 4 feet.

The form of gutter, so universal to-day in the hard sharpness of cast-iron eaves gutter, was rare in early days. The more usual form was the straight parapet type as at Old Palace Yard, Coventry, where the bottom of the gutter rests on the top of the wall. At Dome Alley, however, it is of modern shape, and rests on plain iron brackets.

The Coventry gutter (Fig. 61) has for decoration a singularly fine vine pattern, combining naturalistic treatment of the leaves and tendrils with a conventional composition. It may be attributed to about 1580. A triangular apron similar to that of Winchester occurs at Upton Court, near Reading, and the spouting is dated 1664.

In Mr Lethaby's book is a sketch of lead gutter (Fig. 62), pipe (Fig. 64), and pipe-head (Fig. 63), on a cottage at Bramhall, Cheshire. The cottage has been pulled down, and it was only after much difficulty that the leadwork was found and photographed in a builder's yard. The gutter (a vine pattern of wave outline)

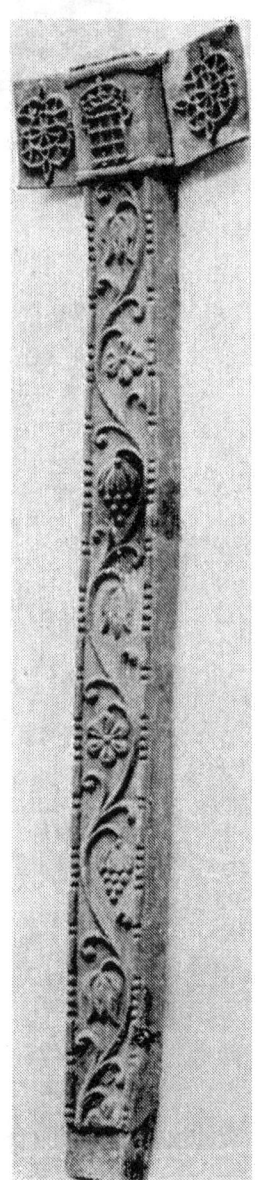

FIG. 64.—Pipe, Bramhall.

and the pipe are particularly beautiful, the head dated 1698 is less noteworthy. It is likely that the pipe and gutter date from about 1600, and that originally the pipe fitted round the gutter outlet without any head being used. As this arrangement would tend to cause overflows, the head was added a century later. The bead and reel ornament on the edges of the pipe is unusual, though it appears on some Anglo-Roman coffins, on an Exeter gutter mentioned below, and on a Durham Castle head of 1699. The vine ornament on the face of the pipe, the socket bearing a crowned portcullis, and the ears covered with a tracery ornament make up the most beautiful pipe in England. The gutter is 9 inches

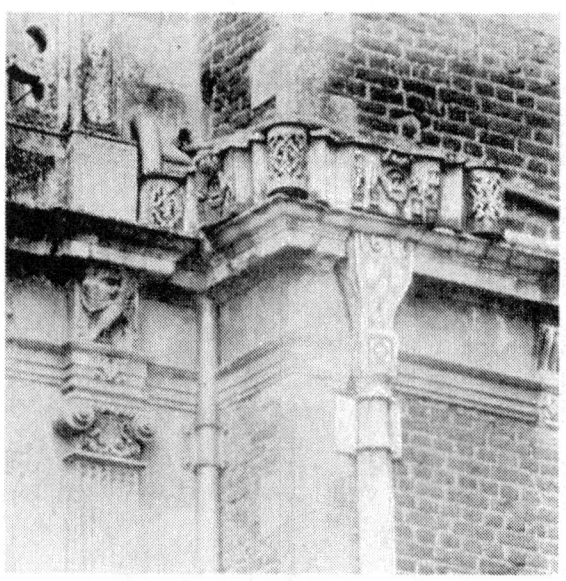

Fig. 65.—Hatfield.

wide by 4 inches deep, the ornamental front being soldered to an L section to form the channel. The pipe is 4 feet 4 inches long, and 4½ by 2¼ inches (external sizes). The ornamental front is a casting soldered to an unornamented channel section to form the pipe.

The head (Fig. 63) has not very much to commend it. The fretty outline of the funnel and the rather meaningless heart ornament suggest the touch of an amateur. It is plainly unworthy of the unique (the word is used advisedly) pipe.

The difference in colour is not due to any legitimate treatment such as tinning or gilding, but to the "picking out" of the pattern in a common welter of oil paint. This head is 22 inches high by 19 inches wide, and its body projects only 4¼ inches.

At Exeter there were on two buildings in North Street, now demolished, fine lead gutters with vine pattern arranged wave fashion, and one had in addition well modelled bead and reel mouldings.

At Leighton Bromswold Church (Fig. 66), a head and two lengths of pipe end

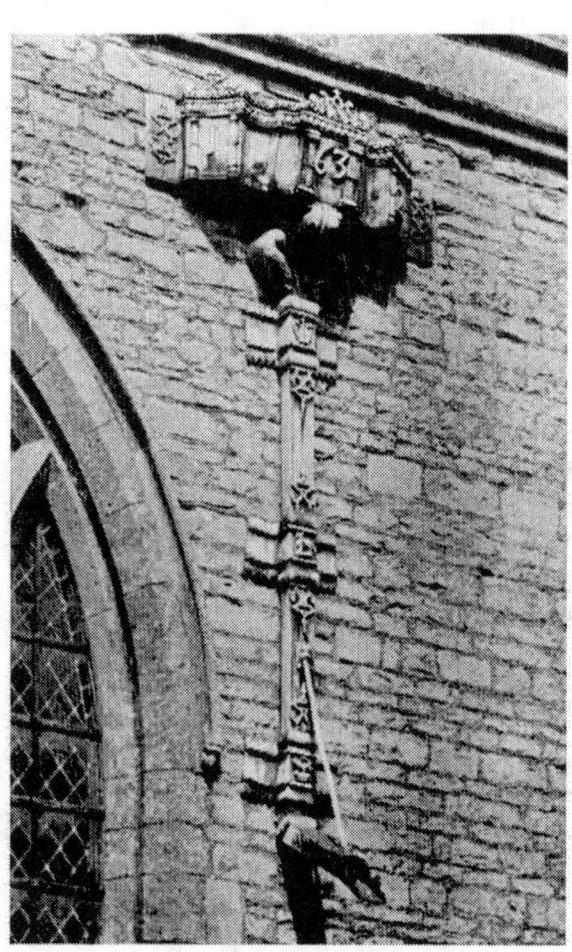

Fig. 66.—Leighton Bromswold.

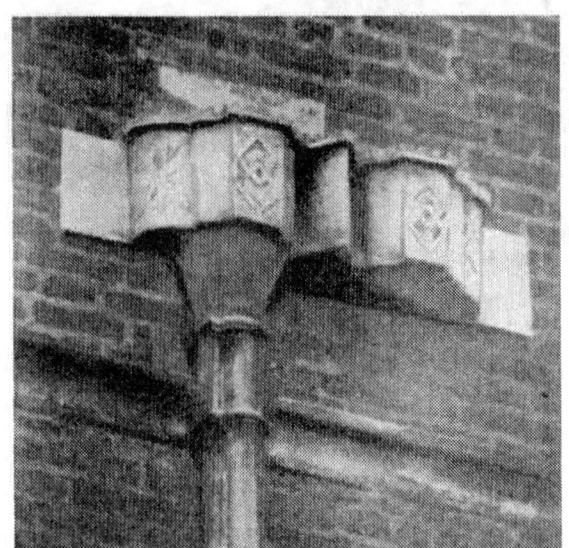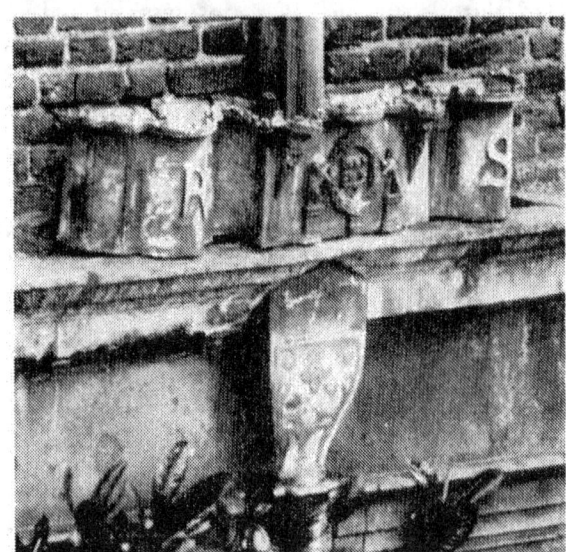

FIGS. 67 and 68.—Hatfield.

unexpectedly in a projecting spout some way from the ground. It is not quite clear why, after using head and pipes, the plumber surrendered the prime use of them by failing to carry the water the whole way in pipes. The projecting spout or shoe is stayed with an iron bar, and the work, apart from its richness and intrinsic value, has a sentimental interest. It is dated 1632, and was fixed on the chancel wall at the restoration done by George Herbert, who was patron of the living. "The Temple" has no poem on "The Church Pipe-Head" to stand by "The Church Porch." It would doubtless have puzzled even the prince of symbolists to have found a spiritual significance in a spout, but the memory of Noah might have provoked his muse.

Great as is the variety in the design and treatment of pipe-heads, it is not surprising, for the positions of gutters and pipes demand irregular, sometimes even bizarre, shapes.

Heads are, in fact, either glorified gutters or glorified funnels; in neither case does water stand in them, they serve simply to direct it to its down-pipe. Irregularity in plan and section is, therefore, no practical disadvantage.

At Hatfield House there is a fine series of heads ranging from 1610. Several are very large, and two of the largest fit round angles of the building and rest on the stone cornice, which is pierced vertically to take the funnel outlet (Fig. 65). They bear the Cecil coat with supporters. On the angles are pierced circular turrets, and an embattled cresting with cable moulding runs round the top edge.

An interesting Jacobean pattern is traced in bright tinning on the front of the funnel. The pipes are rectangular, 5 by 3 inches, with a semicircular bow on the front face projecting $1\frac{1}{2}$ inches.

Some of the heads have simple chevrons and interlaced diamonds (Fig. 67) in bright tinning. They are so like the Knole heads in small details as to tempt the belief that the master plumber who finished working at Knole for the Earl of Dorset about 1608 went on to Hatfield to do the work there in 1610.

R. S. on the head of Fig. 68 is, of course, for Robert Syssil, a spelling which has not survived to support the pronunciation.

At Abbot's Hospital, Guildford, is a series of fourteen pipe-heads and pipes dated from 1627 to 1629. The departure from the early manner becomes here more marked, and frankly classical pilasters appear on the fronts of some of the heads. Two on the High Street front are very elaborate and fit into the corners. One bears the initials G. A., the date and the arms of George Abbot, Archbishop of Canterbury, the founder of the charity. The delicate brattishing on the top is a delightful feature (Fig. 72).

The modelling of the flower ornaments on its fellow (Fig. 69) is capable if a little clumsy.

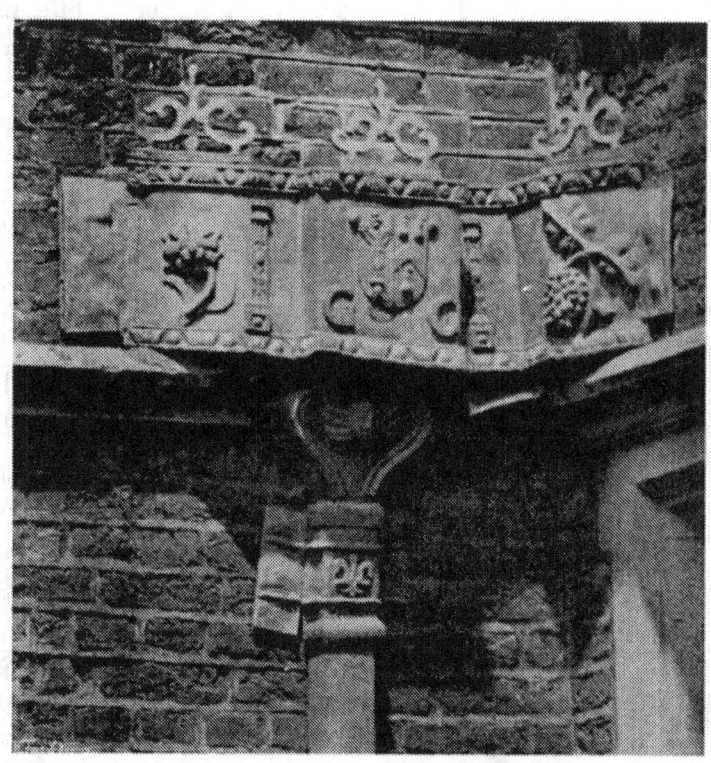

FIG. 69.—Abbot's Hospital, Guildford.

The heads in the quadrangle are smaller and simpler. Fig. 71 shows one with two heavy horizontal bands which perhaps strike the eye as ugly, but they are valuable

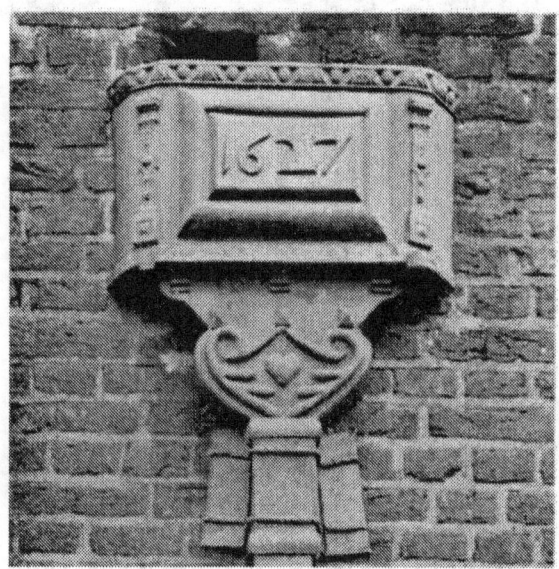
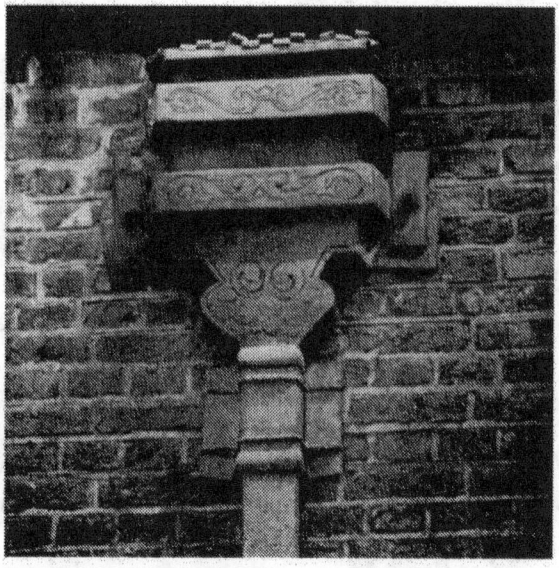

FIGS. 70 and 71.—Guildford.

for the vigorous shadows which they give. The head of Fig. 70 is an example of a rather early head which has lost the early feeling and has not found its way to a satisfactory alternative. The treatment of the funnel is weak and amateurish, and the panel bearing the date has a clumsy moulding. The pilasters are a good example of how not to use architectural detail as mere ornament.

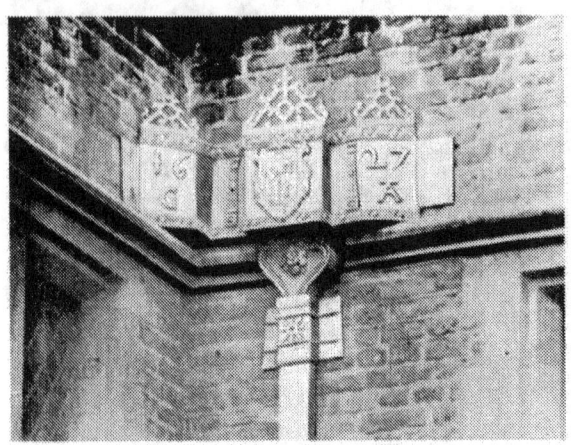

The pipe sockets are really more interesting than the heads, having raised cable bands and ornamental patterns tinned on the face. The pipes have been painted freely, and as the tinning only stands up about one-sixteenth of an inch it is visible only on careful examination (Fig. 73). There are nine patterns in all, including various types of cross and the fleur-de-lys. Another pipe socket, probably of 1750 or later, has a delicate lion's mask enclosed in a beaded pointed oval.

FIG. 72.—Guildford.

At St John's College, Oxford, are four magnificent heads of 1630, the important features of which are the elaborate painting and gilding of the lead. The royal arms and the arms of Archbishop Laud are blazoned in their proper colours, and the turreted face of the heads and the funnel outlets are painted black and white in chevron bands and in many other delightful patterns.

We are indebted to the painstaking care of Mr F. W. Troup for the restoration of this colour work. Mr Troup's measured drawings of the heads are reproduced in Figs. 76 and 77, and photographs of two in Figs. 74 and 75. Fortunately there were sufficient traces of the old colour to make its accurate renewal a certainty and not a speculation. This colour treatment was probably not uncommon in the seventeenth century, but three centuries have weathered most of it away. Two heads on the Bodleian Library retain traces, but apparently only of black and white. Gilt relief was doubtless quite common; the heads at Condover Hall and on the new buildings at Magdalen College, Oxford, are so treated. As Viollet-le-Duc says: "Mediæval lead was wrought like colossal goldsmith's work," and a profusion of gilding would lend actuality to this impression. It is curious in this connection to note (Mr Massé's book is the authority)

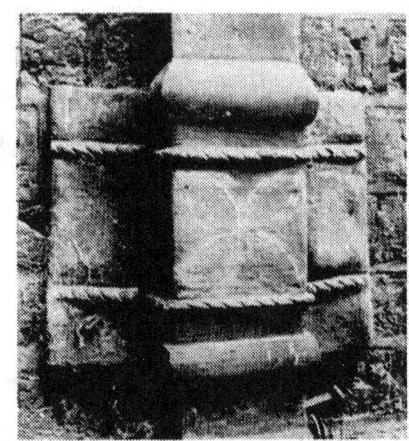

FIG. 73.—Guildford.

that the painting and gilding of pewter were stringently forbidden, and cases are cited where failure to obey the Pewterers' Company resulted in heavy penalties. A plumber's meat was apparently a pewterer's poison.

At St John's College, Cambridge, are also admirable heads dated 1599. By way of leading up to the later work described in the next chapter, a criticism may be ventured of some remarks on lead heads by Mr Reginald Blomfield, A.R.A., in his fine history of Renaissance Architecture. He says that towards the latter part of the seventeenth century the older and simpler treatment of heads gave way to more recondite forms owing to the ambition of the plumber, now become a very dexterous workman, to show his skill. He points to the 1730 head in the Square of Shrewsbury (Fig. 79) as illustrating the change that was destroying English craftsmanship. Mr Blomfield suggests that the workman had long since passed the limitations imposed by technical inexperience, and could not resist the temptation to sacrifice artistic value to mechanical skill. The elaborate work on the heads of Haddon and Knole and Hatfield of the early seventeenth century must, however, have required as full a knowledge of the plumbers' craft in all respects as the later work at Shrewsbury and elsewhere. While the richness of the later work is generally produced merely by applying an excess of separately cast ornaments, the early work is not lacking in an equally rich but withal restrained treatment of applied castings. In addition, we have the delicacy of the pierced work,

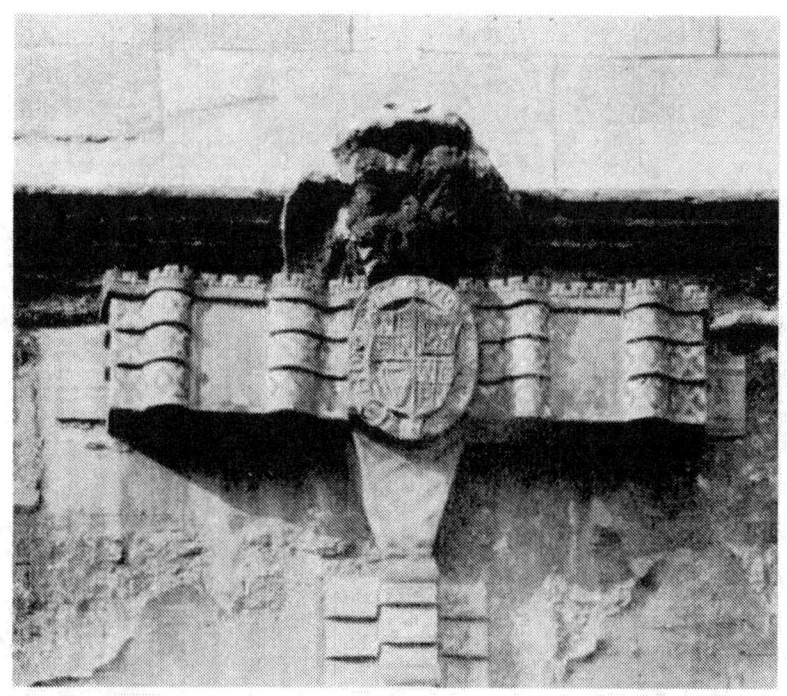

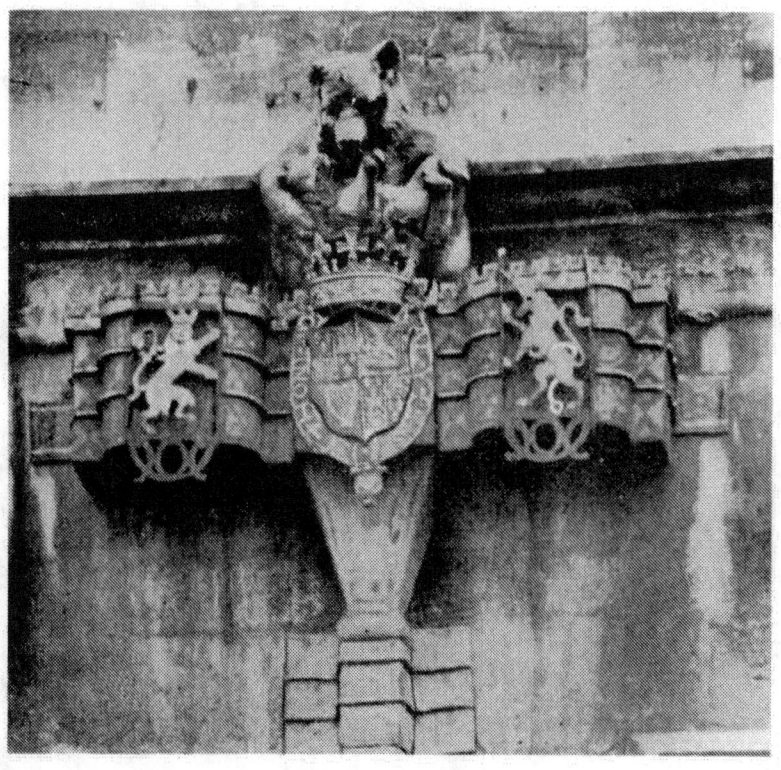

FIGS. 74 and 75.—St John'sCollege, Oxford.

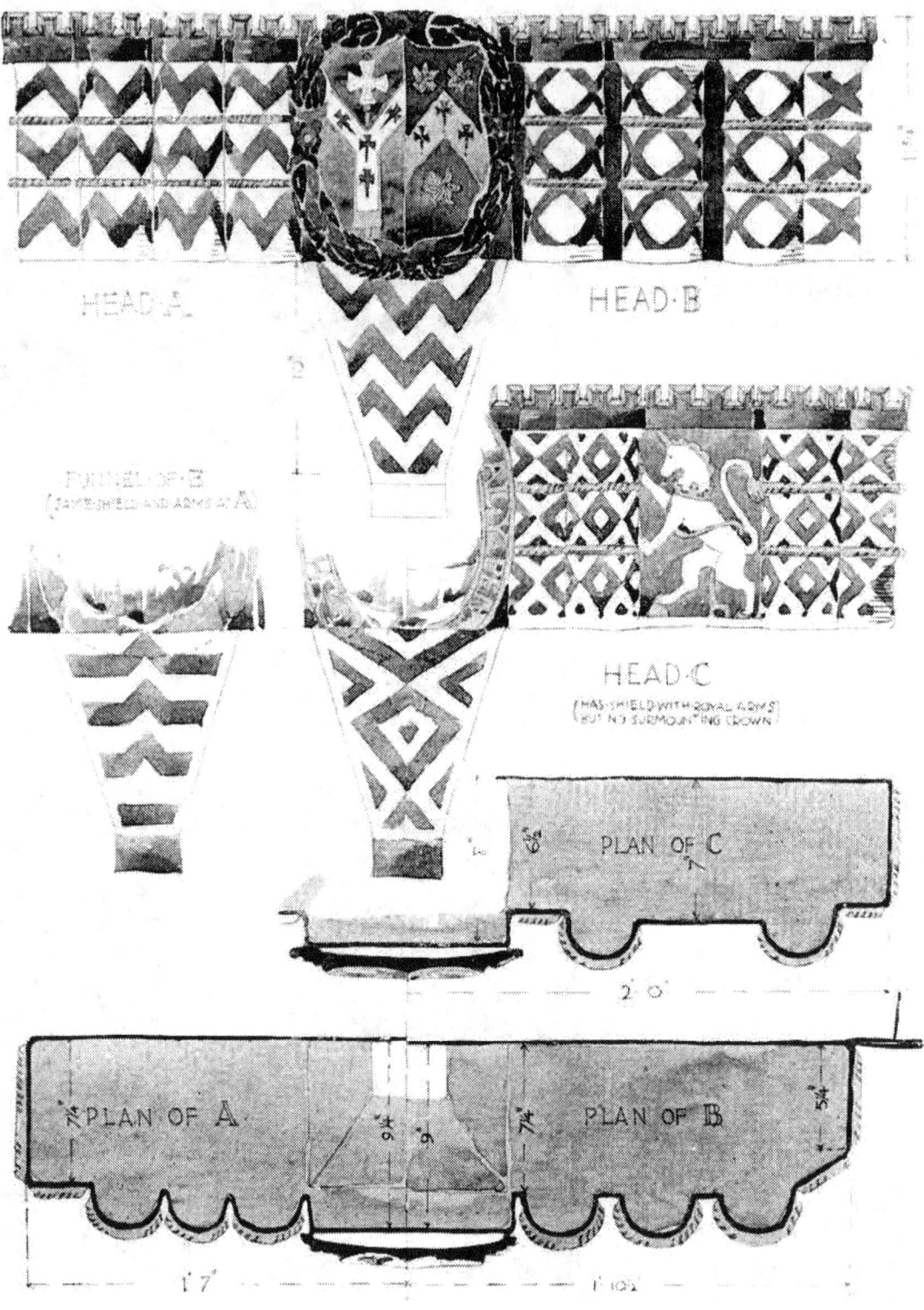

HEAD·A

HEAD·B

FUNNEL OF B
(DAMAGED AND ARMS OF A)

HEAD·C

(HAS SHIELD WITH ROYAL ARMS)
(BUT NO SURMOUNTING CROWN)

PLAN OF C

PLAN OF A

PLAN OF B

FIG. 76.—ST JOHN'S COLLEGE, OXFORD.

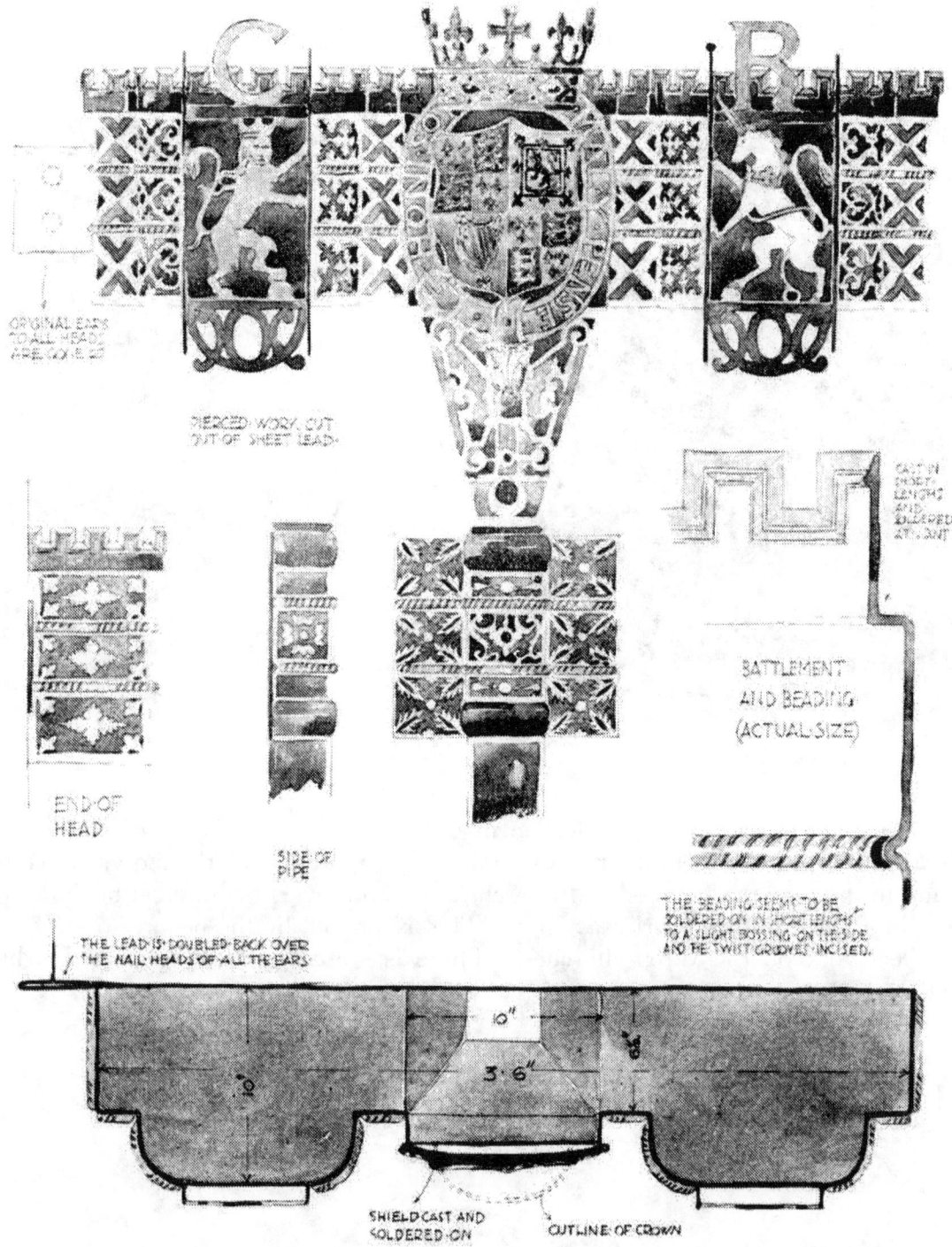

ORIGINAL EARS
TO ALL HEADS
ARE DONE SO

PIERCED WORK CUT
OUT OF SHEET LEAD

CAST IN
SHORT
LENGTHS
AND
SOLDERED
ON JOINT

END OF
HEAD

SIDE OF
PIPE

BATTLEMENT
AND BEADING
(ACTUAL SIZE)

THE BEADING SEEMS TO BE
SOLDERED ON IN SHORT LENGTHS
TO A SLIGHT BOSSING ON THE SIDE
AND THE TWIST GROOVES INCISED

THE LEAD IS DOUBLED BACK OVER
THE NAIL HEADS OF ALL THE EARS

10"

3 6"

10"

6"

SHIELD CAST AND
SOLDERED ON

OUTLINE OF CROWN

PLAN OF HEAD

FIG. 77.—ST JOHN'S COLLEGE, OXFORD.

and the colour treatment of painting, gilding, and tinning, which called for a dexterity as marked as is needed for cast work however elaborate.

With regard to the modelling of the cast ornaments, the lion of 1589 on the Windsor head is at least as good an effort as the acanthus leaves and swags of the later heads, and the most that can be said for the later work is, that in the technique of casting undercut

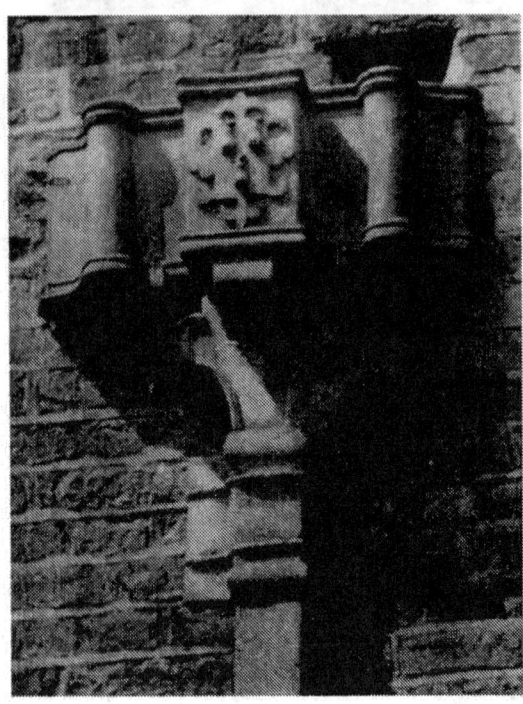

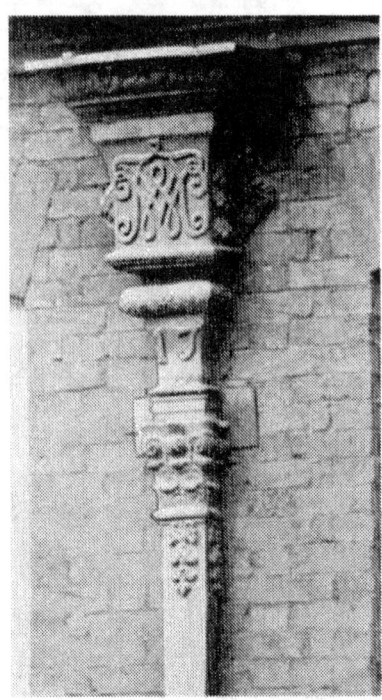

FIG. 78.—Guildford.   FIG. 79.—The Square, Shrewsbury.

work greater skill was shown. The decline in charm which we feel towards the end of the seventeenth century is due rather to the sinking in importance of the individual craftsman owing to the growth of power of the architect. Moreover, the interest taken by the architect in the details of leadwork was faint. This is proved by the poverty of design of the water leadwork on the Wren churches. There is scarcely a head in London which is interesting.

CHAPTER III.

RAIN-WATER PIPE-HEADS (*Continued*).

The Overlapping of Styles—Bolton Hall—Stonyhurst and Bideford—Local Schools of Leadwork—
Shrewsbury, Nottingham, and Aberdeen.

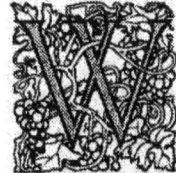 E turn now to the later work, in which the classical detail has become triumphant, and Gothic detail, where it appears, seems uncomfortable and apologetic.

For the sake of convenience the history of pipe-heads has been divided into two chapters.

A division into two periods is naturally much too arbitrary to do more than suggest broadly, that in this kind of leadwork there were two main influences—the mediæval and the Renaissance. Owing to the sporadic working of the new ideas, and the slowness with which they penetrated to the more remote parts of England, there is naturally a great overlapping of styles. A marked instance of this is found at Kendal, Westmorland, where a head of 1711 much resembles in general treatment the Guildford heads of 1627. The applied ornaments are escallop shells and fleurs - de - lys, and there is a parapet of delicate brattished work which is astonishing for 1711. At this date the finish at the top of pipe-heads was almost invariably a heavy and not very interesting cornice. Kendal was remote, and the old manner consequently lingered there.

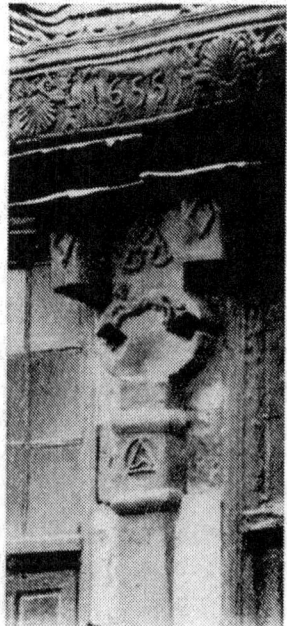

Fig. 80.—Coventry.

Old Palace Yard, Coventry, has a remarkable series of leadwork. Reference

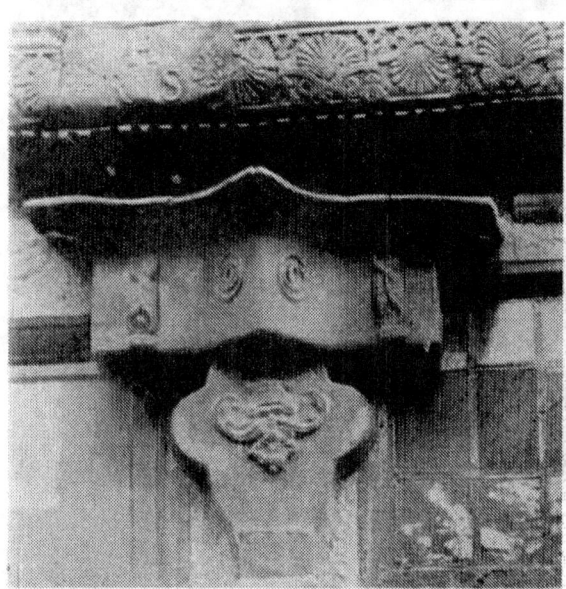

Fig. 81.—Coventry.

has already been made to one gutter of about 1580 (Fig. 61). Seven heads of 1656 and thereabouts receive the discharge from a fine shell-pattern parapet gutter, with dates, initials, and coats of arms interspersed (Fig. 81). Most of the heads have classical cornices of great projection with dentils, but much Gothic detail lingers in battlement and discs of tracery. They alternate with pilasters and arabesque masks. There is a charming disregard of consistency but the pleasantest result. This mingling suggests a Commonwealth plumber adding stock patterns in the new taste to those his father left him, and using one or the other according as they happened to fit the plain lead boxes that called for enrichment of some kind.

The Coventry craftsman evidently did not desire to

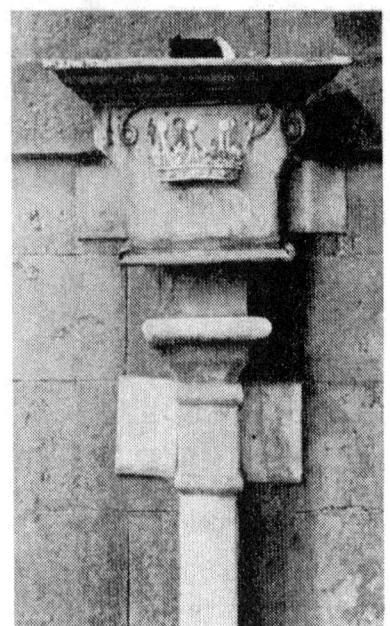

FIG. 82.—Haddon Hall.

deliver his work from the power of the dog. One head (Fig. 80) has a running hound, and a pipe socket has two vigorously modelled spaniels. The building is delightful throughout. Woodwork and plasterwork are full of interest, but dilapidation grows apace, and the little courtyard has a neglected, almost doomed, look, which bodes ill for its survival. A motor car factory of uncompromising utility and vileness has been added recently. One fears that the success of the English Juggernaut may soon claim another victim, and one that Coventry can ill spare.

At Charlton House, Kent, the heads are dated 1659 and

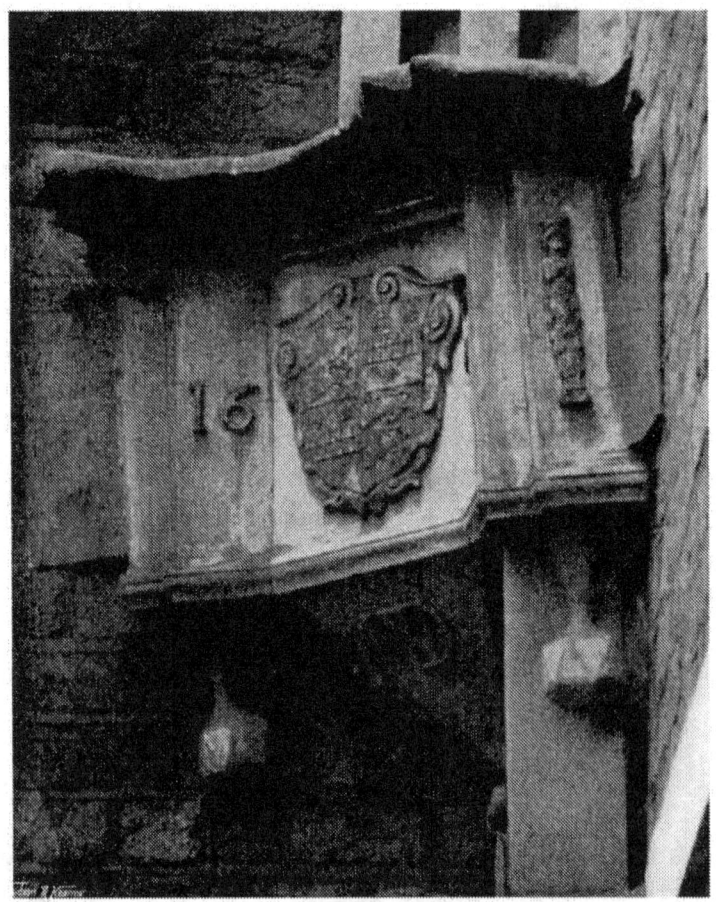

FIG. 83.—Charlton House, Kent.

are, therefore, not so early as the house. The elaborate treatment of the shield of arms (Fig. 83), the pendant knobs and the queer little ornaments suggestive of mummies

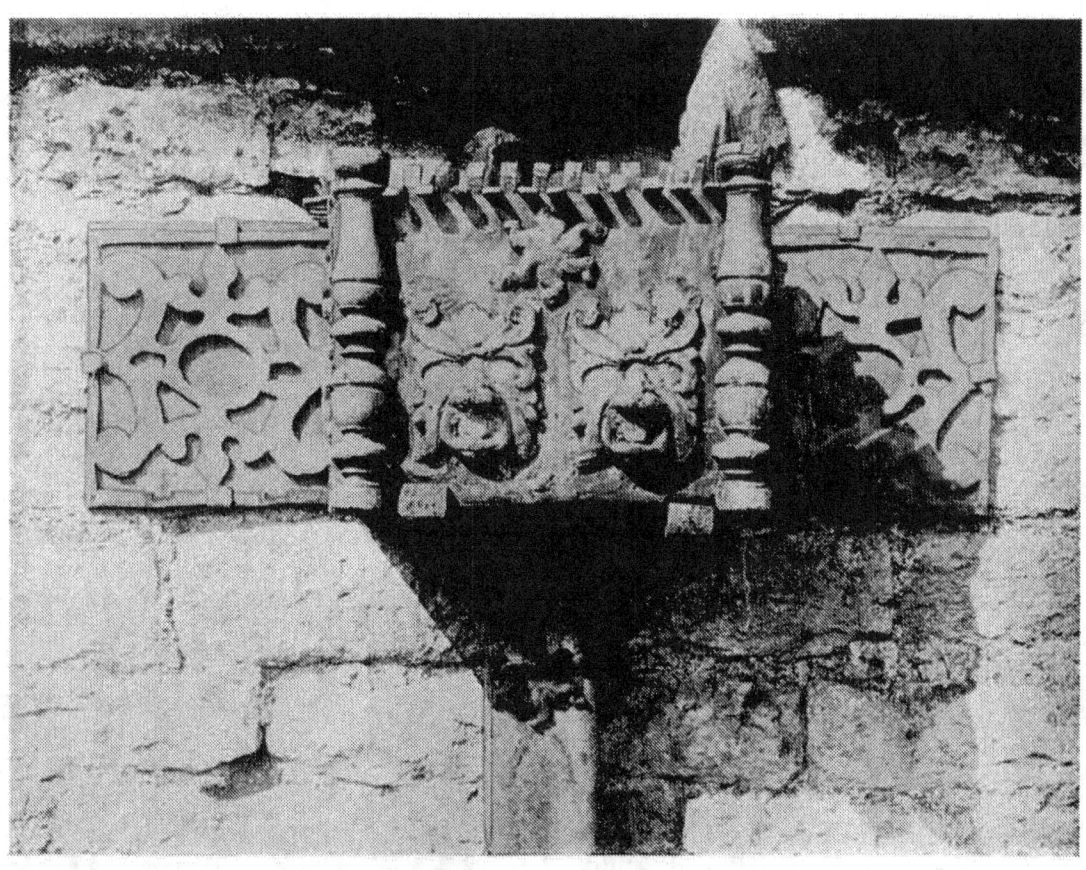

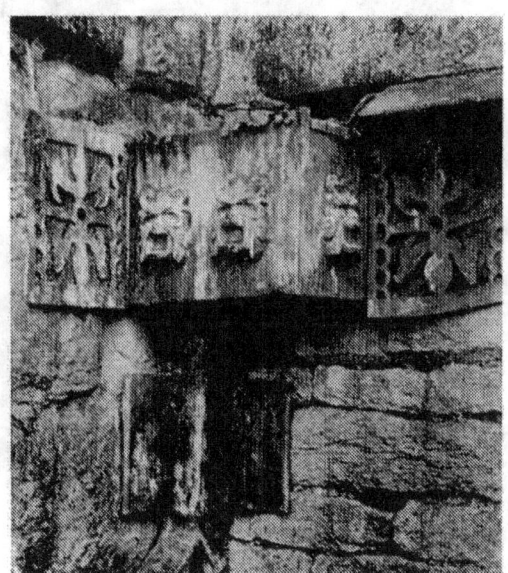
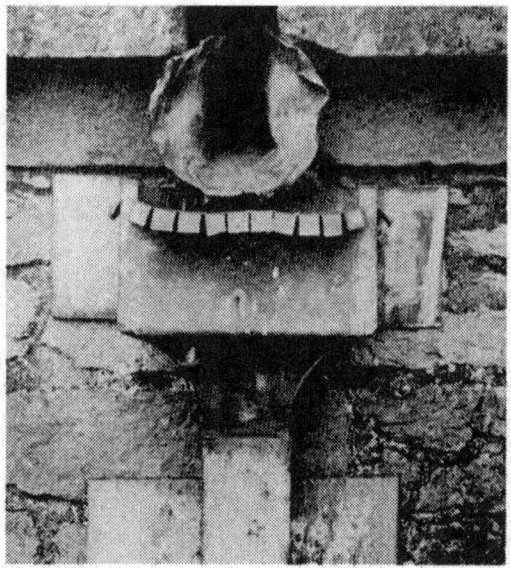

FIGS. 84-86.—LEAD PIPE-HEADS, HADDON HALL.

give the leadwork a pleasant individuality.   Another head (Fig. 88) is notable for the
big shield of arms standing above the bowl.

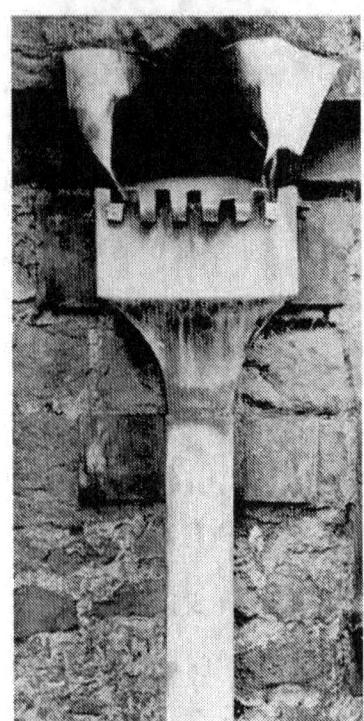

Returning to Haddon Hall, there are some heads in
the Upper Court with rich arabesque masks and balusters
at the corners, which mark a break from the older manner,
and have quite an Italian look.   Even on them a slight
projecting embattled cresting is retained for the delightful
spots of shadow, which it throws on the top edge (Figs. 84
and 85).

There are also several heads (Figs. 86 and 87) of
very simple treatment, which are most difficult to date.
They may be ascribed to about 1670.   There is in the
Guildhall Museum, London, the front only of a head,
dated 1676, the top of which is nicked and bent over in
exactly the same way.   It would be hard to devise heads of
such perfect simplicity which yet should be so entirely successful.   There
is not even a
pipe socket: the
outlet of the
head is made
rather smaller
than the pipe.
This is, of course,
not a thing to

FIG. 87.—Haddon Hall.

imitate, because though the junction of the pipe
and pipe-head is of satisfactory appearance,
there must be trouble at the lower end of
the length of pipe, where it joins to the next
length.   Unless the lower length be fitted with
a socket (though not necessarily ornamented)
it will have a slovenly look, because it must
be worked to a larger opening to take the
upper pipe.   At Hatfield some of the sockets
are of the same size as the pipe, and the
spigot ends of the pipes above are worked
to a smaller size to make the joint.   This,
however, besides looking a lazy piece of work,
has the practical disadvantage that the bore
and, therefore, the water-carrying capacity of
the pipe is reduced.

In practical points such as this it is not
always safe to follow the older work, which
sometimes shows strange lapses.   Jerry-building is not a purely modern vice: it is as
old as laziness.

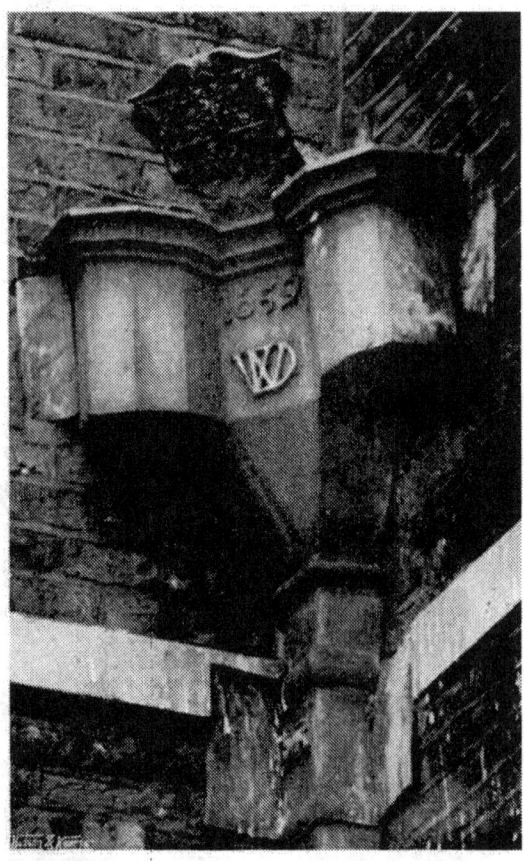

FIG. 88.—Charlton House, Kent.

The long vase-shaped head (Fig. 82) is illustrated not so much for its intrinsic merit (it is rather dull) but because it was a common form throughout England for a century later. This type frequently has a lion's mask on the face, as at Hampton Court (Fig. 93), and can be seen in scores in London on the Inns of Court and the city churches. Some at Hampton Court have the flat front covered with a very intricate monogram of George II.

From 1700 onward one finds that a building has generally only one type of head. The applied ornaments vary somewhat, but fancy was dying, and the wealth of invention we find at Haddon and Knole about 1600 had become ancient history.

At Poundisford Park, near Taunton, there is a very complete system of rain-water lead-work (Fig. 90). From the valleys at each side of a high-pitched roof the water descends through heads and pipes (obviously recent) into a pretty horizontal gutter with ornamental top edge. The outlet from this gutter conducts the water into a turreted head (Fig. 89) with pipe discharging into a handsome lead cistern. The "castle" treatment of the head is so distinct from the stiff feeling of the pots of flowers which, with the date 1671, decorate the cistern, that one is tempted to think the head is earlier. As, however, the Durham head of 1699 (Fig. 95) combines the same "castle" motive with a markedly classical cornice, we may take the Poundisford Park head as probably contemporary with the cistern (which is

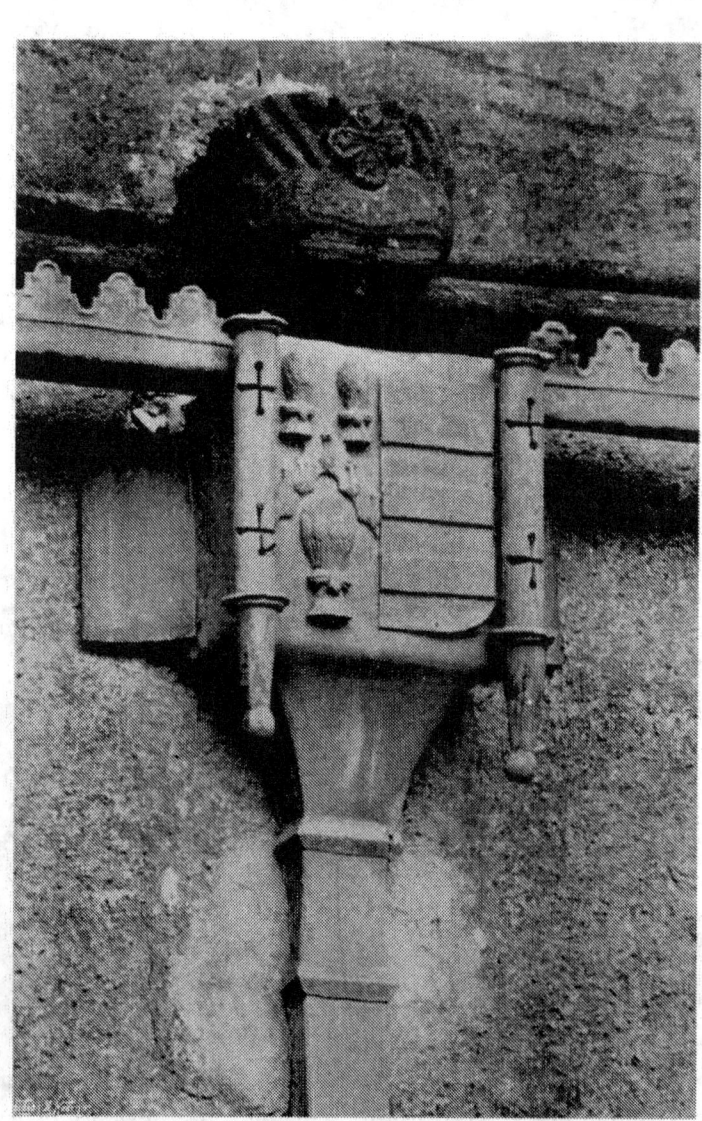

FIG. 89.—Poundisford Park.

illustrated in the next chapter). We have here a parallel in leadwork to the mingling of the two manners in stonework which appears on the Salisbury Chantry at Christchurch and elsewhere. The gutter is notable; the same pattern, but doubled, appears on another house at Taunton, and in the Devizes Museum there is a similar gutter, which came from the Bear Hotel, Devizes. At East Quantock's Head there is a head with a parapet of the same outline, which was evidently a peculiarity of the Somerset-

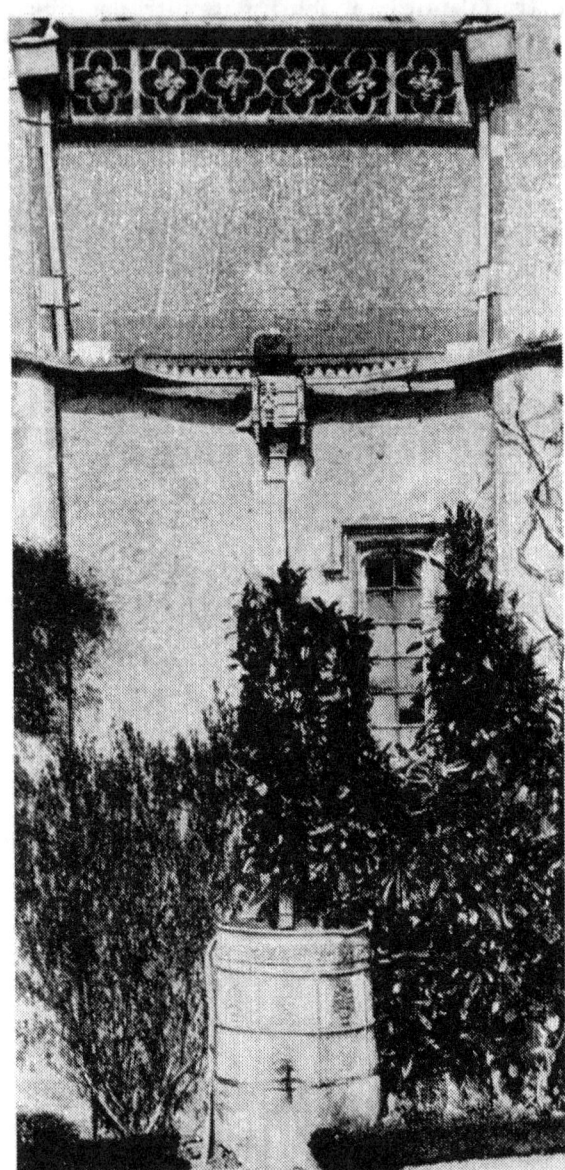

FIG. 90.—Poundisford Park.

shire plumber. The same outline but in a
feeble variation is found at Stanwick, Yorkshire. A head not unlike that at Poundisford Park is on Torrington Church, Devonshire (Fig. 91). The corner turrets are less
actively warlike than those of Poundisford
Park, as becomes the peaceful nature of their
home, and the vine decoration which strug-
gles round the little parapet has a soft and
pleasant air. The formal flower ornament
on the pipe socket has a peculiar interest,
as it amounts almost to a trade mark of the
west country plumber. Either at Taunton
or Exeter there was apparently an eminent
family of leadworkers, who did the best of
the ornamental work of the two counties over
a long period of years. This flower orna-

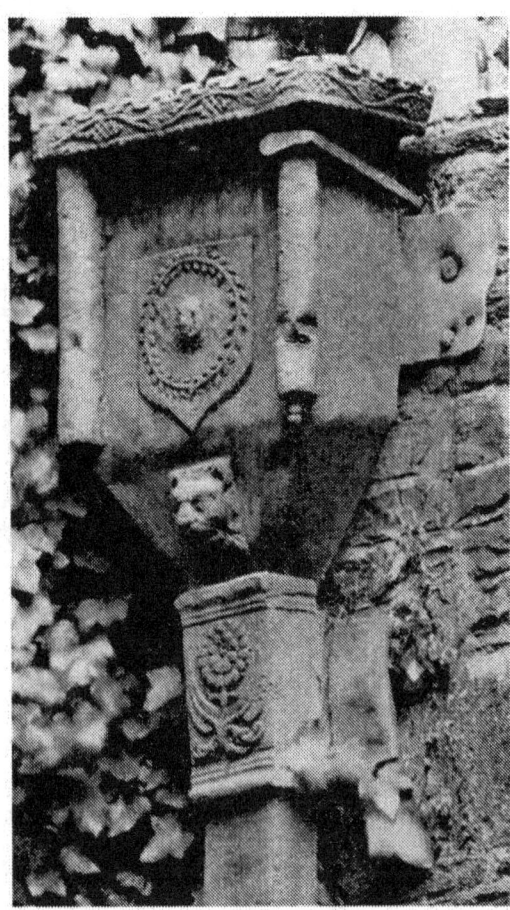

FIG. 91.—Torrington Church.

ment crops up continually. It will be noticed
on some of the cisterns illustrated in Chapter
IV. The head at Petworth, Sussex, dated 1654,
is rather uninteresting, but it has a certain
dignity. The Durham Castle heads have an
especial value historically, as showing the pains
taken that heraldry should tell its story accurately. A head of 1661 fixed to the south wall
of the chapel bears a shield with the arms of the
See of Durham alone, which was then vacant.

Bishop Cosins' Correspondence (Surtees, 1870-71, vol. 55, p. 341) gives under "Durham Repaires," 8th May 1666: "Paid Alderman Myres, plummer, for 13 stone of lead covering the ovell of the fountain, mending the gallery leads, and a quarter's wages for keeping the pipes, £2. 6s. 3d." Very possibly this important citizen was the author of the head dated 1661.

The example of Fig. 95 bears on the richly mantled round shield the arms in pale both of the

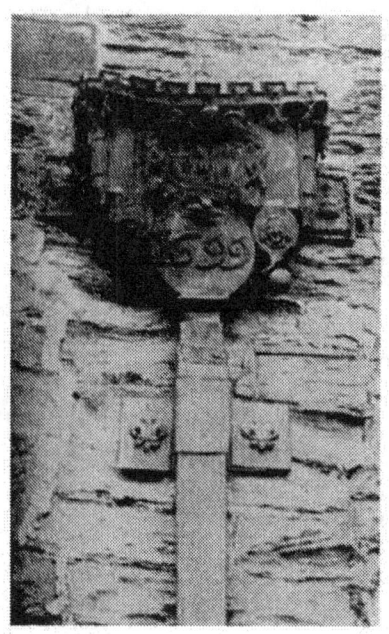

FIG. 92.—Durham Castle.

see and of Bishop Crewe. As Crewe was a baron in his own right, we have as his personal mark the baron's coronet as well as the prince-bishop's coroneted mitre which indicated his office. The tasselled labels of the mitre stand clear of the flat surface of the head, and are unusually narrow. The lower member of the cornice is a delicate bead and reel moulding, the upper an ogee with a rich but shallow classical pattern worked on the face.

The baron's coronet

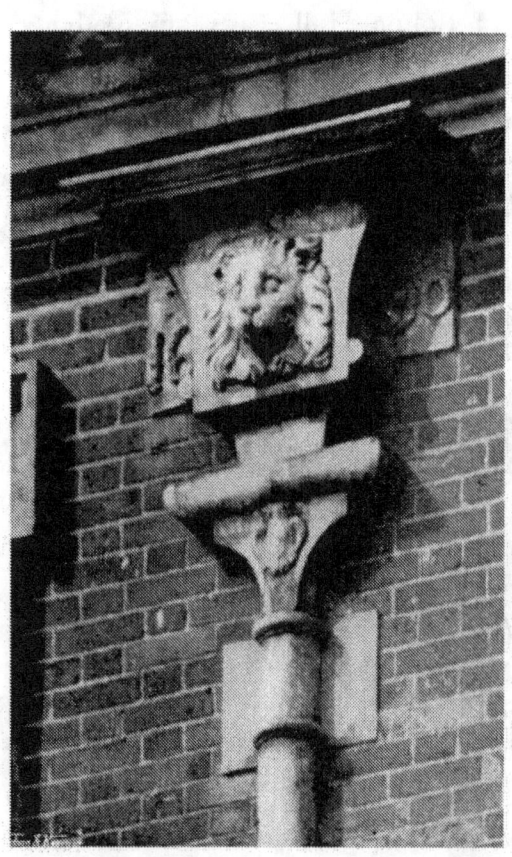

FIG. 93.—Hampton Court.

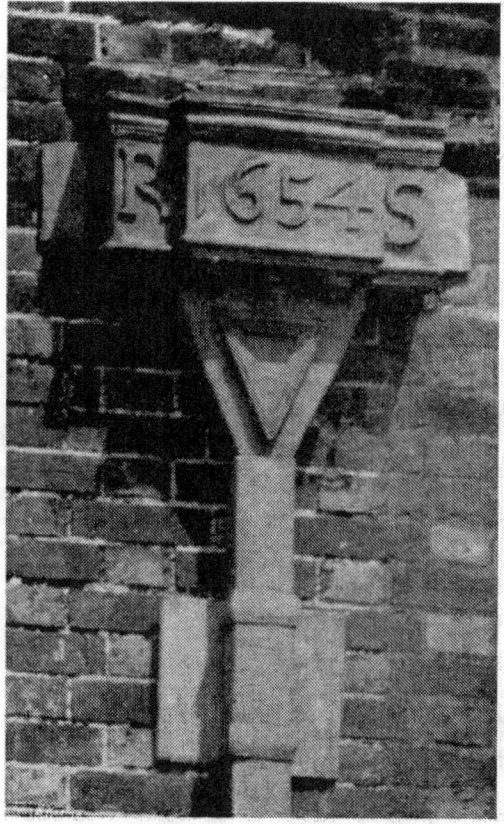

FIG. 94.—Petworth.

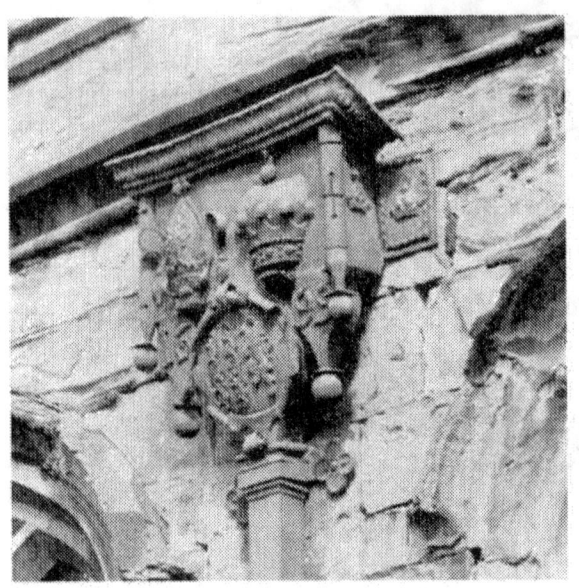

FIG. 95.—Durham Castle.

recurs both on the side of the head and on the ear. In the latter case it is enclosed by a moulding which looks like the cast cable which is so pleasant and constant a feature in the old work, but is actually a flat ribbon closely twisted. Unhappily, the original lead pipes have been abolished, and iron substituted. The altogether odious cast-iron ear, which fastens the socket to the wall, seems a needless barbarity. Of all the offences of cast-iron pipe, surely the band ear of this type is the greatest. If it serves no other purpose, though, it is a commentary vigorous enough on the distance we have travelled since 1699.

Another head of 1699 (Fig. 92) has battlements with a pierced valance of Tudor ornament instead of the classical cornice. The attempt to remain Gothic must have amused the plumber vastly. He has perpetuated his sense of humour in two bewigged and laughing faces on the lower part of the head.

Very similar to the Durham heads are those of Bolton Hall (Figs. 96-99) though here the Renaissance Rubicon has been finally crossed. The only suggestion of mediæval parentage which remains is in the pierced fronts of the cylinders. The variation of heraldic ornaments gives great historic interest to the heads. The arms are those of Charles, sixth Marquis of Winchester, afterwards Duke of Bolton, and of his second wife, Mary Scrope. The design is somewhat over rich, but the modelling of the Paulet hinds and of

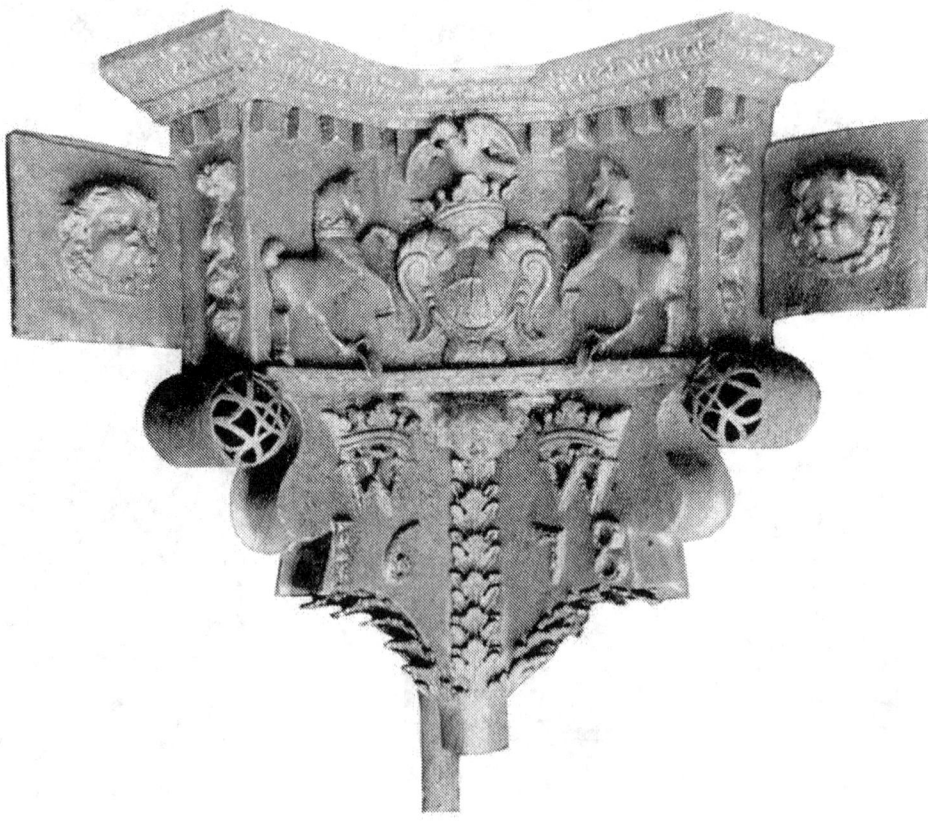

FIG. 96.—Bolton Hall.

the Scrope choughs which support the shields is
especially vigorous.   In one head the Paulet coat
is supported by the Scrope choughs, a hybrid
arrangement due, doubtless, to the Scrope shield
having been lost, and the gap filled by a plumber
who was a Gallio in heraldry.   The cherubs are
podgy in the best gravestone manner.   The date
deserves a word.   The simple, clear figures of
the Windsor and Knole heads are left behind for
a pretentious, husky type, which accords with the
general treatment of the head, but is not very

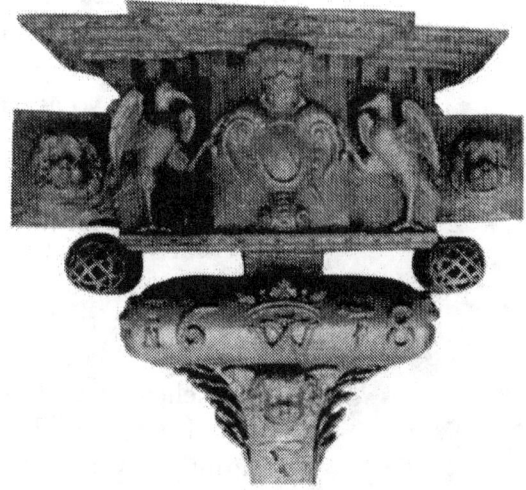

FIG. 97.—Bolton Hall.

admirable in its own right.   A head
on Winchester College has similar
numerals.   About 1700 they were
common.

Bolton Hall was burnt down in
1902, but the leadwork escaped practically unhurt, and Fig. 99 shows the
heads as in place before the fire.   The
roof was covered with lead, which
melted and cascaded into the heads
and down the pipes.   However, no
harm was done, as the melted lead
settled down in the bottom length
of pipe, whence it was removed by
the simple method of splitting the
pipe up the back.

A technical word may be added as to the
making of these heads, which applies, more or less,
to all heads of the late seventeenth century.   The
main box part is made of cast sheet-lead beaten to
the shape and soldered up.   The cornice has been
cast in lengths, mitred, and soldered on.   The
dentils and all other ornaments are separate castings
soldered on.   The substance of lead averages 10 lbs.
to the foot, but varies between 7 and 12 lbs.   The
method of fixing, viz., simply soldering on from the
front instead of also pinning through to the back, is
slovenly and unlike the best work at Haddon;
hence the dropping off of ornaments, and muddled

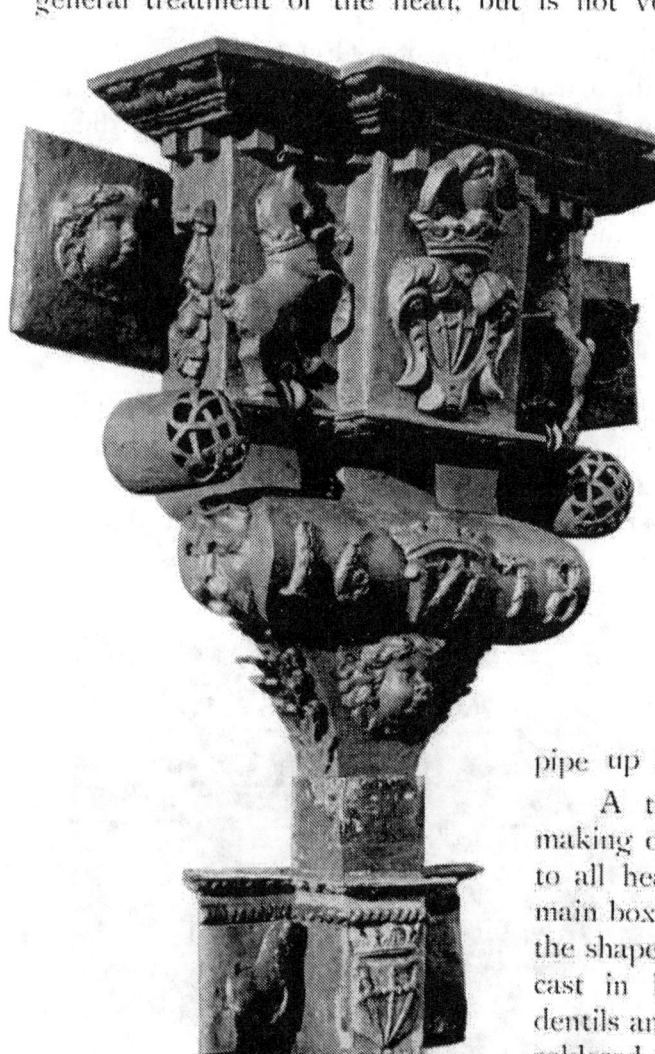

FIG. 98.—Bolton Hall.

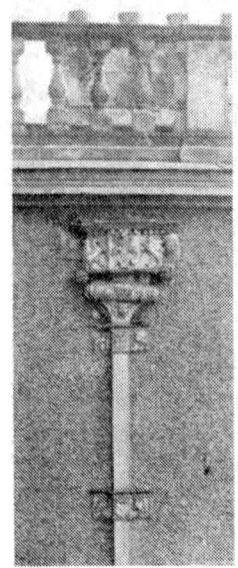

FIG. 99.——Bolton Hall.

refixing. The overlapping acanthus leaves at the bottom of the head are characteristic of the period, and while giving an undeniable richness, do so at the price of troubling the general effect. In 1678 there has ceased to be much reticence in the use of applied decoration. There are no traces of gilding, colour, or bright tinning. The pipe sockets and ears have cable-moulded bands, and are also decorated with the heraldic devices. The pipes used with the flat heads are rectangular (5¼ inches by 3½ inches), and with the angle heads are circular (4½ inches). The flat heads are 2 feet 11 inches wide by 2 feet 10 inches high over all, and the angle heads 2 feet 2½ inches from angles to edge of ears.

The Hatfield Park head, dated 1680 (Fig. 100), is a very dignified work. Like the earlier heads of 1610, it rests on the stone cornice. There are few heads that accord so fitly with their architectural setting. The lead cornice is of a strong yet graceful moulding that matches the stone cornice. The two semicircular projections on the face of the head are taken up on the face of the pipe, and there is an economy in the applied ornament which is refreshing at this date. The whole effect, if a little stiff, is eminently scholarly. If there is a weakness, it is in the rather hard line of the horizontal projection on the funnel, which catches the light a little harshly.

In this head one seems to see the hand of an architect behind the plumber. The earlier leadwork, save in one notable exception at Knole (Fig. 56), seems to have been done with little reference to the general treatment of the building. The plumber was probably told to provide the required number of stack pipes and heads, and the design was left to his own fancy. There was a lack of co-ordination, which produces results

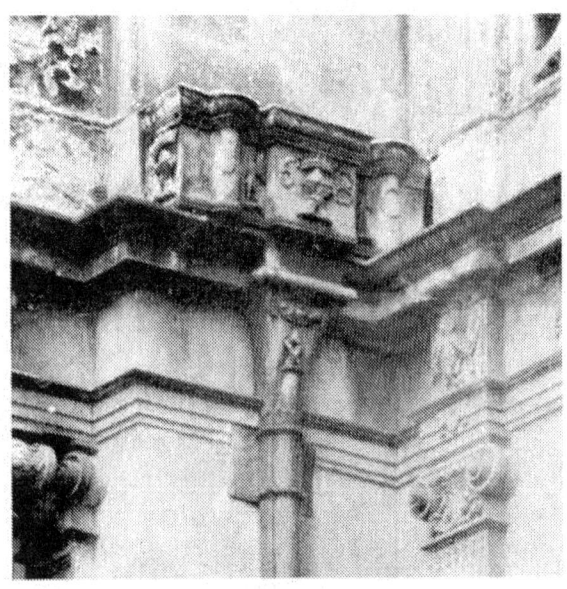

FIG. 100.——Hatfield.

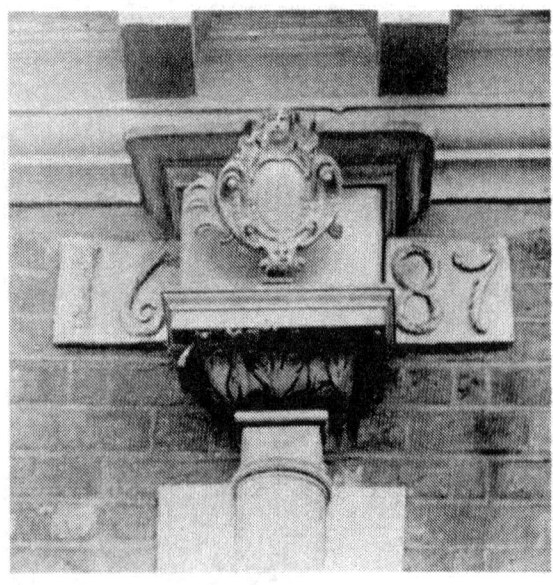

FIG. 101.——Winchester.

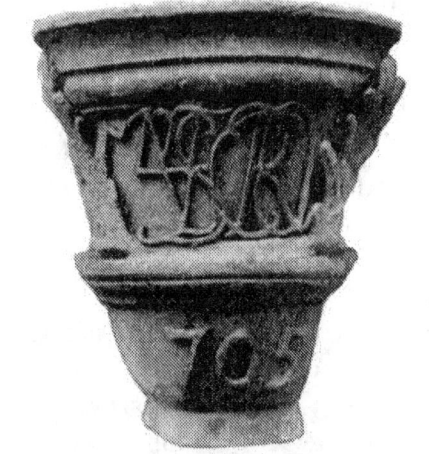

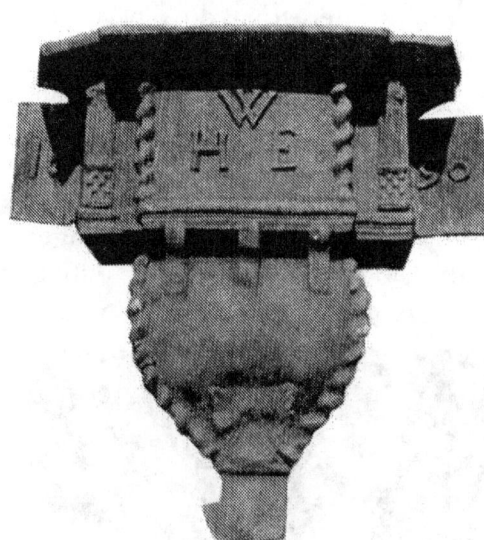

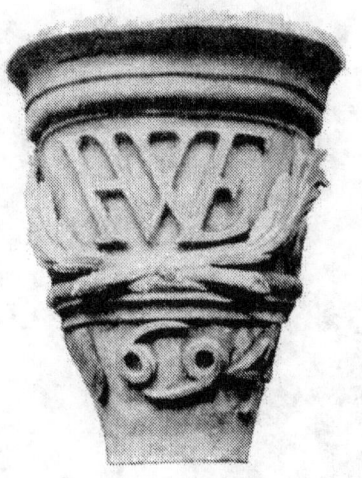

delightful enough, but diverse enough to prevent any unity in detail, even if it existed in the general scheme of the building. One cannot think of Inigo Jones allowing a plumber any voice in the design of his leadwork : Wren was certainly less careful. The early Palladian work with elevations in the grand manner did not admit of the careful proportions of its stonework being disturbed by streaks of lead pipe. The thought of a down-pipe on the front of the Banqueting Hall verges on profanity. Palladianism was the death of leadwork. There are down-pipes and heads on the side elevations of Wren's work at Hampton Court. The heads are large and ornamental, but they are not very interesting. On the Judge's Lodgings at Winchester is a head dated 1687 (Fig. 101). It is interesting that the shield, which was probably painted with a coat of arms, is fixed to the head only at the top and the bottom, and stands quite clear between. The pipe-heads on St Laurence Jewry have shields standing free in the same way. At South Kensington Museum there are on loan seven heads from the Old Manor House of Bucklebury, Berkshire, long since destroyed. They are of two main types, one rather pretentiously architectural, the other of the funnel shape, which in its simpler and undecorated form is so common on late eighteenth-century buildings. One of the latter (Fig. 102) is redeemed from banality by the two antler-like ornaments and the undecipherable monogram. It is altogether a rather slovenly piece of work, and seems to be an amateurish copy made in 1705 of the head dated 1694 (Fig. 104), which has ornaments of great simplicity and distinction.

The larger head (Fig. 103) is an excellent example of 1690 : the twisted edging is not only rich, but its softness seems peculiarly suitable to the material. The pilasters are unusually treated. They are fluted, with Ionic capitals, and have a dado of chequers, which lighten the design with a pleasant spottiness. The three connections between the bowl and the funnel are also rare : they give the general effect of trusses, but are only thin straps. The lettering is admirable, and

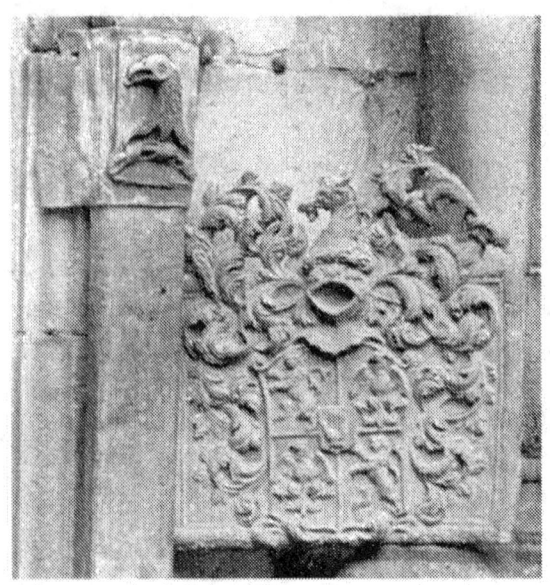

FIG. 105.—Stonyhurst.

stands for Sir Henry Winchcombe and Elizabeth, one of his two wives of this name. On the Lay Vicar's House and the Custom House, Exeter, and also on the Stone House, Topsham, are simple heads of the end of the seventeenth century, semicircular on plan, and edged with a bold egg and tongue moulding.

At Dartmouth, on St Saviour's Church, a pipe socket is entirely covered by a large mask. With every desire to escape being gibbeted as a blind Gothic enthusiast, it is difficult to avoid the conclusion that the further we move from mediæval into classical treatment, the less interesting do pipe-heads become. Not only is classical detail substituted for mediæval, but the change seems often to have destroyed the craftsman's sense of material. Of this perversion the Stonyhurst and Bideford heads (Figs. 105 and 106) are fair instances.

The Stonyhurst head, shown in the photograph with a pipe by its side, is no longer in position, but four others, two exactly as the photograph, and two with funnel outlets added, still serve their original purpose. This work can be dated from the heraldic charges as being between 1689 and 1717, and is notable for many reasons. It is the only highly decorated head, the front of which is cast in one piece, apparently from a carved wood pattern. It looks more like a Sussex iron fire-back than a lead head. The sharp modelling

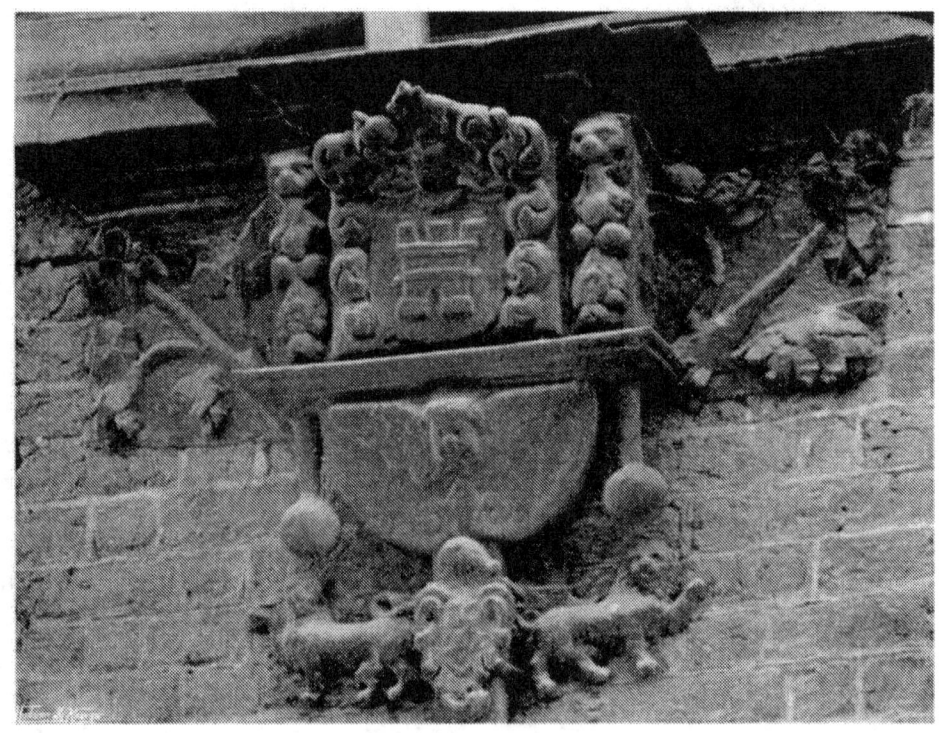

FIG. 106.—Bideford.

shows that the plumber had abdicated his control, and was content to reproduce in lead what another had carved in an alien material. It is not suggested that no carved wood patterns were used in the earlier work, but at Stonyhurst the feeling of the pattern material dominates the finished lead instead of being subordinate to it. As an example of the richest possible heraldic treatment it is admirable. There is scarcely an inch of surface not covered either by the coat, crest, or mantling, and yet, owing to the unity of treatment, and the absence of dates, cherubs, initials, &c., there is no suggestion of overcrowding.

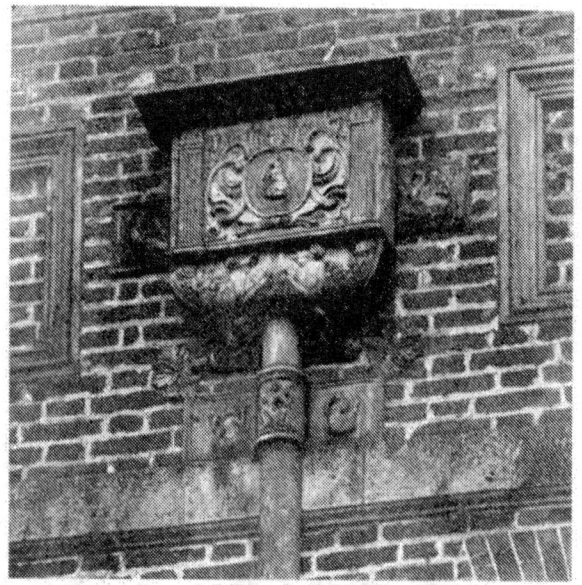

FIG. 107.—Frampton Manor House.

The Bideford head (Fig. 106), which is also of about 1700, suggests a nervous horror of plain surfaces. It is a plaster-work rather than a leadwork design. It shows not only an almost wanton luxuriance of ornament but also a lack of economy in material. The designer seems to have thought in trowelfuls of plaster rather than in weight of rather costly metal. The treatment has, however, one advantage over the Stonyhurst work in that the surfaces are rounded and easy, as becomes the nature of lead, and the general design is at least vernacular. Even if it

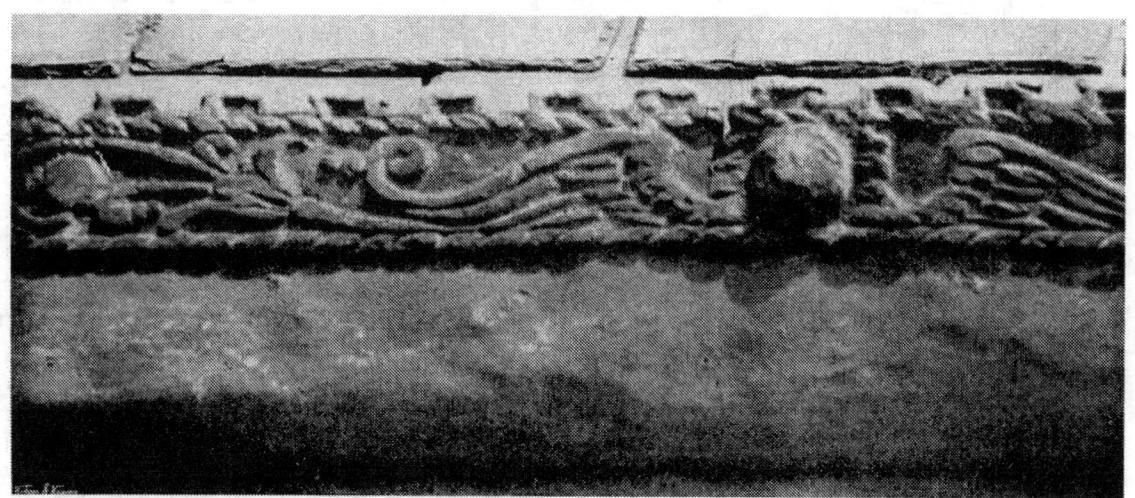

FIG. 108.—Gutter, Barnstaple.

is a plaster design it is English and not foreign. The later English plumber may have rather blundered with his material, but he at least never borrowed ideas from such ingenious gentlemen as Artari and Bagutti. One does not often find the pendant knobs, and there is something very naïve about the two leopards who are prancing away from the pipe along the brick wall. The modelling of the stalks and leaves to the right and left of the bowl is in a naturalistic manner, quite foreign to the fat stiff ornament which flanks the shield. The cherub is the most ordinary touch on a quite extraordinary composition, which shows the riotous ease with which the plumber played with his material. This head is but one of a pair; the second is similar, but hardly as rich. At Barnstaple there is a lead gutter with toy battlements and a rope moulding enclosing ornament, which is a medley of vague flowers and wings (Fig. 108).

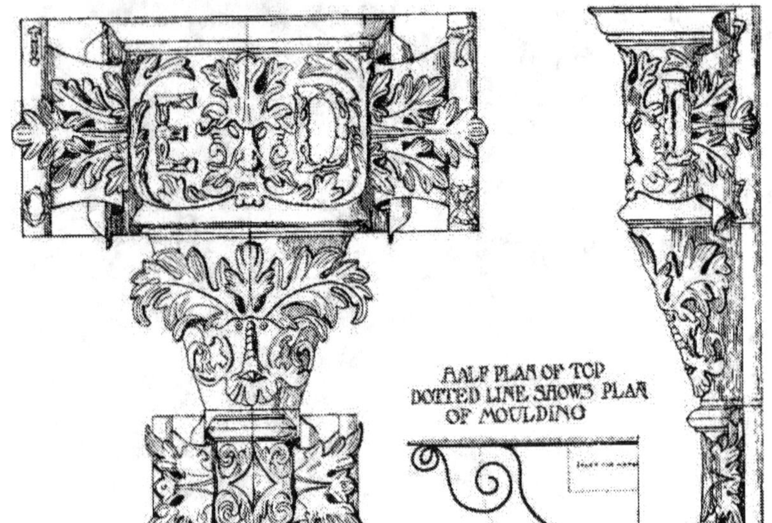

Very architectural are the heads at Frampton Manor House, Boston, Lincolnshire (Fig. 107). The fluted pilasters, the flourishes round the central panel, and the rich modelling of the lower part of the head give it a distinctly baroque effect. Altogether it is quite foreign in feeling. The pipe ears and the side wings of the head itself have delicately moulded watery creatures—swans and mermaids. There are leaves on each side of the lower part of the bowl, connected with it by stems, and fixed to the wall—most unreasonable leaves that do nothing. This head is very characteristic of the early eighteenth century, and is certainly one of the finest existing of its type. At Mel-

HALF PLAN OF TOP
DOTTED LINE SHOWS PLAN
OF MOULDING

PLAN OF UPPER BANDS

PLAN OF UPPER BANDS

FRONT   SCALE   SIDE

FIG. 109.—Canons Ashby.

Traced by permission of Mr A. Hartshorne, F.S.A., from his Plate in the "Spring Gardens Sketch Book."

bourne, Derbyshire, there are several heads obviously cast from the same patterns. This is another case of the peripatetic habits of plumbers, for Melbourne is a long way from Boston. There is another, very similar, but less worried, on Sawley Church, Derbyshire. On a late and ugly head at Kendal there are creatures of a dragon sort, modelled like the Frampton swans with needless delicacy. At Llanelly on the estate offices are a very rich head and pipe (figured in "Arch. Cambrensis," fifth series, vol. xvii., p. 236). At Raby Castle there is a very refined example consisting of a plain box with delicate balusters at the corners and a cornice. It is dated 1712.

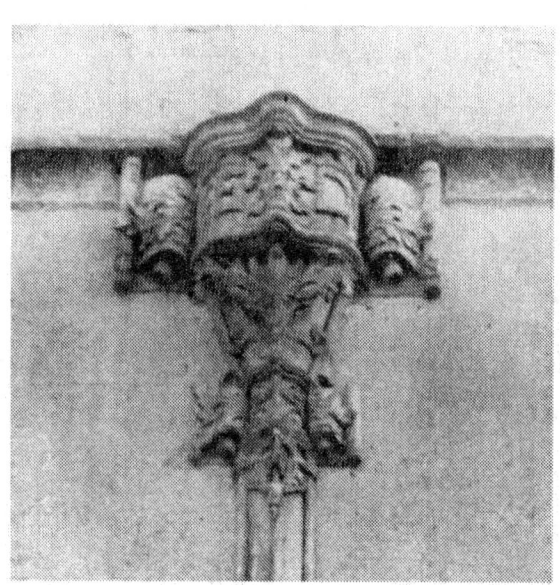

FIG. 110.—Canons Ashby.

Figs. 109 and 110 show by photograph and measured drawing what is perhaps the most rococo of English heads. It is from Canons Ashby. The rich sweeping curve of the curled ears is its most interesting feature, and one that deserves repetition in a less exuberant key. At the Architectural Museum, Tufton Street, there are

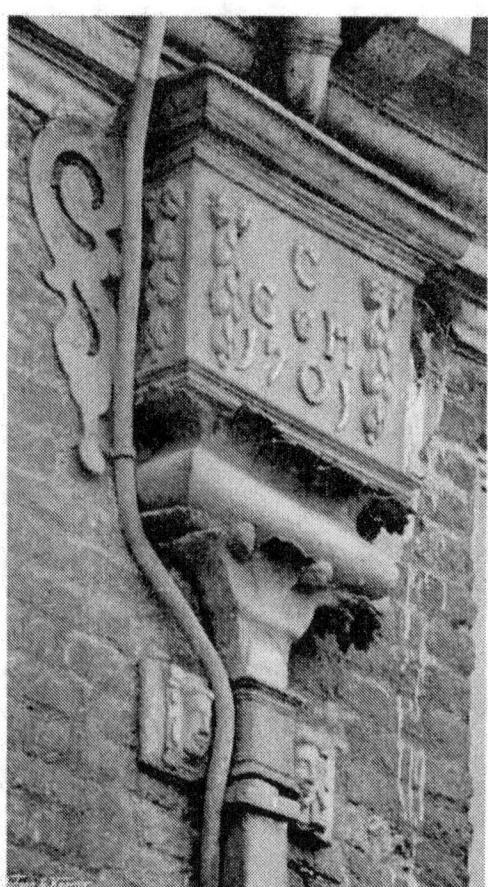
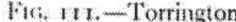

FIG. 111.—Torrington.

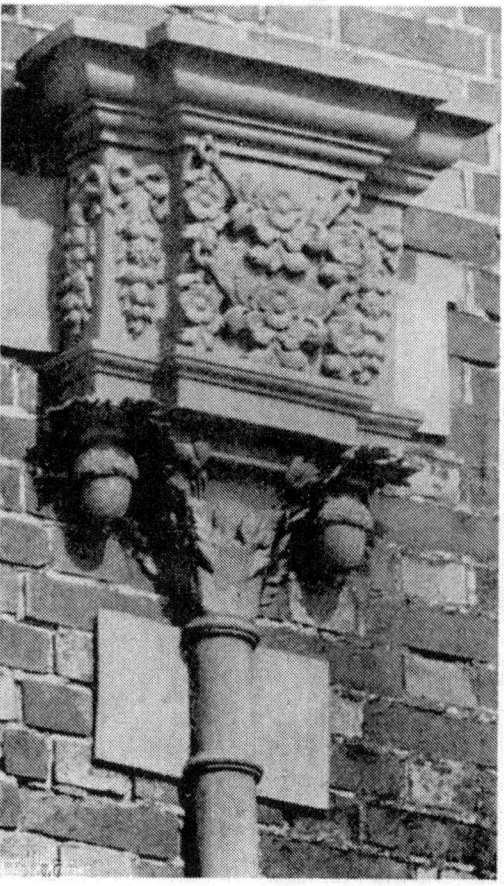

FIG. 112.—Petworth.

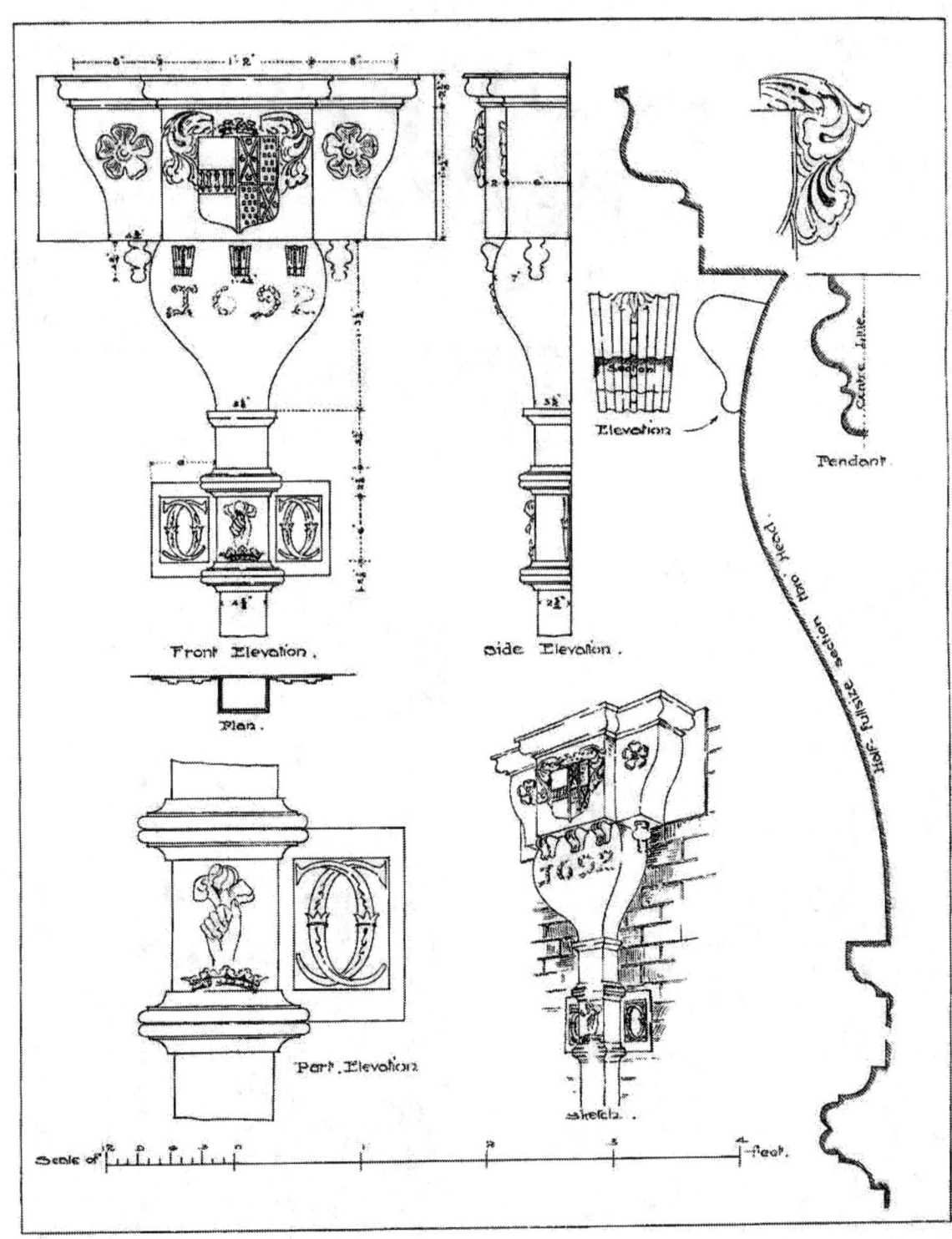

FIG. 113.—LYDNEY PARK, GLOUCESTER.

four old lead heads, the most important of which, the example from Lydney Park, Gloucestershire, is here represented by Mr Erskine Cumming's measured drawings (Fig. 113).

At Levens Hall rank was something more than the guinea's stamp. It was writ very large on the pipe-head. But for this the head is uninteresting (Fig. 117). Petworth, Sussex, provides a head (Fig. 112) which is a veritable museum of lead flowers strung and festooned over the bowl. It gives a rich effect and is very orderly and balanced. The piercing of flat sheet lead, as in the head of 1701 at Torrington, North Devon, is unusual, and gives a papery effect. Note also the rather smirking masks on the ears of the pipe socket (Fig. 111). Shrewsbury has only one early seventeenth-century head. It is dated 1610, has an embattled gutter running into one side, and raised chevron ornament on front. It is very similar in character to the Knole and Hatfield heads of the same period. The building on which it is fixed is very ruinous and is likely to disappear soon. In the eighteenth century local schools of plumbing seem to have taken shape, and to have influenced the craft

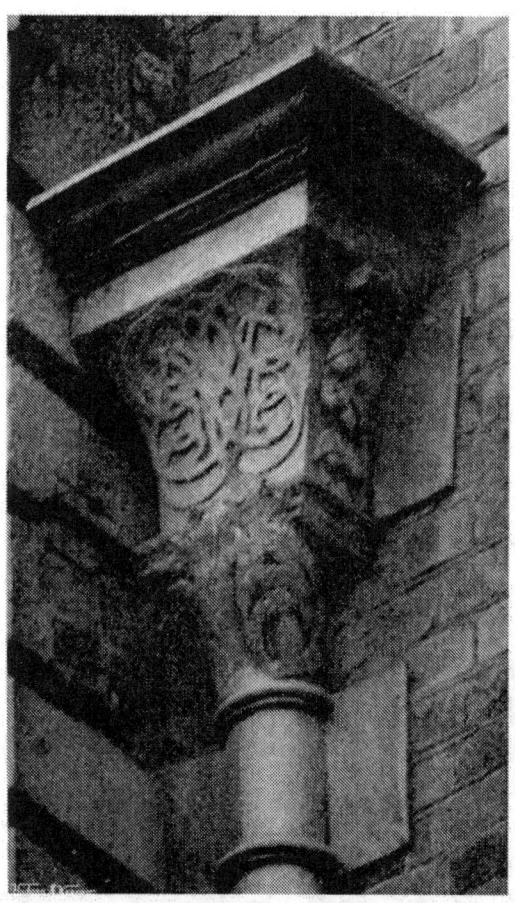

Fig. 114.—Shrewsbury.

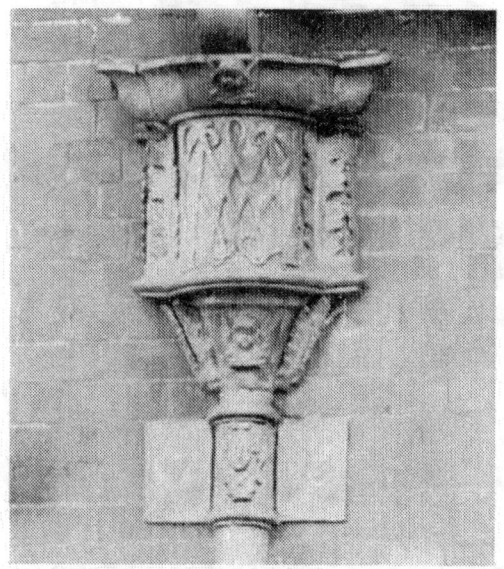

Fig. 115.—The Constabulary Offices, Shrewsbury.

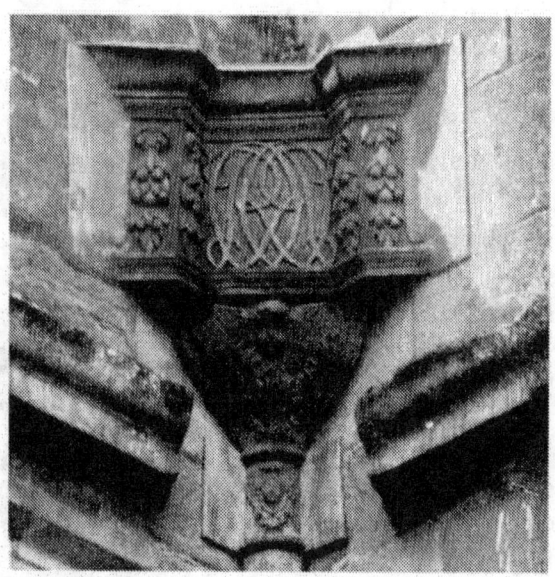

Fig. 116.—Condover Hall.

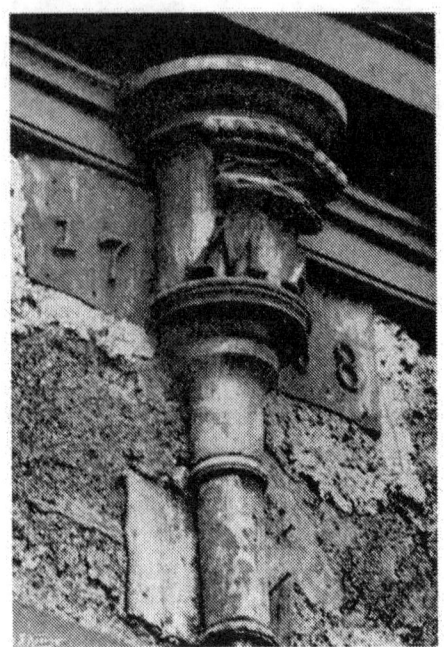

FIG. 117.—Levens Hall.

of a large district. The recurrence of the same ornaments on cisterns shows this to have been the case in Devonshire. Shrewsbury affords another notable instance. There are many heads of the type of Figs. 115 and 118 with simple cornices and very elaborate monograms, and many bear the municipal leopard's mask. They show great technical capacity, and give a note of gaiety to the bald brick and stucco elevations. Reference has been made in the last chapter to the two heads in The Square, dated 1731 (Fig. 79), which show the rich and fretful methods of this period at their best. The example of Fig. 115 was fixed in 1715 on a now

demolished building, and is at present in use at the constabulary offices.

The most attractive shape is that of Fig. 118; the head of Fig. 114 has the same elaboration of monogram and acanthus ornament, but the shape is not good.

There continued in the district a definite tradition in this manner until 1800, producing designs generally lame and unhappy, but not without a certain dexterity. At all events they showed an appreciation of past merits, and even about 1800 we find pipes with semicircular front projection like the early seventeenth-century pipes of Hatfield. The same pipe occurs at Warrington, dated 1740. Shrewsbury pipe sockets sometimes take the form of Corinthian capitals (Fig. 79), a superfluity of architectural naughtiness which is not un-

FIG. 118.—Shrewsbury.

amusing. Condover Hall, near Shrewsbury, has an angle head in the distinctive Shropshire manner (Fig. 116). The cornice mouldings are of careful proportion, and the strings of flowers are excellent of their kind, if a little too suggestive of plaster. The woman's head on the pipe socket is another common feature of the local work. There remains the gilt relief, which lightens the general effect. This Shropshire school stretches down to Ludlow, where there are several late heads of merit. Another local school is that of Nottingham. The work remained interesting until a late date. There is considerable refinement in the head of Fig. 119, though the double-headed eagle is a tame enough bird and poorly executed. The very late example of Fig. 120 is of a happy simplicity, if somewhat amorphous.

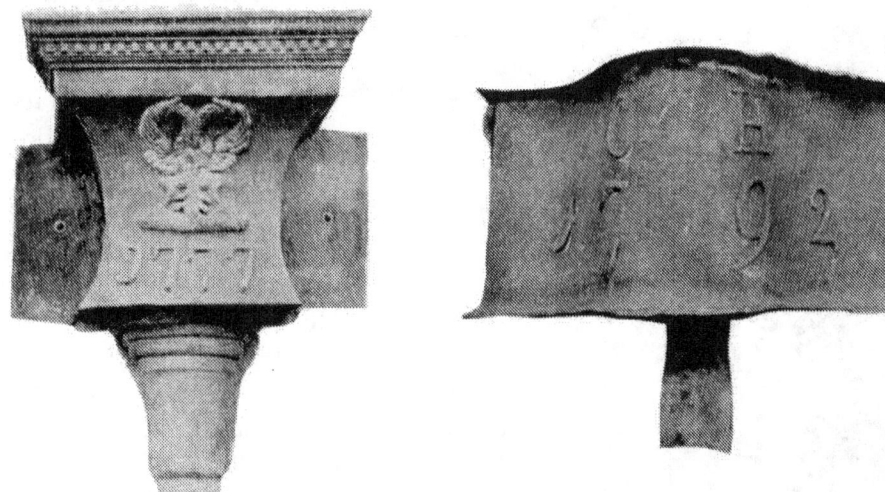

Figs. 119 and 120.—Nottingham Museum.

The last examples of local peculiarities are taken from Aberdeen. The head of Fig. 121 is in the possession of Mr William Kelly, to whose acute and sympathetic observation the author is indebted for

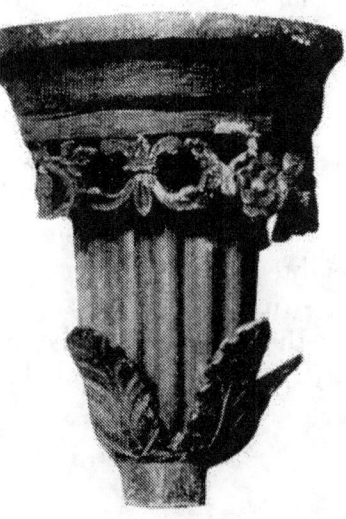

Fig. 121.—Aberdeen.

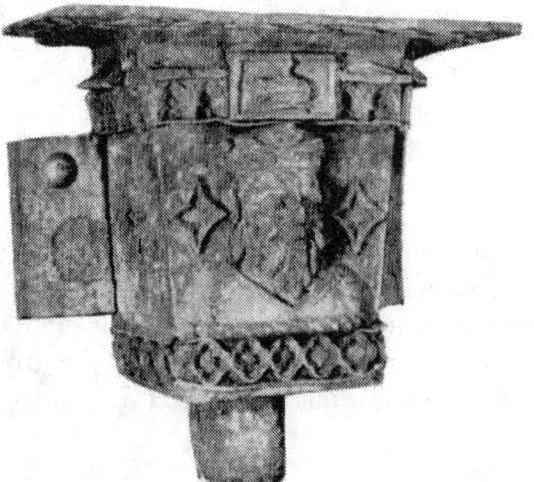

Fig. 122.—Plumbers' Company Museum.

much valuable information anent the Aberdeen leadwork. It is one of a type that occurs all over the town, though some are even more elaborate. The three large leaves, with modelled faces and serrated edges, are full of vigour, and the cast open-work valance, composed of a rose separated from the thistles on either side by fleurs-de-lys, is a striking feature. It will be noted that these ornaments are inverted. The top mouldings are perhaps rather too heavy, but the whole composition is eminently successful. As the date is probably about 1750, this head contrasts pleasantly with the far less spirited work of like date in England.

The example of Fig. 123 is quite characteristic of the general Adam feeling which pervades the leadwork. On others of plain funnel shape there are delicate swags. The Aberdeen heads repay study the more, in that Scotland generally is rather weak in leadwork.

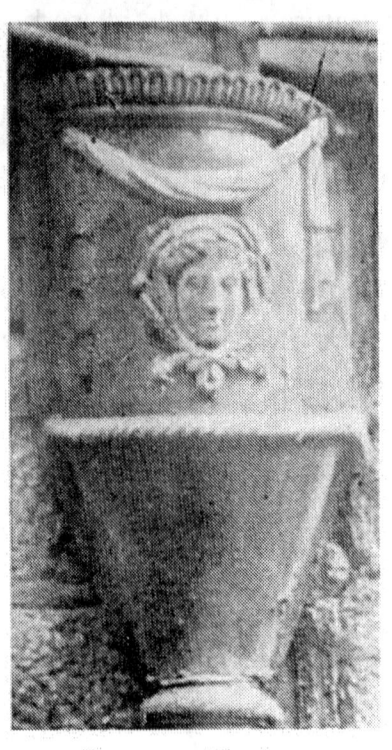

Fig. 123.—Aberdeen.

The example of Fig. 122 is an echo of Strawberry Hill. Carpenters' Gothic one knows, here is plumbers' Gothic. The head is now at King's College, London, and is the property of the Worshipful Company of Plumbers. It came from Grimsthorpe, a house of the Earl of Ancaster, but it is impossible to trace its precise date. The Saracen's head and coronet were probably stock enrichments, for a facsimile head came from the demolished Christ's Hospital. Surely Gothic tracery was never put to odder use. The two quatrefoils which line with the Saracen's nose have a particularly forlorn look, but how this head would have pleased Horace Walpole. At Wollaton Hall, near Nottingham, the Saracen's head appears again on pipe-heads and sockets, dated 1746, but here the general design is of the ordinary classic sort of that date. As for pipe-heads in Ireland, as far as early work is concerned, their place is in the chapter which the snakes occupy in the traditional history, but this may be "another injustice." In Dublin there are some heads of the type of Fig. 93, but they do not call for separate illustration.

To the symbolist on the prowl rain-water heads will be a disappointment. It would be only reasonable to look for some decorative motive suggesting water, but search has so far been vain, if we except the horizontal zigzag bands that are fairly common. As however, zigzags as symbolic of water are archaic, the symbolism, if it can be claimed, is probably quite unconscious. There are eighteenth-century cisterns which bear frogs and such like on their fronts, a commentary grim enough on the fauna of eighteenth-century drinking water, but hardly fit food for the symbolist's meditation. One looks in vain for bands of wavy lines on the front of a head, or some modification of the wave scroll. One would be grateful even for a fylfot.

CHAPTER IV.

CISTERNS.

Possibilities of Decorative Treatment—The Great Tank at St Fagan's—Methods of Making—West Country and London Cisterns Compared—Detailed Descriptions of Examples Illustrated.

AIN-WATER cisterns have so obvious a connection with pipe-heads that we may consider them next, though they are related in form to fonts.

Their decorative problems are altogether different from those of lead pipe-heads. Pipe-heads are generally out of reach. They admit of a delicacy of treatment in piercing and modelling the lead that makes for gaiety, and even allows frivolity. It would be difficult, however, to be frivolous on the front of a cistern. Such ornament as is used must necessarily be in low relief. Anything like the outstanding detail which is permissible on a font would be, on a cistern, in grave danger of harsh treatment from the domestic can and bucket. Yet even so, there is a notable variety of treatment.

The limitations of form are of necessity considerable. Cisterns can only take simple shapes. They may be rectangular, polygonal, circular, or segmental on plan, but variety ends there. For practical reasons their sides vertically should be straight. Their top edges must be strictly horizontal and unrelieved by parapets or any like finishes, such as give an unending variety to rain-water heads. Decoratively the aim is, suitably to ornament a flat surface of regular outline, and speaking broadly, there are four main ways of doing this.

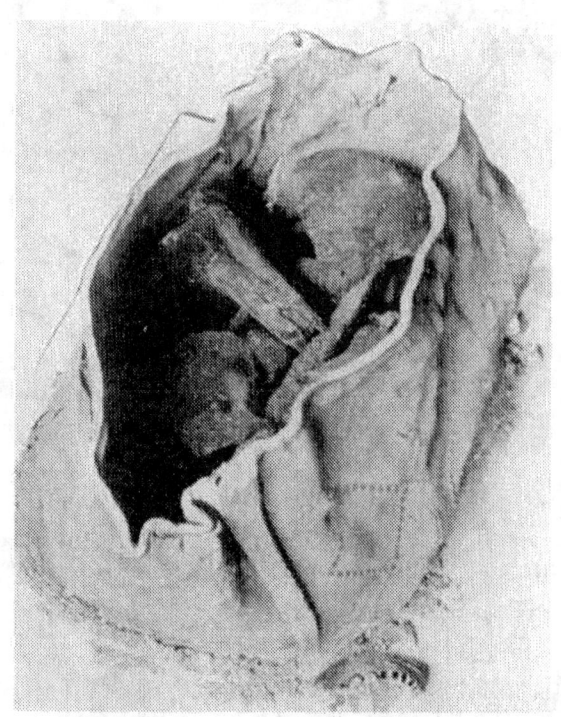

Fig. 124.—St Eanswith's, Folkestone.

1. To treat the surface with some unobtrusive recurring ornament in the same way that a mediæval mason diapered a wall, a method entirely and unfortunately neglected.

2. To panel the face by applying moulded ribs, and further to diversify the surface

by spotting it with small ornaments such as dates, small figures and heraldic charges, the ordinary method of the seventeenth and eighteenth centuries.

3. To model a considerable part of the surface in low relief, so as to produce a certain unity of effect not obtained by simple panel treatment. This method obtains only in rich work, like the most elaborate example at Lincoln's Inn (Fig. 147).

4. To make a moulded frieze the dominant decoration, e.g., the jardinière at Charlton (Fig. 151).

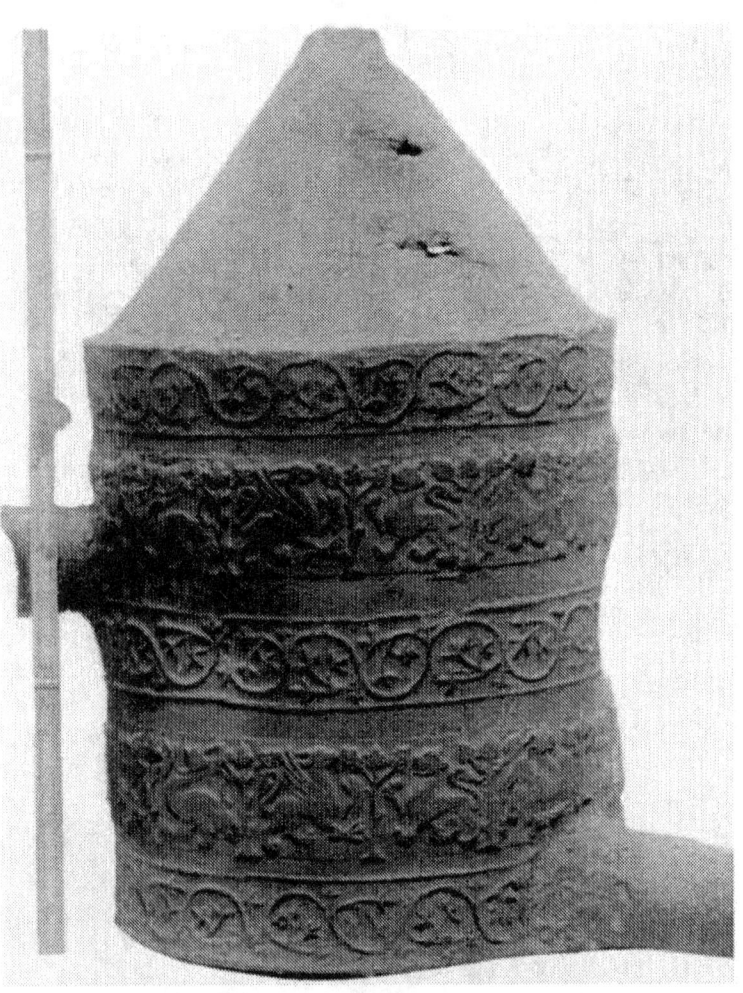

FIG. 125.—Italian Tank, British Museum.

To deal with them in order, apologies are needful for the inclusion in this chapter instead of in the later chapter on sepulchral leadwork of the gruesome example given in Fig. 124. Moreover, it is a reliquary, and not a cistern. Decoratively, however, the two things are the same.

The example is from St Eanswith's, Folkestone, the bones are probably those of the saint. We may put aside, however, the ecclesiastical significance of this lead box and its contents. Decoratively the idea is excellent. The surface is covered with a network of dots (one lozenge of which has been emphasised by the engraver for the sake of clearness). Each dot is lozenge shaped, and near the top of the box the lozenge pattern is crossed by a horizontal line of the same dots. Whether this reticulation is intended actually to suggest a net, or is merely a pleasant combination of dots and lines, seems not to be material. It is illustrated mainly as showing a type of decoration which might well be adopted for relieving flat surfaces in modern leadwork, and is in fact the only example that at all fits the first type classified above. The box has a rough cover (not fitted to it) which apparently was originally part of a Roman coffin. It has at one end, on the underside, five parallel cable mouldings. The reliquary itself seems to be (for historical reasons too long to be set forth here) of the twelfth century.

In Fig. 125 is illustrated the exquisite lead cistern which the British Museum possesses, but it is of Italian origin. The nature of the ornament suggests that it may be of the late fifteenth century, but it is obvious that the disfiguring inlet and outlet pipes are the addition of the Philistine. The conical top also seems to be no part of the original. The second and fourth bands of ornament are particularly interesting owing to their similarity in character to the frieze of the Bovey Tracey tank, and the remaining three bands are of the same family as the frieze of the Lincoln Cathedral example. These parallels are worthy of mention as showing that the decoration of the English leadwork of Renaissance times not only has roots in the earlier work, but is also allied to foreign examples. Two French cisterns at South Kensington, and one at the Cluny Museum, Paris, are also treated with horizontal bands covering the whole surface, a very delightful method which seems to have found no favour in England.

There is one distressing feature in the attempt to trace the development of the design of flat surfaces in leadwork. No English rain-water cistern of ordinary type exists that can positively be dated as being of the sixteenth century or earlier.

The *Builder* of 23rd August 1862, gives a sketch of a cistern dated 15——. The artist found it in the merciless hands of a dealer in building material, who doubtless made unrighteous haste to convert it into saleable goods. It bore the initials E. R. in quatrefoils, and the royal arms with supporters and somewhat elaborate mantling. Except for the Gothic touch in the quatrefoils, it apparently did not differ much from the later ribbed examples. Parts of the front and ends were divided by ribs into square panels, having spots of ornament not now decipherable on the sketch. It had, however, two unusual features in moulded plinth and cornice.

The earliest dated example known to the author is illustrated in Figs. 126-129.

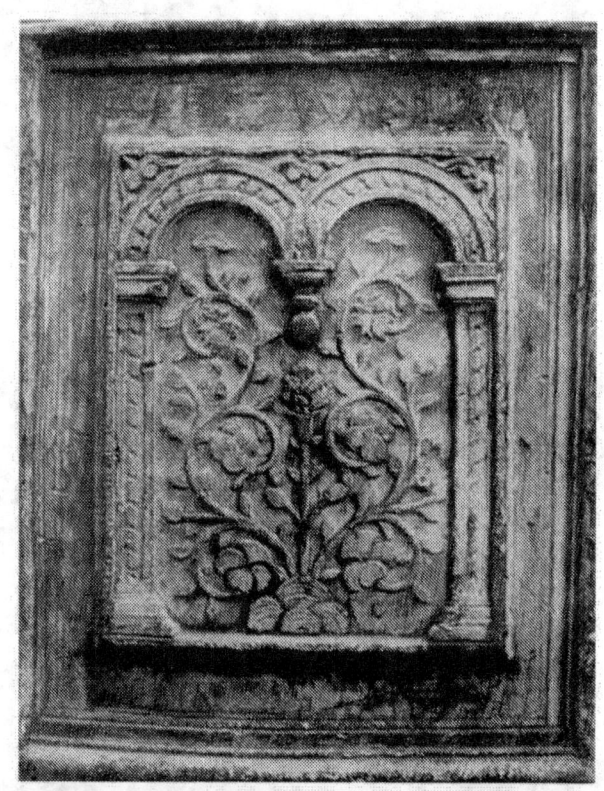

FIG. 126.—St Fagan's: Detail of Recurring Panel.

The Earl of Plymouth is the fortunate possessor, at St Fagan's Castle, Cardiff, of this magnificent example of English, or rather Welsh, water leadwork.

It is a delightful feature on its stepped stone base in the middle of a round garden, between the main entrance to the Castle and the drive. Save for the battery of time it is quite circular. The dimensions are—height, 44 inches; circumference about 240 inches. Each of the panels is $18\frac{1}{2}$ inches by $14\frac{3}{4}$ inches, and the frieze is $6\frac{1}{2}$ inches in depth. The latter was not made in uniform lengths, but joined at irregular distances with a view, apparently, to interfere as little as possible with the more important features of the design. Weight of metal has not been spared. The cistern is as much as half an inch thick on the top edge, to which wise extravagance its per-

Fig. 127.—St Fagan's: Detail of Frieze.

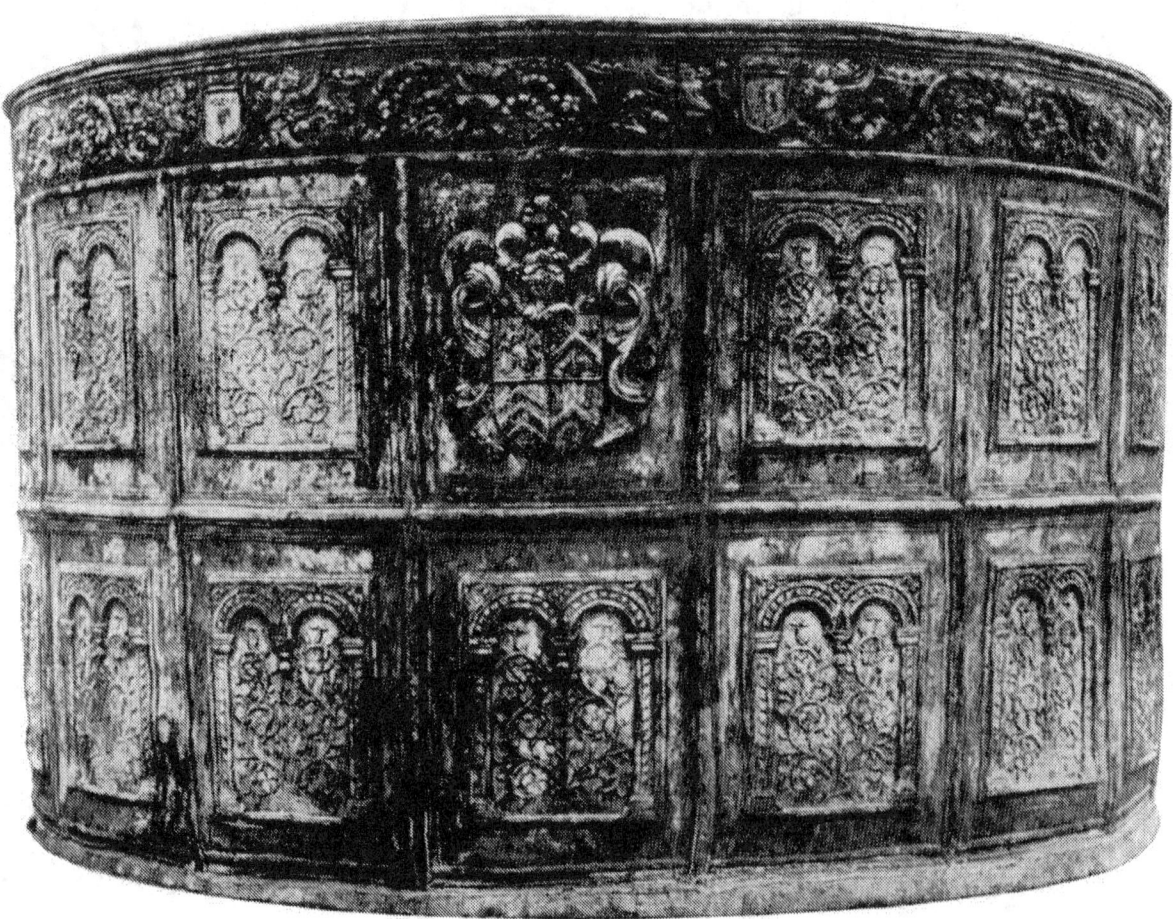

Fig. 128.—Round Cistern, St Fagan's, Cardiff.

manence is largely due. Nowhere is it less than a quarter of an inch thick, as far as can be judged without the aid of calipers. The relief is slight on the repeating panels, about a quarter of an inch, increasing a little on the royal panel, and jumping to about three-quarters of an inch on the panel containing the Lewis arms. Thirty out of the thirty-two panels into which it is divided are cast from the same pattern, which is shown large in Fig. 126. The remaining two give respectively the royal arms, with the date 1620, and the arms of Sir Edward Lewis of Van, St Fagan's, Penmark Place, and Llantrithyd. This knight of many places bought the manor of St Fagan's from Sir William Herbert in 1615-16. The tank would, therefore, seem to be one of the things with which he beautified his new estate, unless indeed he brought it from Van, a place near Caerphilly and some six miles from Cardiff. There remains at Van some Tudor work and a large round dovecot. The date does not necessarily deny this, as it may indicate the setting of the tank in its new place, but the nature of the ornament makes it likely that 1620 was the

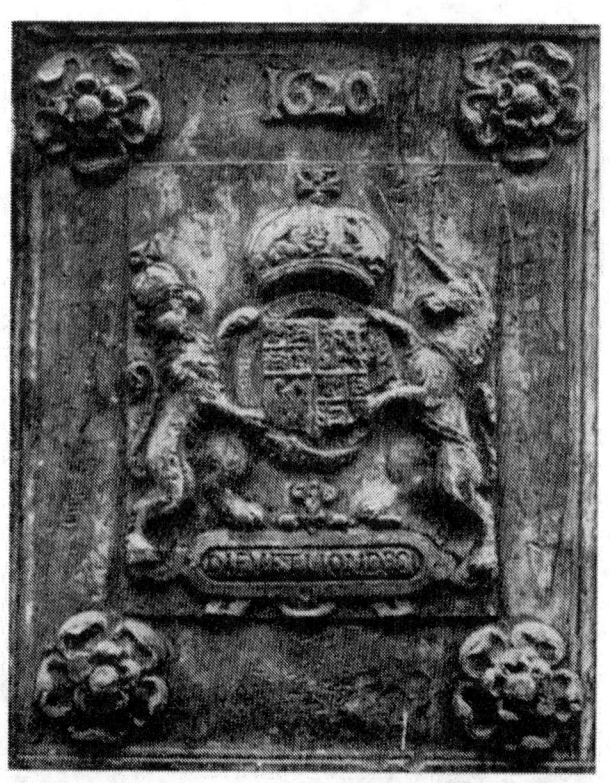

FIG. 129.—St Fagan's: Detail of Royal Panel.

date of its making. As, however, the panel with the Lewis arms was obviously (from its treatment and from the seams on the inside of the cistern) inserted after the main part of the cistern was made, a pleasant taste of doubt remains.

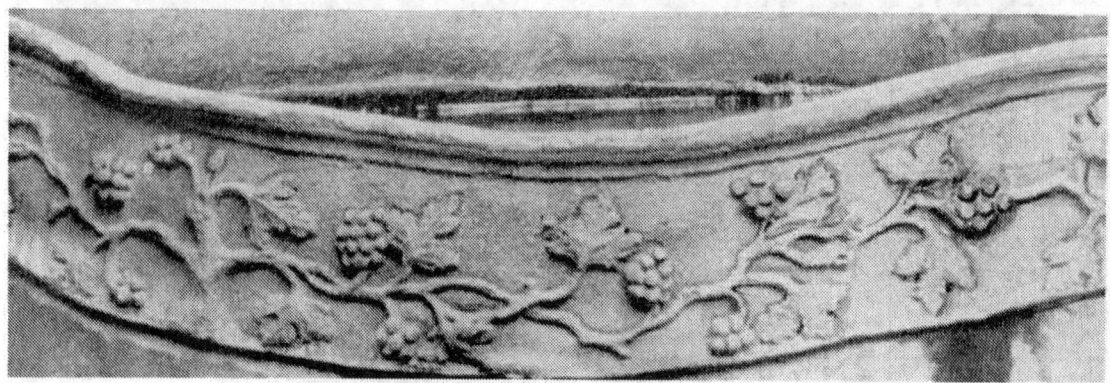

FIG. 130.—Cistern, Lincoln Cathedral: Detail of Frieze.

It is likely that the cistern as it stands now is not complete. Probably a fountain stood in it originally, with some conceit like a cupid or nymph spouting water. If it was a local production it is a feather in the cap (unhappy metaphor) of the Welsh plumber of the seventeenth century. Speaking generally, the main impression

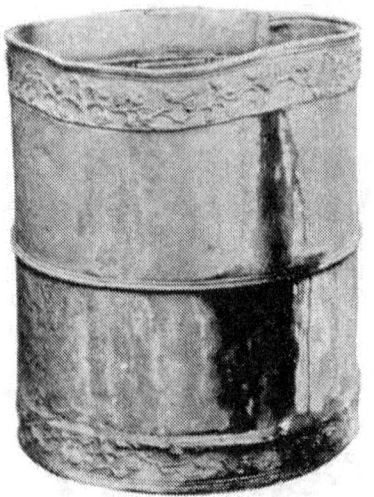

it gives is of a curious likeness in general treatment to the arcaded Norman fonts, of which there are six in Gloucestershire. The comparative nearness of these fonts makes it a not too flighty suggestion that they may have influenced the design.

At Kempston Hall, Dorsetshire, is an angle cistern with curved front divided by mouldings into six panels, ornamented with the date 1633, lions rampant, a fleur-de-lys, and the initials H. A.

It would be unwise to dogmatise as to the date of the example of Fig. 131, which is at Lincoln Cathedral. It looks very early, indeed the ornament has a flavour of the fourteenth century, but is probably as late as 1650. Though plain it is full of interest. The running bands of ornament are unlike the usual formal treatment of flower motives (Fig. 130). The three vine patterns on gutters

FIG. 131.—Lincoln Cathedral.

(illustrated in Figs. 58, 61, and 62) all repeat, and have a definite composition. But these Lincoln flowers meander round their native tub in a pleasantly casual fashion, which is

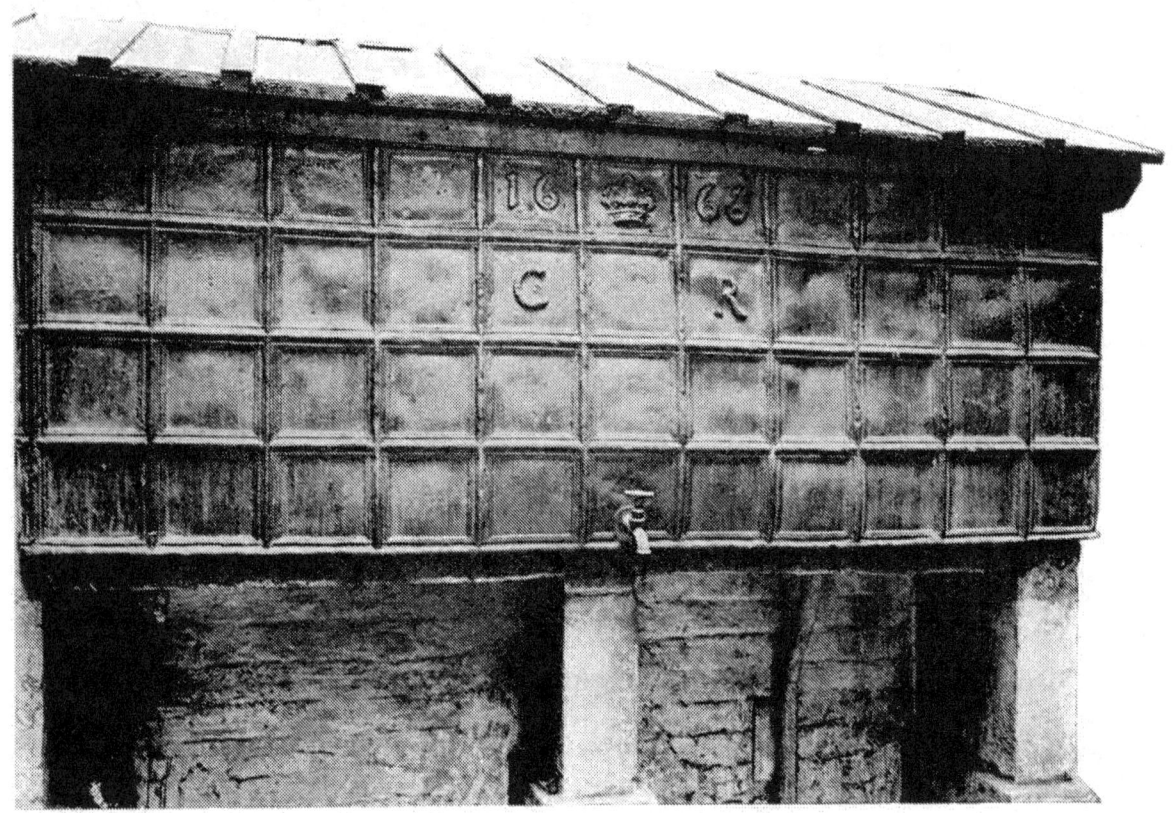

FIG. 132.—No. 10 Downing Street.

foreign to the usual primness of leadwork. On the west country cisterns of the seventeenth century the top and bottom bands of ornament have their ingenious little woodland scenes modelled in the same irregular way, but figures almost necessarily import a freer treatment. The Lincoln ornament is naïve to the point of being amateurish, and there is no effort to give the line of stalk a distinctive sweep, which would pull the design together.

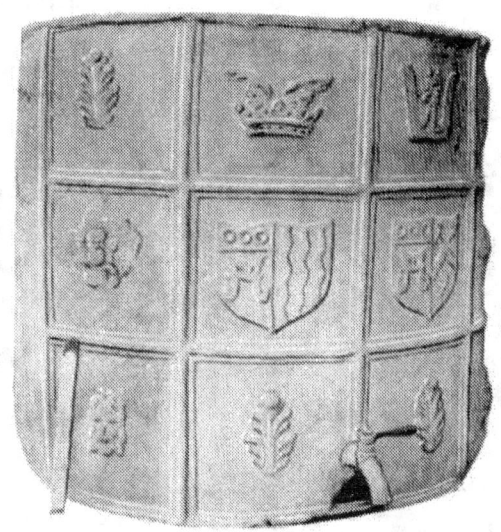

At No. 10 Downing Street, Westminster, there is a plain panelled cistern dated 1666. It is very sparingly enriched, as only five of the forty-four panels, into which the ribs divide it, bear ornaments, which are the date, a crown, and C. R.

At Ayscoughfee Hall, Spalding, Lincolnshire, there is a fine cistern almost circular (Fig. 133) and about 3 feet in height. The winged coronet is an interesting ornament. It is rather unusual to find no frieze round the top of the cistern, such as we have in the Bovey Tracey and Poundisford Park circular examples, which are similarly divided into square panels.

Fig. 133.—Ayscoughfee Hall, Spalding.

This is but one of many pleasant things at Ayscoughfee Hall, which, under municipal care, has a somewhat neglected look.

Bolton Hall, Yorkshire, has a fine series of lead cisterns, which are of the same period as the pipe-heads illustrated in the last chapter. They stood originally at

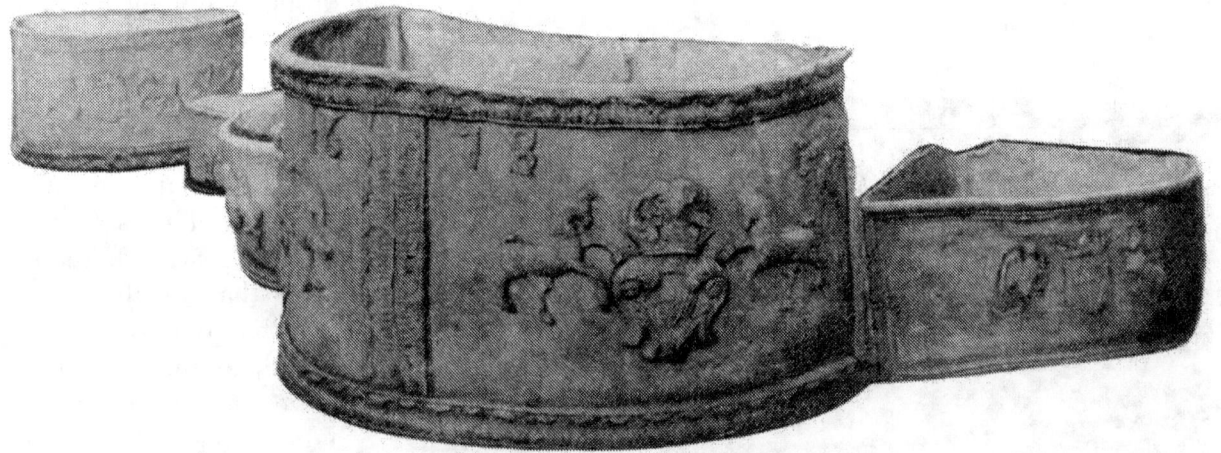

Fig. 134.—Bolton Hall, Yorkshire.

the foot of the stack pipes, and it will be noted that the cistern at the right of the group in Fig. 134 is angled on plan to suit the angle pipe-head already mentioned. The semicircular plan of the larger ones is unusual, and a pleasant variant of the ordinary rectangular form. The simplicity of their treatment is in contrast with the rather crowded ornament of the pipe-heads. There is no attempt to panel the fronts with ribs. On the larger cisterns the classical leaf moulding which runs round the top and bottom divides the semicircular front vertically with a double band. For the rest they

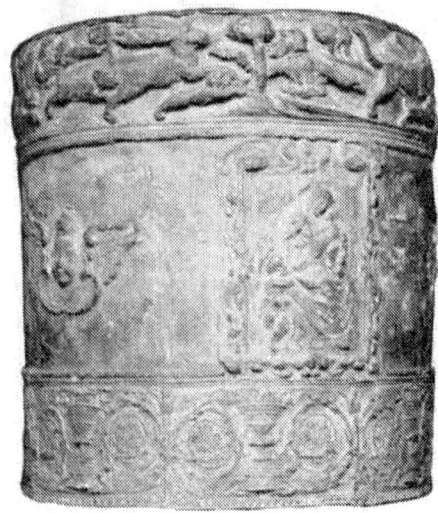

Fig. 135.—French Cistern,
South Kensington Museum.

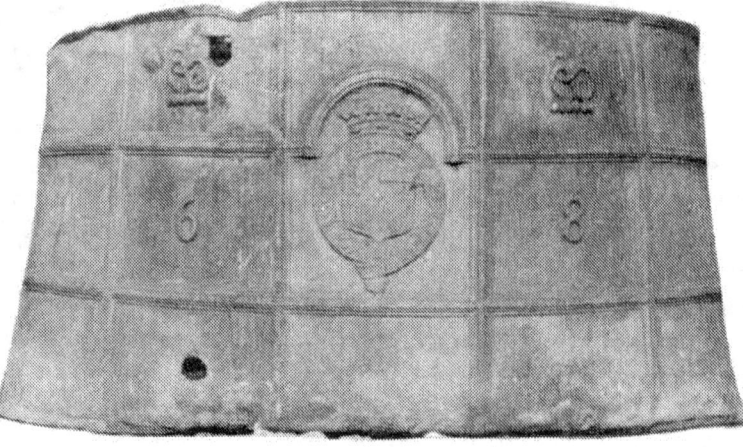

Fig. 136.—Nottingham Castle.

were content simply to apply the coat of arms of the Paulet and Scrope families, with their supporters. On the small angle cistern the Scrope choughs support the Paulet shield, due probably to muddled refixing at some time when a number of the heraldic ornaments had

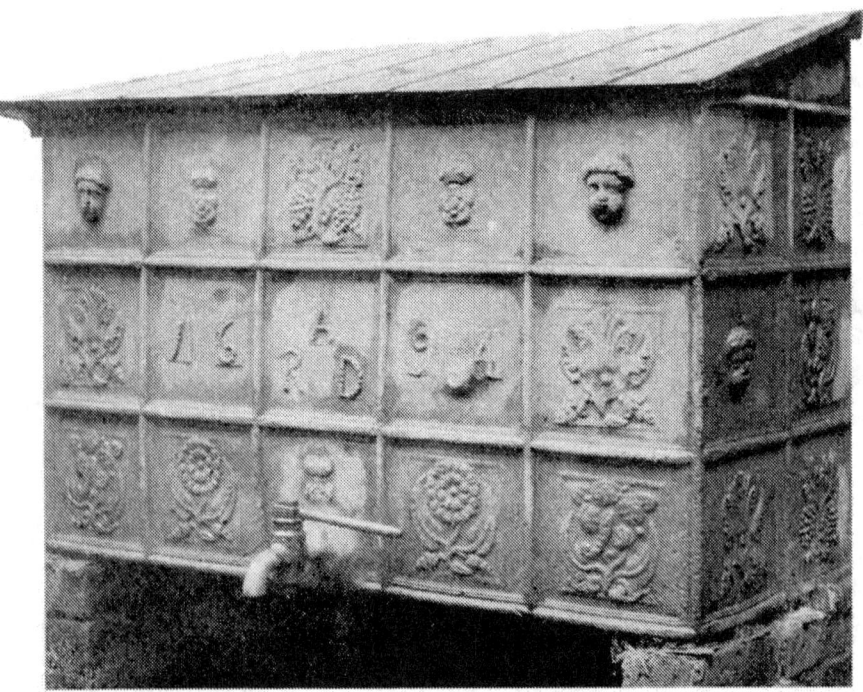

Fig. 137.—Exeter, 1694.

dropped off, owing to bad work when the cisterns were first made. There are more applied ornaments missing from late seventeenth and eighteenth century leadwork than from that of the sixteenth and early seventeenth centuries. The later men were more intent on piling on enrichments than in seeing that those they applied were firmly fixed. Although cherubs are plentiful on the pipe-heads, the Bolton cisterns lack their celestial presence.

They are more plentiful on cisterns than on fonts. The Slimbridge font (see Chapter I.) dated 1664 might almost, except for its size, be a rain-water butt. It has four cherubs, but seventeenth-century cherubs did not discriminate between spiritual and secular tubs, and took up their abode as readily on the latter as on the former. It is worth recording that we do not find English cisterns decorated with religious emblems, if we except cherubs, which are as often profane amorini as heavenly products. On a French cistern at the South Kensington Museum, illustrated here by way of comparison (Fig. 135), there is a panel of the Virgin and Child. Very lean and strenuous dogs are coursing round the frieze. The round tank, dated 1681, at Nottingham Castle is an admirable example of the plainer sort (Fig. 136). The arms are those of Henry Cavendish, K.G., and the "serpent nowed" is the Cavendish crest. The outward slope of the sides, from the top downwards, adds

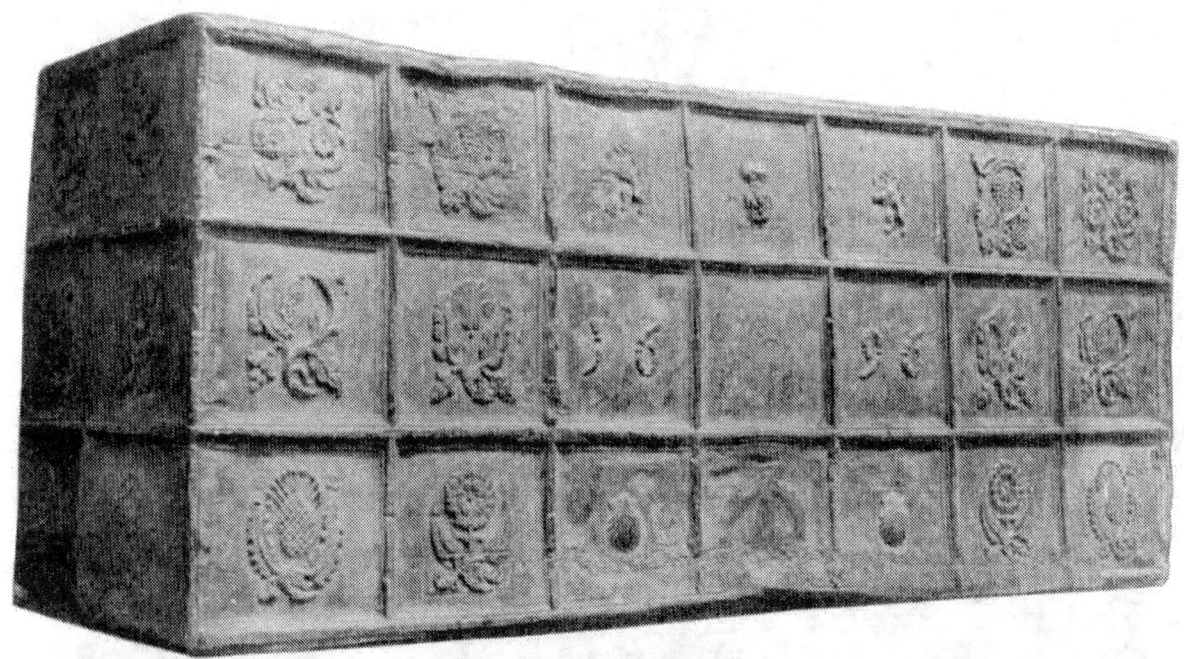

FIG. 138.—Exeter, 1696.

decorative interest to the tank, but makes it less practical when it comes to cleansing it. After all, if one drinks water from a lead cistern, a few bacteria more or less are not of much account, and seventeenth-century courage was undisturbed by those pleasant creatures whose names make a point of ending in *coccus*.

There is a vigour about the decoration of Devonshire and Somersetshire cisterns of the late seventeenth and early eighteenth centuries which cannot be claimed for the London work of the same date.

The Exeter examples dated 1694, 1696, 1708, 1715, and 1724, and the tanks at Poundisford Park and Bovey Tracey all have a delightful variety of flower and animal ornaments which are freshly amusing. Probably they were made by the same plumber. Some of the ornaments which are seen on the tank of 1694 (Fig. 137) are repeated on that of 1724. They obviously are cast from the same or duplicate patterns. There is a delightful disregard of scale. In a sporting scene on the 1724 cistern the huntsman is but little larger than the dogs, and the stag has a quiescent air which does not quite match with the violent activity of the three dogs (one high in the air) which are after him. But it makes a quite dramatic picture.

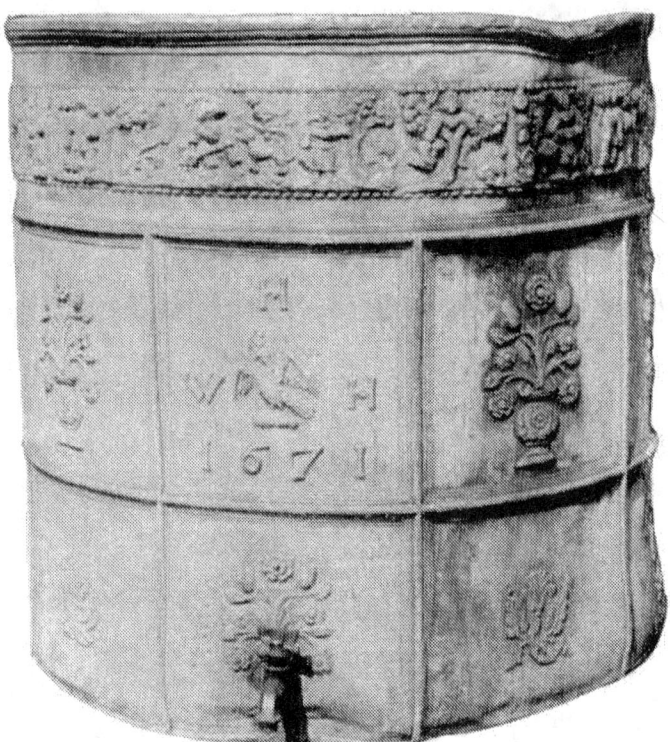

FIG. 139.—Poundisford Park.

The Deanery at Exeter possesses two very much alike, dated 1694 and 1708. The former is illustrated in Fig. 137, and the admirable modelling of the vine pattern in the middle of the top tier of panels is worthy of note.

The cistern of Fig. 138, in the possession of Mr Harry Hems, at Exeter, is a particularly good example of simple panelling. It is dated 1696, and probably had all panels filled with devices, though two have gone. The six ornaments repeating at the right and left of the front are especially interesting. Perhaps the second from the right-hand top corner is the happiest, the vine pattern being employed most successfully. The return ends are decorated with the same six ornaments. It will be noted that there are square outlines round these ornaments, which suggest that the ornaments were cast separately and applied. This is not so, however.

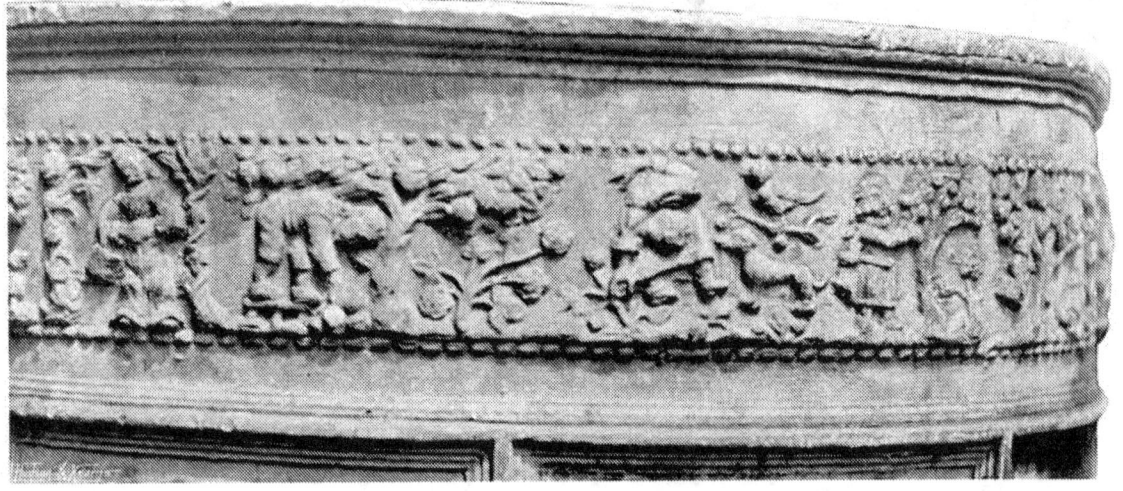

FIG. 140.—Frieze of Cistern, Poundisford Park.

The outline merely marks the edge of the loose pattern, where it was pressed into the casting sand. A word may be added here as to the method of making this cistern, which applies to most of this type. It was similar to that employed for Sussex iron fire-backs. The various ornament models were either temporarily fixed to the main pattern before it was pressed into the flat bed of sand, or they were separately impressed after the main pattern had been employed. Never, however, do we find in lead-work such freakish ornament as in one early fire-back, where the ornament is the impress of the moulder's hand, a trick amusing enough, but scarcely art. The front and sides of the cistern (Fig. 138) were cast in one flat sheet, which was bent at the front angles, and also at the back, returning 3½ inches. The return pieces are soldered to a sheet-lead backing. Two stays of sheet lead 13 inches deep divide the inside into equal distances; they reach to within 6 inches of the top, and stand clear of the bottom. In the middle, tying the front and back, is a circular solid bar of lead 1½ inches in diameter. Other dimensions

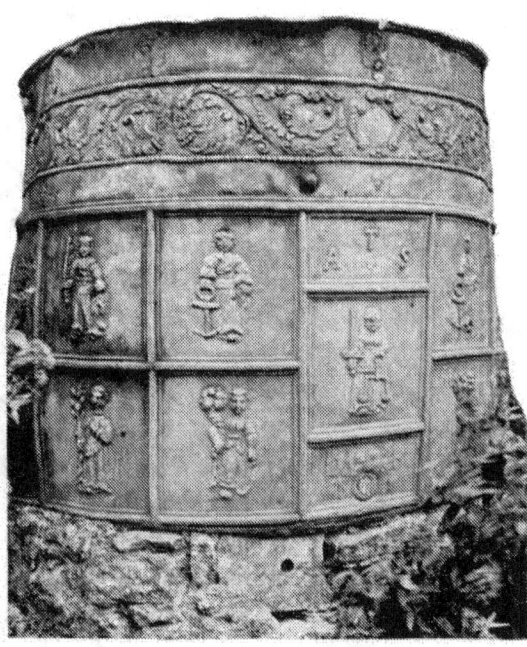

FIG. 141.—Bovey Tracey.

are: length, 6 feet; height, 2 feet 4 inches; width, 2 feet; greatest thickness, ¼ inch.

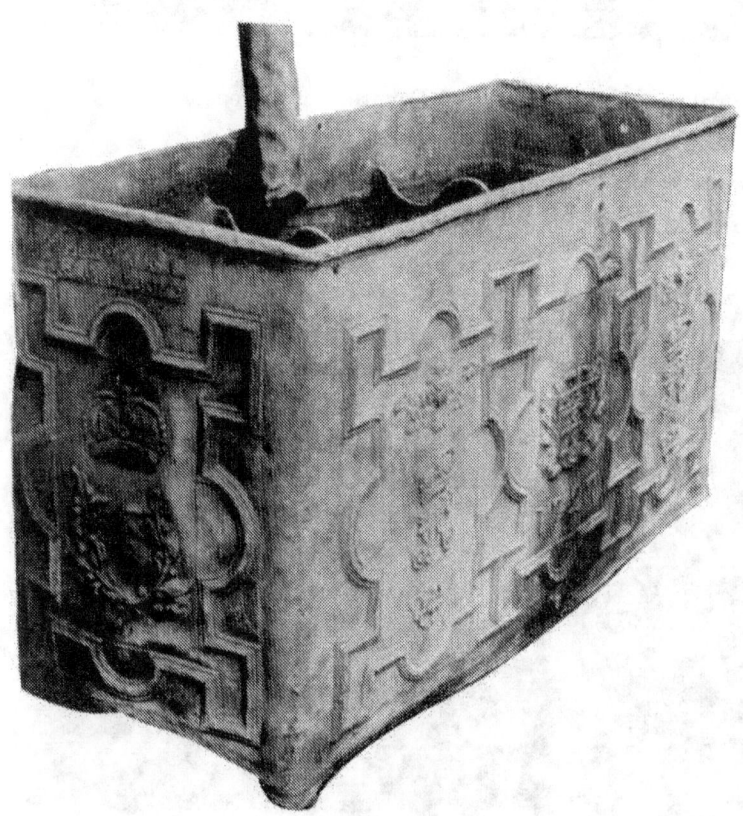

FIG. 142.—St Mary's, Scilly.

The cistern at Poundisford Park, Taunton (Fig. 139), is shown in sequence to the illustration of the rain-water head in Fig. 89. It is dated 1671. The arrangement of the pots of flowers in the panels is formal enough, but fancy has been given rein in the little frieze that surrounds the top. The scenes, as is befitting, have a garden atmosphere. One pleasant-faced urchin is apparently about to help himself from a fruit tree, while another is contemplating a rather weedy dog. Trees mingle with flowers, and altogether the composition is delightfully casual. The decoration of the Bovey Tracey tank (Fig. 141) is rather stiffer, and the frieze, though of a graceful arabesque, has not the vernacular charm of the Poundisford Park example. The little figures in

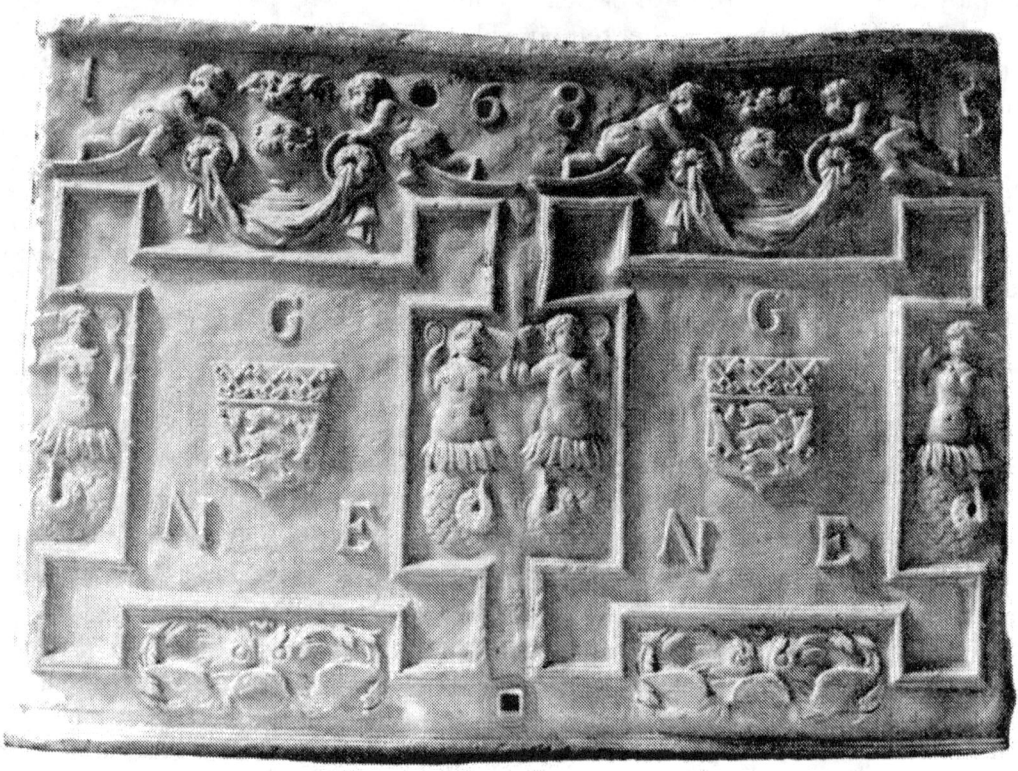

FIG. 143.—Cistern with Arms of the Fishmongers' Company, at Inwood.

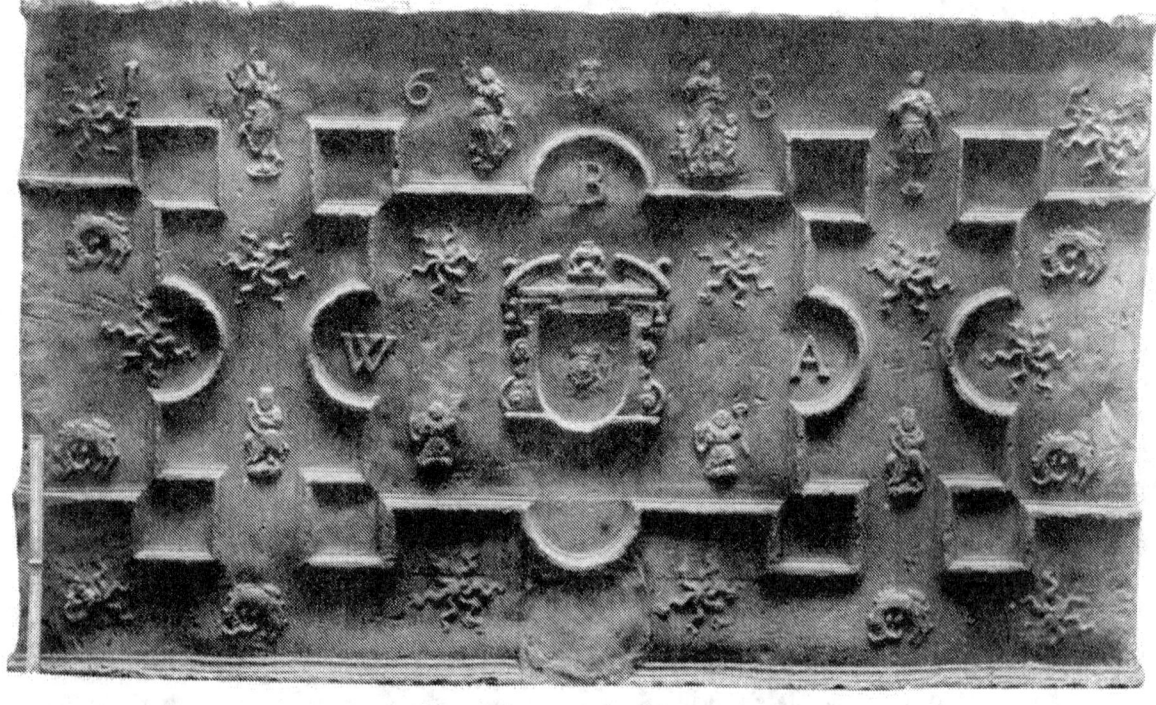

FIG. 144.—Child's Bank, Fleet Street, 1685.

FIG. 145.—Child's Bank, Fleet Street, 1757.

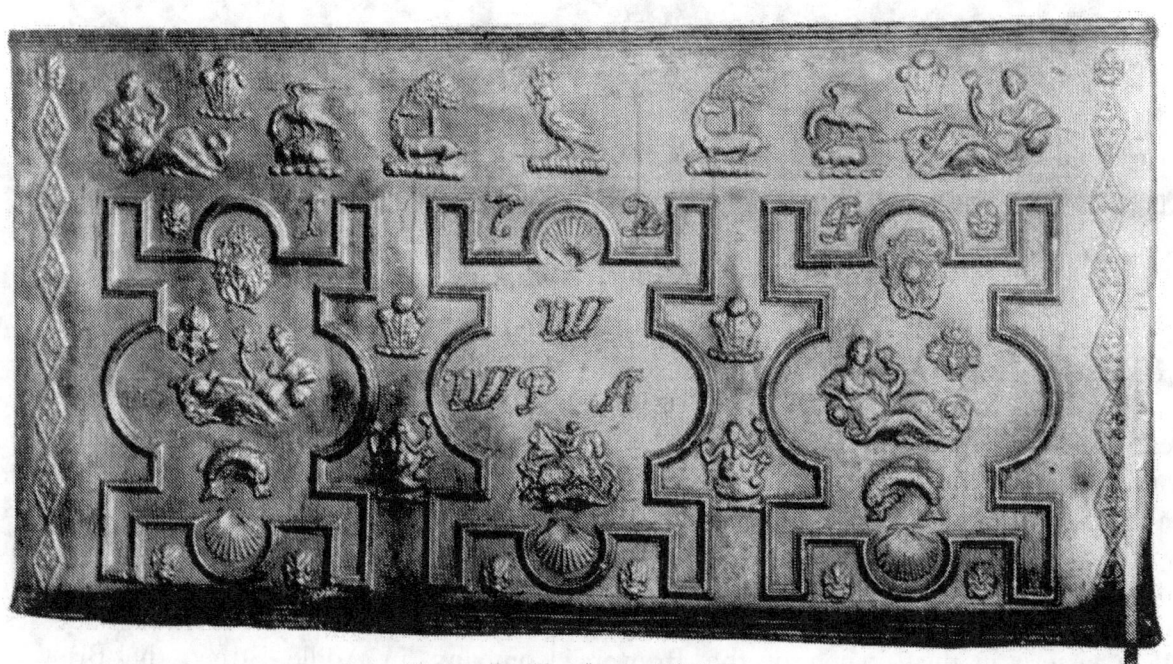

FIG. 146.—The Record Office.

the panels are charming. Justice with sword and scales has forgotten to bandage her eyes, and the lady with the cornucopia has rather the air of one of Miss Honeyman's Sallies. Hope holds her anchor with impressive stolidity, and the other little people have engaging characters of their own.

At St Mary's, Scilly, one expects something rather unusual. One may be forgiven the vague hope of finding some graceful convention of daffodils on the leadwork that would accord with the subtropical atmosphere of the Isles. But London throws its influence afar. The cistern of Fig. 142 is not only of the ordinary London type, but even bears, which is unusual, the name of the maker, "Walker, London," a name one seems to have heard before. It is a royal cistern, and bears the initials and crown of

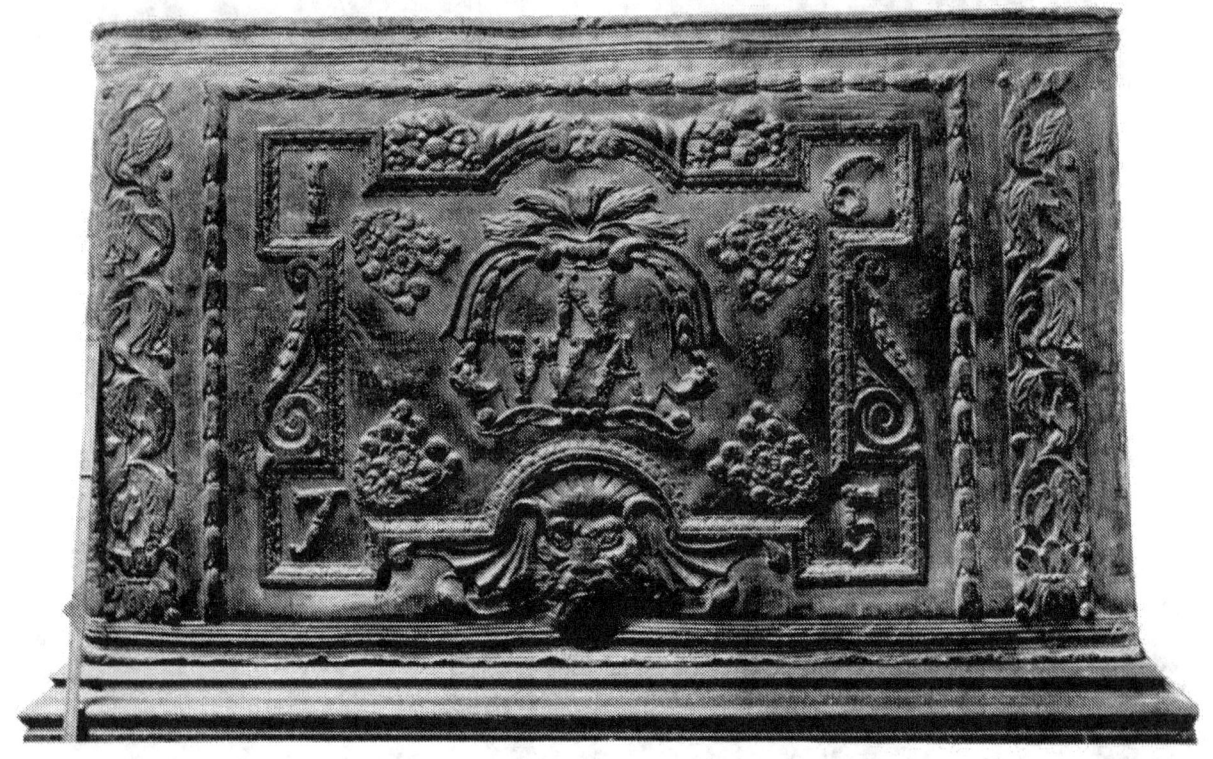

Fig. 147.—Lincoln's Inn.

George I. or II. The cherubs are very fully bewinged, and the arms of the central panel are those of H.M. Ordnance Office, which controlled the Castle at St Mary's.

In all the tanks of this type, and there are still scores in London, the ingenuity of the designer was busiest in the treatment of the ribs. There seems to be no end to the combinations of half circles and straight lines. This sort of design is an affair of set-square and compass, and frankly is not difficult. The London work is not rich in fancy. There is not in the modelling of the applied ornaments anything like the gaiety we find in the enrichment of work of similar date in the West of England. London plumbers dotted the faces of their cisterns rather mechanically with shells and stars and stiff little goddesses. On a cistern in the kitchen of the Brewers' Company, in Addle Street, the Brewers' coat of arms is repeated thirteen times, surely a little too often. For the rest it has

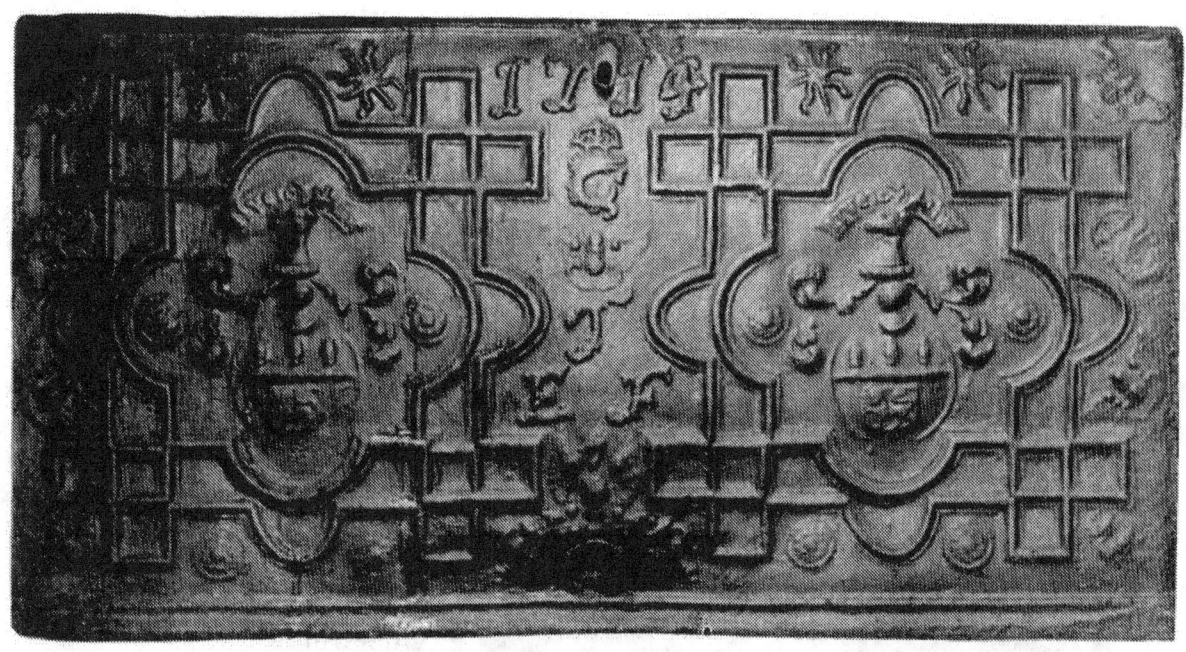

FIG. 148.—4 Queen Square, Bloomsbury.

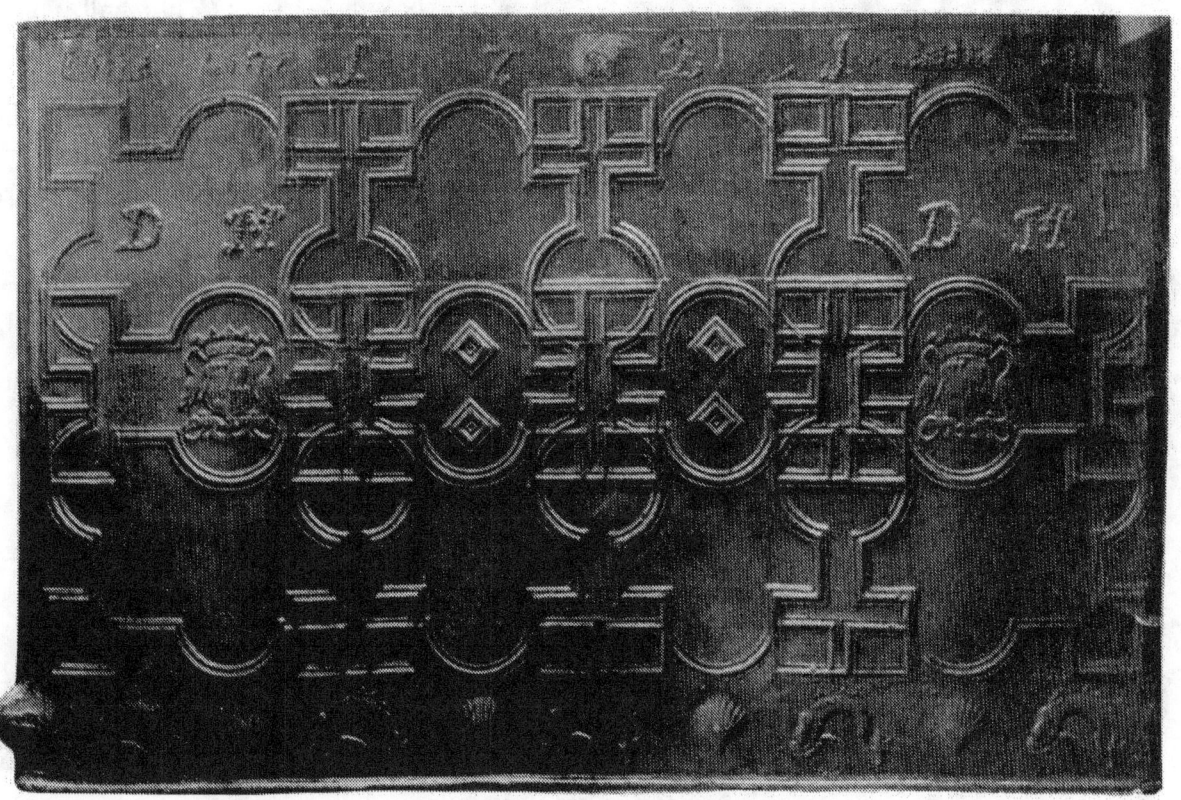

FIG. 149.—20 Hanover Square.

stars and shells between the ribs. A swag or two, however, gives it a little variety. It is singular that swags are so little used in leadwork, seeing that they were such usual

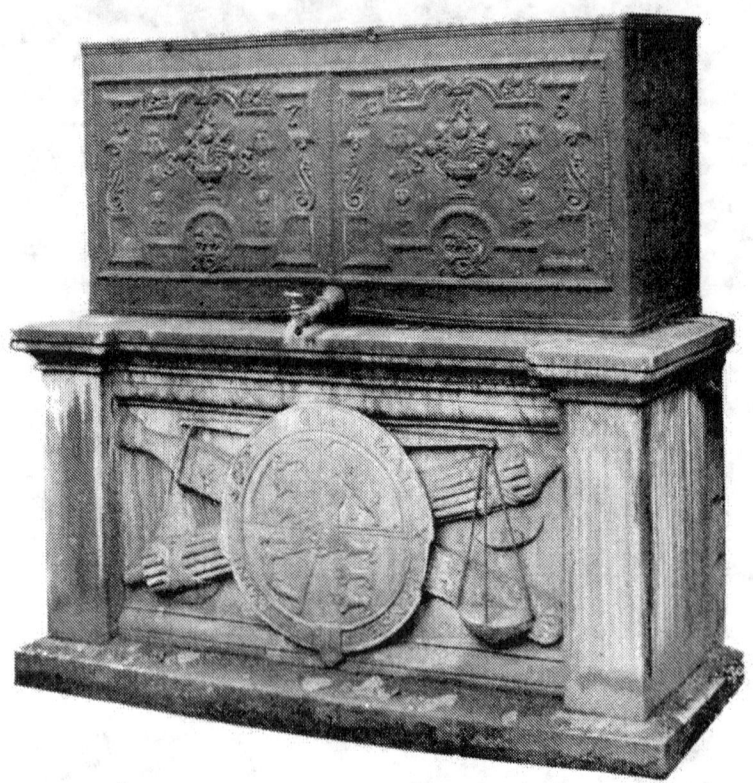

enrichments in the allied craft of plasterwork. The City Companies are rich in cisterns. There is one at the Bakers' Company dated 1720.

At Inwood there is a London cistern dated 1685, which bears the arms of the Fishmongers' Company (Fig. 143). The modelling is distinctly better than the average, and Mr Starkie Gardner regards this tank as an example of the degree of relief that may properly be applied to panelled leadwork. There are several examples of merit in the Guildhall Museum, London.

Child's Bank, Fleet Street, has three to its credit. Fig. 144 shows one of the best in London. It is dated 1685. The half panels return round the sides, and in this show a pleasant disregard of the prevailing practices. The

FIG. 150.   44 Great Ormond Street.

ornaments are admirable. The stars are gay and curly, and there is an echo of history in the very small bust of King Charles I. between the 6 and the 8. The little figures are

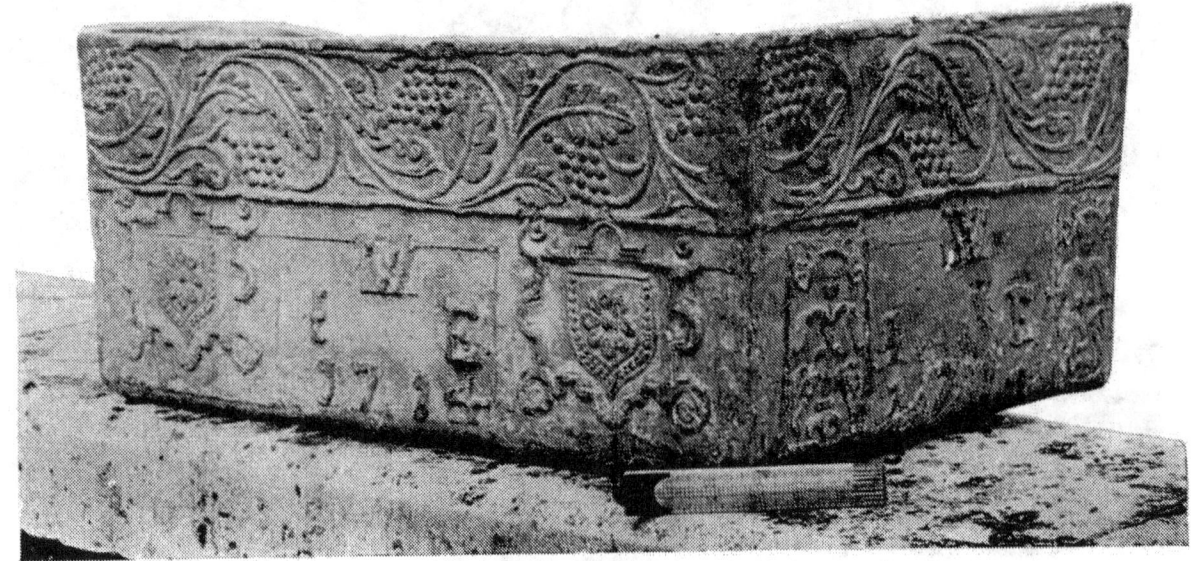

FIG. 151.—Charlton House, Kent.

vigorous and interesting.   Those at the right and left of the lower tier may be taken to be King David harping on his harp.   As to the remaining ornament, which occurs six times, it is difficult to dogmatise.   It suggests an exasperated prawn, or perhaps a fresh-water relative inhabiting London cisterns—anyhow a watery creature.

A second cistern at the same Bank is dated 1679, and retains a little Gothic feeling in the fleur-de-lys, but some Tudor roses are very feebly modelled.

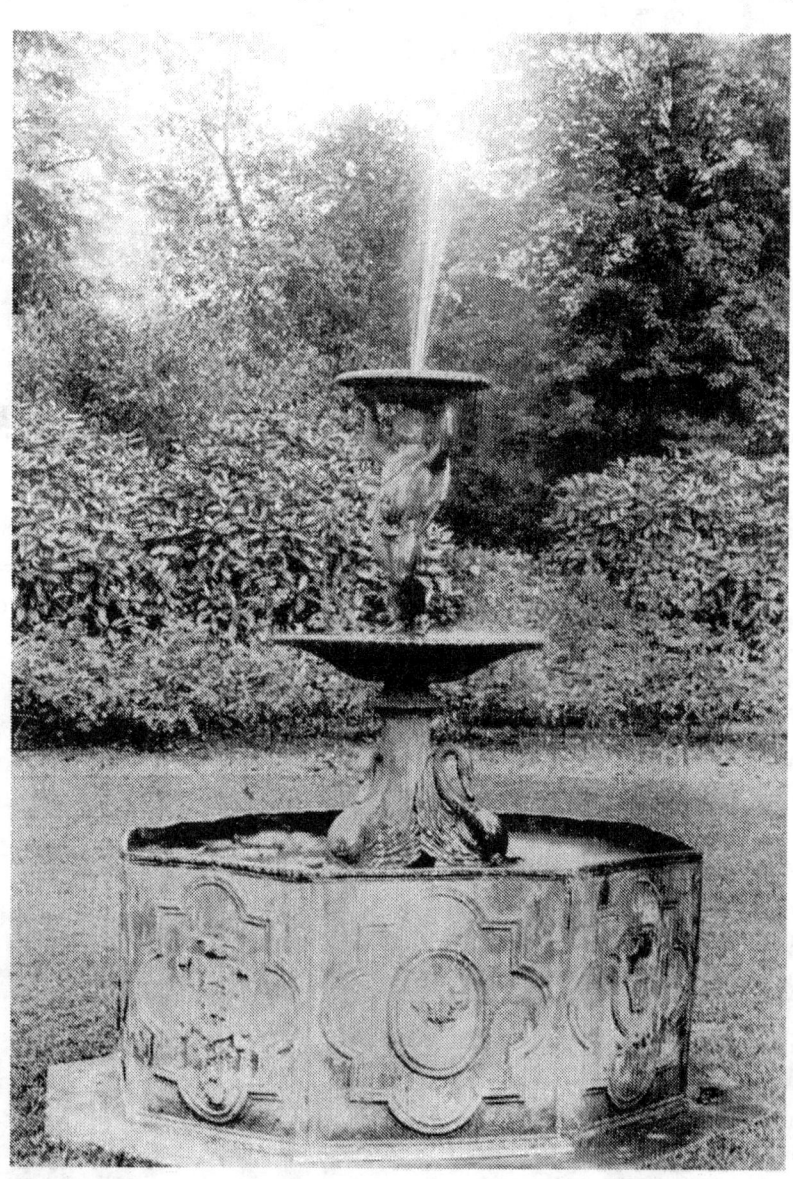

The tank of 1757 (Fig. 145) is the third of the series, and is a good example of the formalism of the later eighteenth-century work.   The some-what excessively whis-kered lions of the oval panels are amusing though, and the strips of rather aimless ornament down the side lighten the gene-ral effect.

At the Record Office, in Chancery Lane, near the doorway of the Rolls Chapel, are four eigh-teenth-century cisterns, one of which is shown in Fig. 146.   This surely reaches the zenith of the marine store style of de-coration.   The plumber has made the front of his tank a museum of his pat-terns.   He must have suffered from an acute horror of plain surfaces. It is an entertaining pro-duction, but one is grate-ful that it does not always happen.

FIG. 152.—Charlton House, Kent.

Mr Max Clarke has at his house in Queen Square a good example (Fig. 148), which yet has some technical failings.   The patterns seem to have been carelessly used, with the result that the alignment of the ribs is very irregular.   The star ornaments are poor compared with those on the tank of Fig. 144, and the lettering is straggling and forlorn.   The treatment of the coats of arms is rather more ambitious than successful.

At 20 Han-over Square ("the common lodging-house of learned societies"), which shelters those who are wise in everything from obstetrics to Irish folk songs, there is a tank in the area, visible from the door-way (Fig. 149). If the Record Office example was a study in spotty orna-ment, this is a

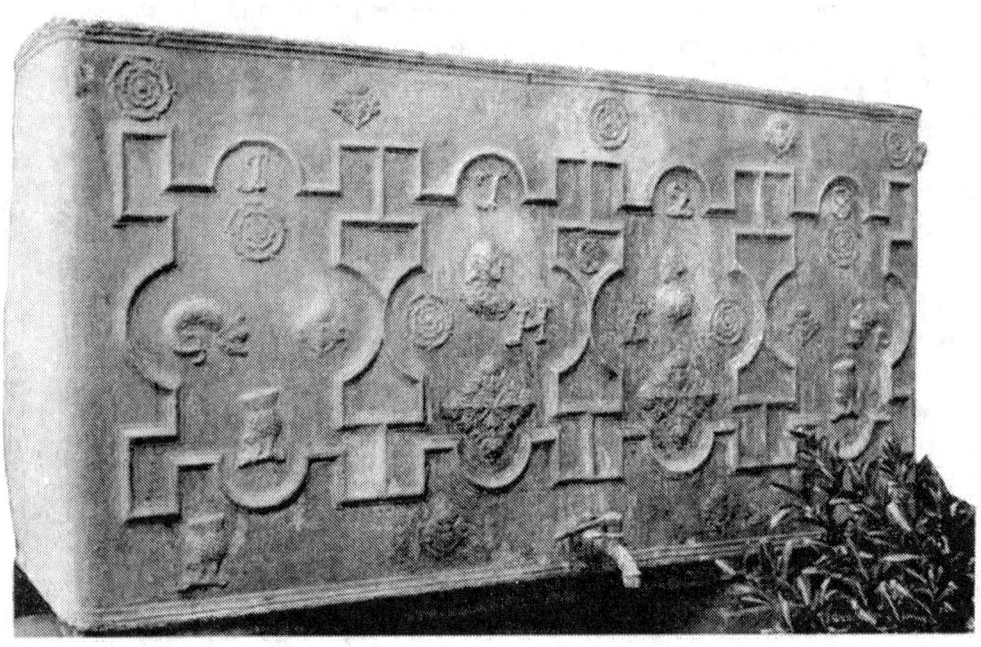

FIG. 153.—Ealing.

liberal education in the interlacing of ribs, almost Runic in complexity.

Lincoln's Inn has three excellent cisterns. One is very plain, divided into two panels with simple ribs, and altogether lacking further ornament. The second (illustrated in Fig. 147) is one of the most elaborate in England, and shows some scholarship in its design. Though the outline of the ribbing is not unusual, the ribs them-selves are richly modelled, and the trusses at the sides give a strong architectural flavour. The trophy of fruits at the top and the mask are admirable of their kind.

The vertical strips of ornament at the ends, while good in themselves,

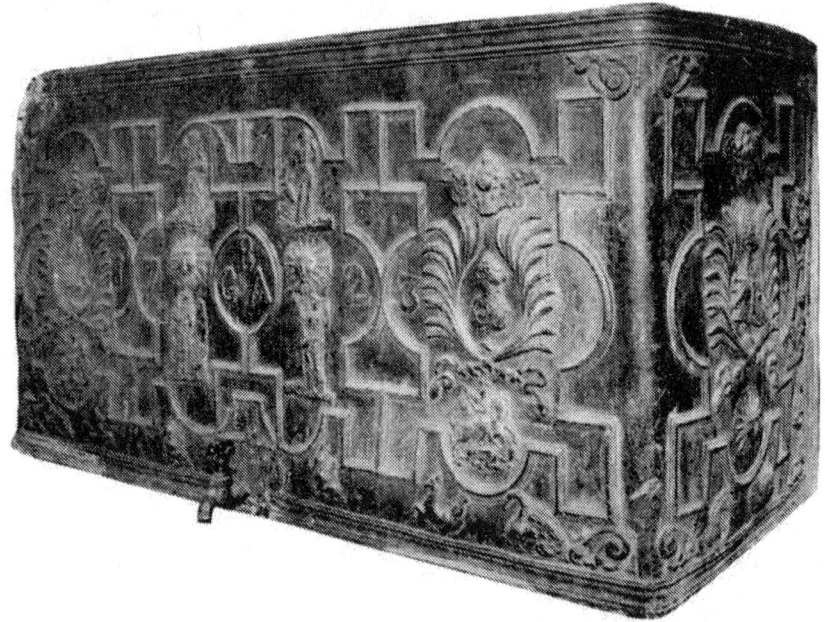

FIG. 154.—Bedford Row.

seem rather a mistake. One feels that the cistern would have been better if it had stopped short of these strips, and finished outside the very good framing of husks. While the pro-portion of the tank would not have been so good, decoratively there would have been a

unity which now it rather
misses. The third cistern in
the Inn is dated a few years
later than the last, and was
evidently inspired by it, as the
ribs and some of the enrichments are the same. Probably
the same patterns were used.

Near by, in Great Ormond
Street, at the Nurses' Home of
the Children's Hospital, there
was a cistern dated 1745 (Fig.
150), evidently made from the
same patterns as the two best
examples at Lincoln's Inn.
The stone pedestal on which it
stands is a modern addition,
set up by Mr Frederick Warre.
He found the tank stowed
away in a cellar, and as Lord
Thurlow once lived in the

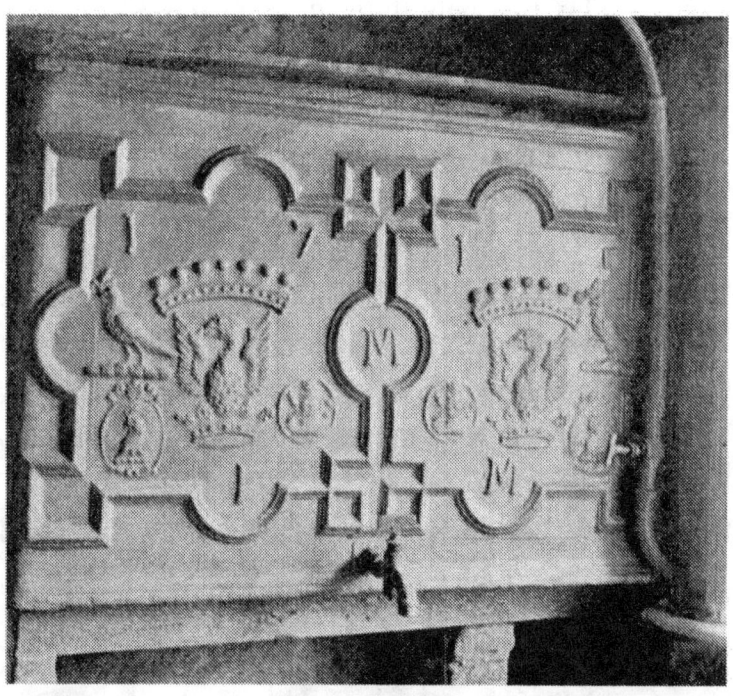

FIG. 155.—Richmond.

house, the scales of justice and the lictors' rods are appropriate emblems of the great

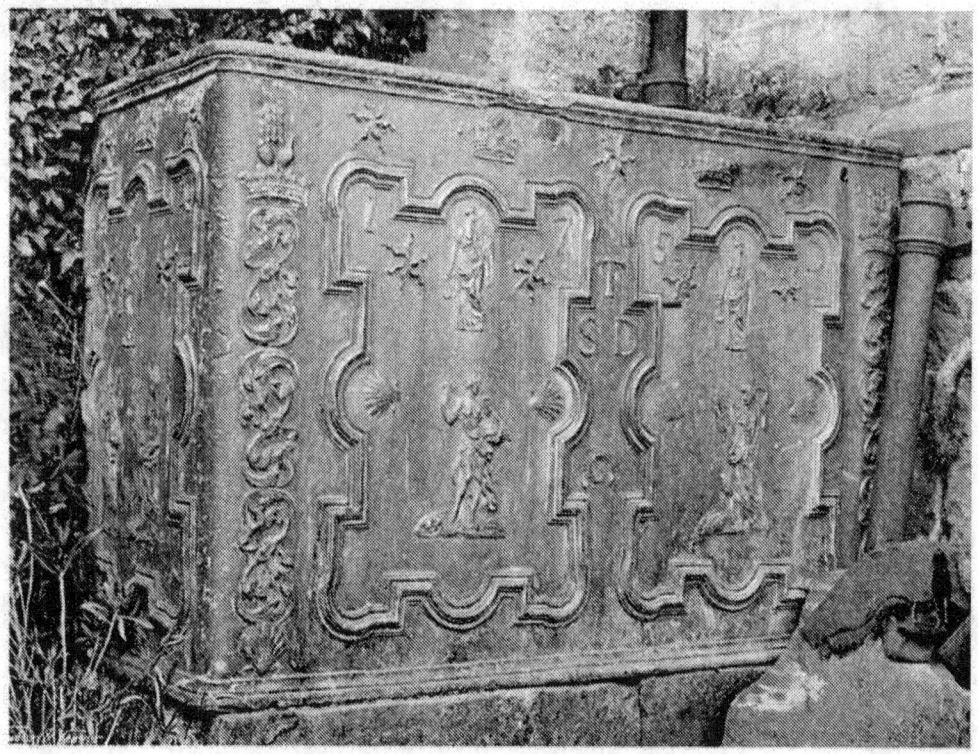

FIG. 156.—Sackville College, East Grinstead.

judge. He was only thirteen years old when the tank was made, so must be acquitted of having any hand in its design.

Very delightful is the little tank of Fig. 151, which Sir Spencer Maryon Wilson of Eastborne has at Charlton House, Kent. It is not strictly a cistern (being only about 24 inches long and 11 inches high), but rather a jardinière. The decoration is more natural than is ordinarily found in 1714, and were it undated, fifty years earlier would be a reasonable attribution. Its great charm is in its colour. It is almost purely white, and might indeed have come from Blakesware, where Elia wrote of the "flower-pots, now of palest lead, save that a spot here and there, saved from the elements, bespeak their pristine state to have been gilt and glittering." At Charlton no gilt survives, if it were ever there.

As far as possible the illustrations for this book are made strictly *ad hoc* by the omission of the surroundings of the leadwork; but the octagonal cistern at Charlton House

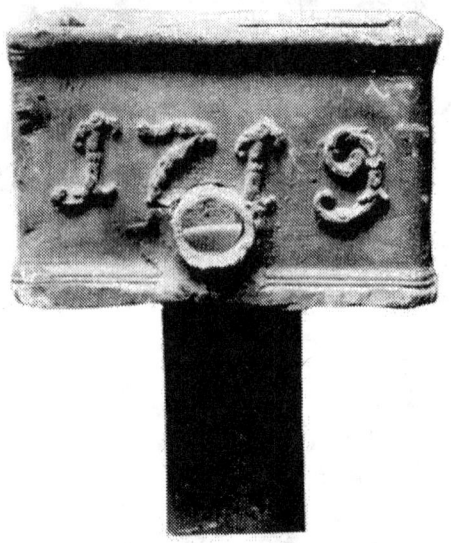

Fig. 157.—Lead Pump-head.

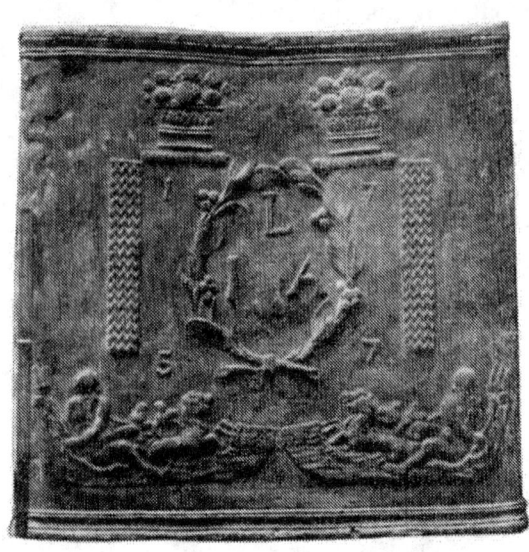

Fig. 158.—Tenterden Street.

(Fig. 152) would lose half its charm if divorced from its charming setting. It stands filled with water-lilies, and is a centre of spouting freshness in a rose garden framed in trees. Each face of the octagon is about 2 feet long, and the tank is a particularly happy example of the panelled type. It was perhaps made in the time of Sir William Langhorne, as the initials W. L. appear on the tank. Originally it was probably, as it is now, the base of a fountain. The upper part is an addition, and was but recently acquired. It is "antique" (precious word), and not old, but the swans and cupid make with the tank a most agreeable composition. There are two more cisterns at Charlton House with ribbing. They are dated 1774, but the octagonal one is probably of the seventeenth century. The cistern at Ealing (Fig. 153) is another injustice to Ireland. The rose and thistle occur several times, but the shamrock is not to be found. There are also two notable square patches of ornament that look like rich embroidery, and have an almost Gothic feeling. The dolphins give the needful watery touch. In a Bedford Row cellar is a cistern of the same date, 1723, and probably by the same hand (Fig. 154).

The outlines of the ribs are identical, and both tanks bear a pair of small busts, which perhaps indicate George I. and his consort. The crossed palm branches are very decorative, and there are several figures, including a George and the Dragon. The lead tank of Fig. 155 is in the kitchen of a delightful house on Richmond Green. It is English enough in all but its ornament, and it has been suggested that the double-headed eagle is an indication that the house was in 1715 a residence of the Austrian ambassador of that date.

The very interesting little cistern of Fig. 158 was taken from a demolished house in Tenterden Street, W., by Messrs Cowtan & Son. It is dated 1757, not a very fruitful period for symbolism, but the strips of zigzag may be there for a purpose. The same ornaments have not been found elsewhere, and, regarded simply as decoration, they are rather a harsh addition to an otherwise pleasant arrangement. The Neptunes are driving their teams in very spirited fashion, and the wreath is quite graceful, if a little attenuated. The baskets of flowers seem rather a mistake. At Sackville College, East Grinstead (Fig. 156), the panelling has a curiously halting but refined outline, and the enrichments are admirable and sparingly used. Fig. 159 shows four delightful low reliefs in the possession of Mr Herbert Batsford. They probably formed part originally of a cistern, and are good typical work of the first half

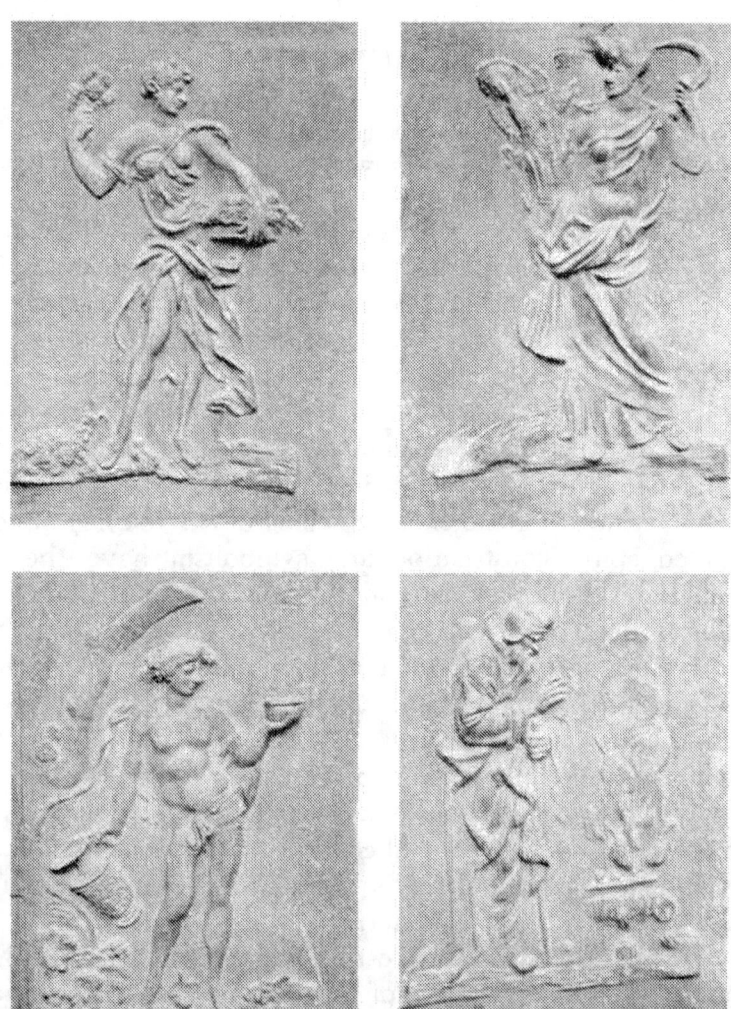

FIG. 159.—Panels of the Four Seasons.

of the eighteenth century. The same reliefs appear on a cistern at the Guildhall Museum, London, which bears the date 1795 and the name of Sir John Cass.

Pump-heads are less common than cisterns, but they are not very interesting. One of normal type is illustrated (Fig. 157), which is rather early in date (for a pump-head). Others bear the stock cistern enrichments, such as shells, stars, and lions' masks.

CHAPTER V.

MEDIÆVAL LEADED SPIRES.

The Character of Spires—Classification—"Collar" and "Broach"—Destroyed Cathedral Spires—Existing Leaded Spires—Scots Leadworkers—St Nicholas, Aberdeen—Old Saint Paul's—Chesterfield.

AMONG the debts of gratitude which architecture owes to lead, there is none more weighty than its use in roofing. The roof may be said to be the second need of architecture, as the wall is the first. The wall gives privacy, the roof brings protection. The spire is the supreme form of the roof; it is the roof spiritualised. In its relation to the Gothic spirit it has a character all its own. In its essence it is the roof of a tower, but it intends more. It is a constructed symbol of aspiration, and its building is one of the greatest concessions to constructed beauty and symbolism which Gothic art has made.

Since lead is the most efficient of all roofing materials, it is fair to say that, in the leaded spire, construction and symbolism have their perfect meeting. Among spires generally, those that are leaded take a small and rather forgotten but still honoured place. The leaded spire has a character all its own, and maintains its character of a spiritualised roof more intelligibly than a stone spire can do. The white, almost glistening, patina which comes with age on lead, where air is not befouled with city smoke, makes the spire stand like a frosted spear against the sky; and the slight twists, which almost every timber spire has taken, give a peculiar sense of life. These are "refinements" which do not fit any theories, but result from the sun sporting with a slender timber structure, made more sensitive by its metal coat. A shingled spire is apt to twist (Cleobury Mortimer is an example), but there is none shingled that compares with the inebriate vagaries of the leaded spire of Chesterfield.

One of the most interesting points that arises with leaded spires, as indeed with all subjects, is the question of origins, and in this connection shingled as well as leaded timber spires must be mentioned. Mr Francis Bond in "Gothic Architecture in England," took some pains to classify spires of all types. He divided them broadly into Pathless and Parapetted. A fresh classification is now offered, on the same lines, but amended.

Pathless—
1. Collar-type, *e.g.*, Ryton.
2. Broach-type, *e.g.*, Braunton, Barnstaple, Godalming, Ickleton, Swymbridge, Almondsbury.
3. Pinnacled type, *e.g.*, Long Sutton, and St Nicholas, Aberdeen.

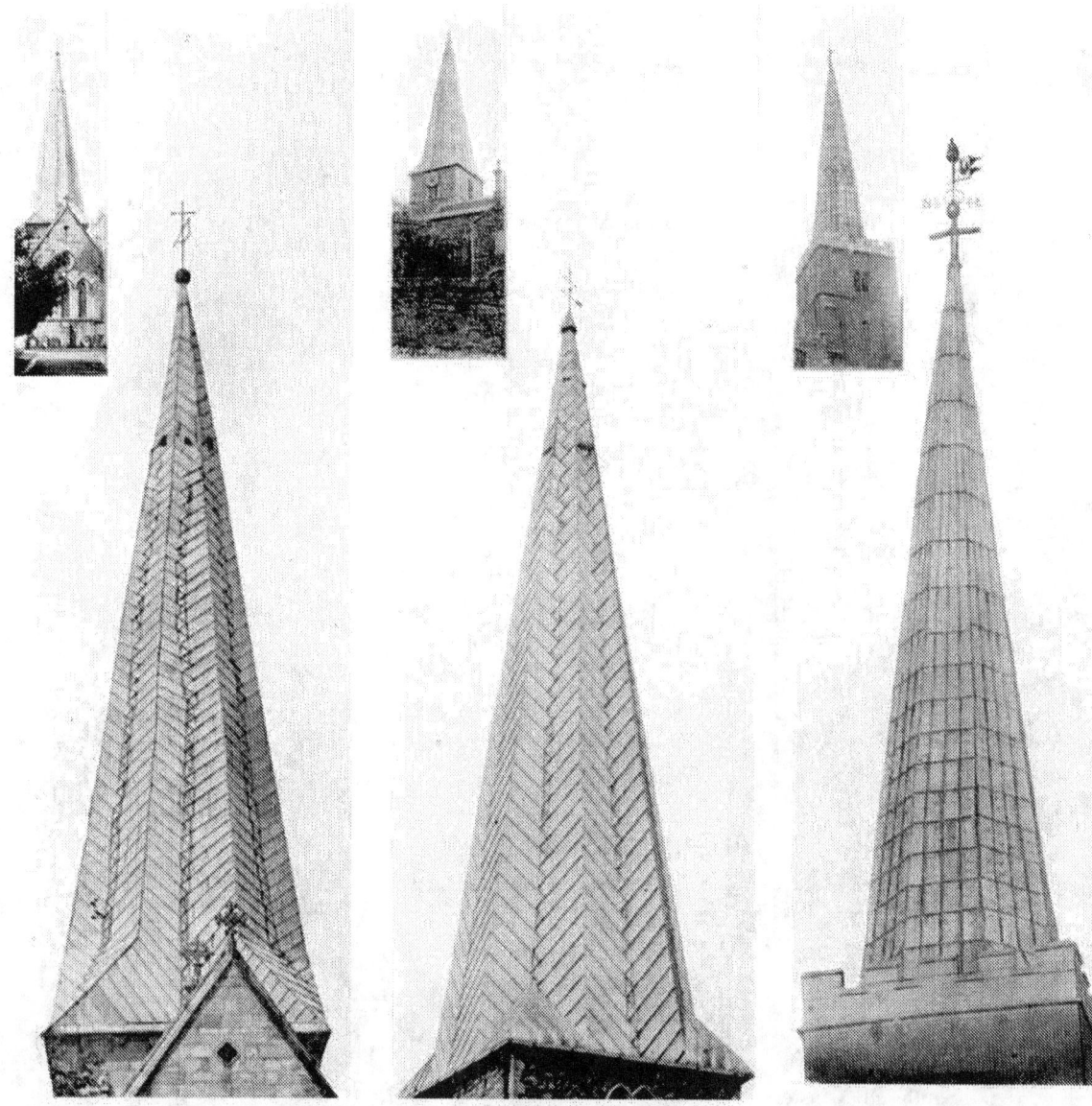

FIG. 160.—Ryton, Northumberland.
(*Pathless Collar-type.*)

FIG. 161.—Almondsbury, Glos.
(*Pathless Broach.*)

FIG. 162.—Harrow, Middlesex.
(*Parapetted Straight-sided.*)

THREE TYPICAL LEADED SPIRES.

Parapetted—

1. Collar-type, *e.g.*, St John's, Perth, the tower of which has a heavy over-sailing parapet within which the spire stands.
2. Broach-type, *e.g.*, Hemel Hempstead.
3. Straight-sided type, *e.g.*, Harrow, Chesterfield, Minster, Great Baddow, Much Wenlock, Wickham Market.
4. Spirelets, *e.g.*, East Harling, Wenden Ambo, Swaffham, Hitchin, Sawbridge-worth, and Ash, Kent.

The pathless collar-type and broach-type can best be considered together, for some confusion has arisen in the definition of leaded spires owing to the somewhat loose use of the word "broach." The spires now described as "collar-type" are sometimes called "broach." The shingled spires (*e.g.*, Shere, Tangmere, Merstham, Newhaven, and Plumpton) are all of collar-type, and may be taken as the first remove from spires square on plan, which are simply lofty roofs. The spires of Southwell Minster have been restored in their original form as pictured in Dugdale (Fig. 163), and Hexham Abbey had a pyramidal roof on the way to being a spire (Fig. 164).

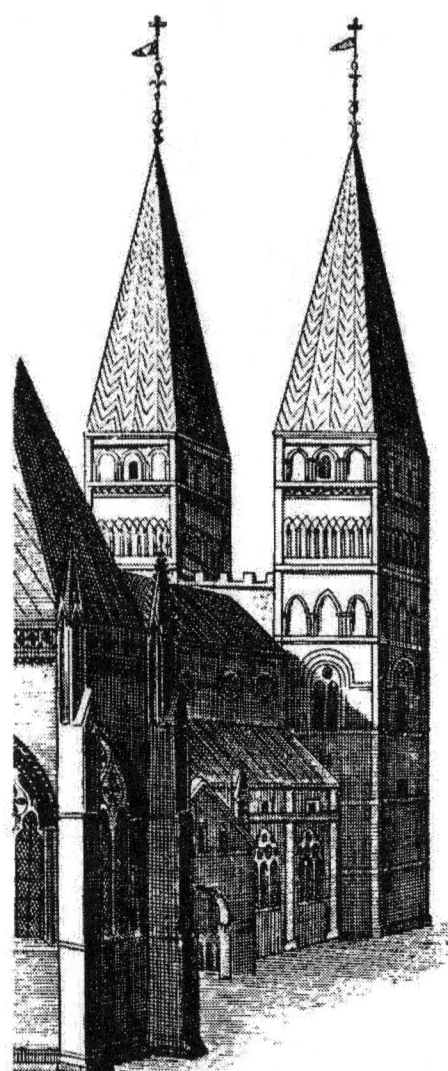

FIG. 163.—Southwell Minster.
(*From Dugdale.*)

The engraving in Dugdale is somewhat mysterious. It was drawn by S. Anderton and engraved by David King. Some corner turrets are surmounted by queer pinnacles, shaped like bulging carrots. These pinnacles look as though they might have been leaded. In collar-type spires the upper portion is octagonal, and the diagonal sides spread and bend outwards to the corners of the tower which they meet in a point. The vertical timbers of the octagon are framed in a collar which is supported by the timbers of the lower part. The collar-type is probably an earlier form of the timber spire than the broach-type.

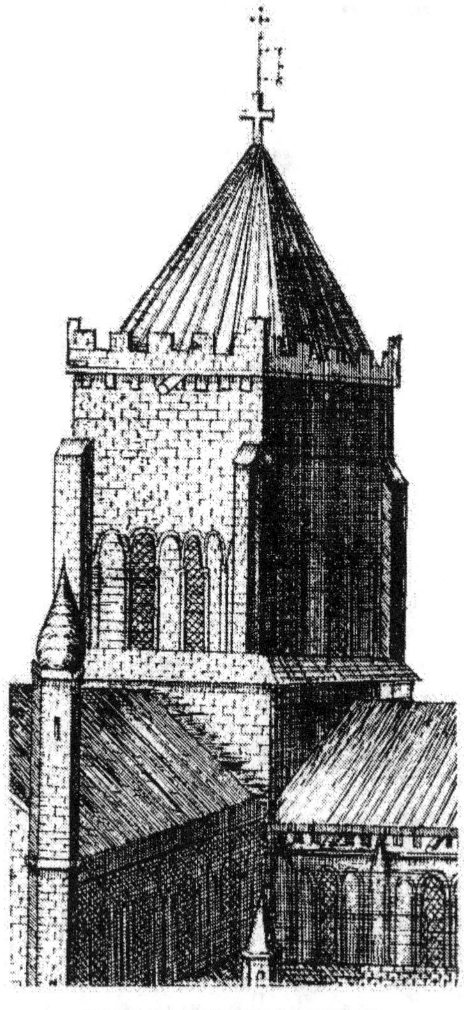

Ryton has a leaded spire of strict collar-type, but in general proportions it is more like the lofty broach of Almondsbury than the squat, shingled collar-type spires. The diagonal ribs meet in a very irregular line on the faces of the octagon (Fig. 160).

FIG. 164.—Hexham Abbey.
(*From Dugdale's "Monasticon Anglicanum."*)

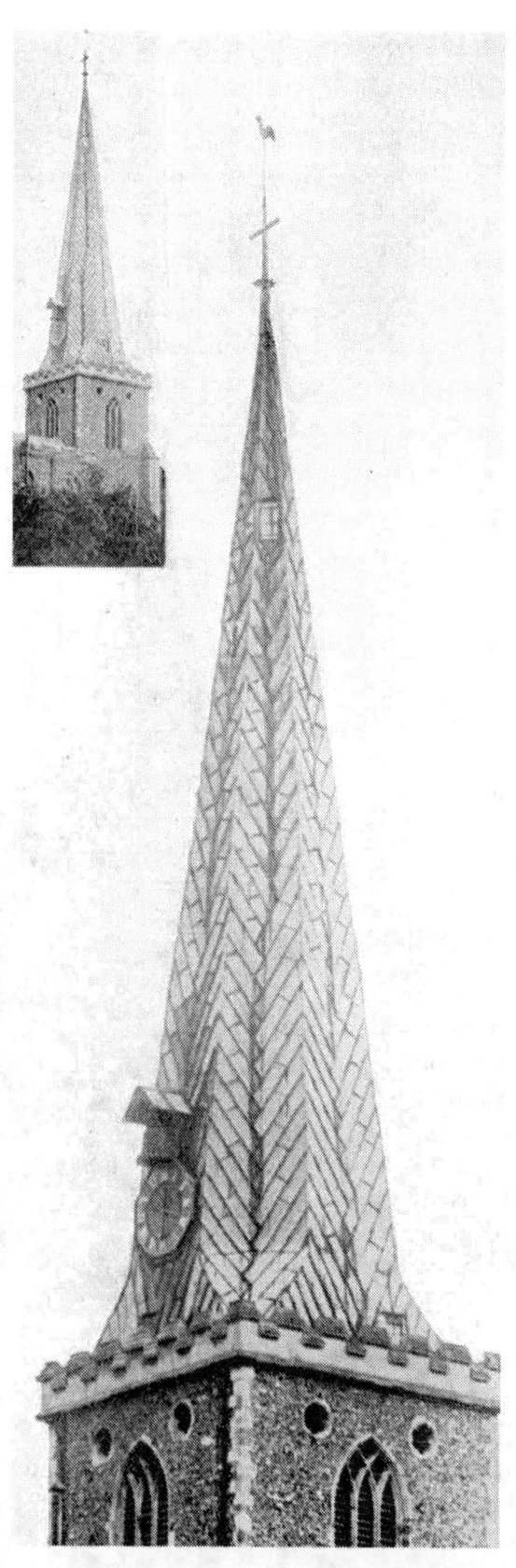

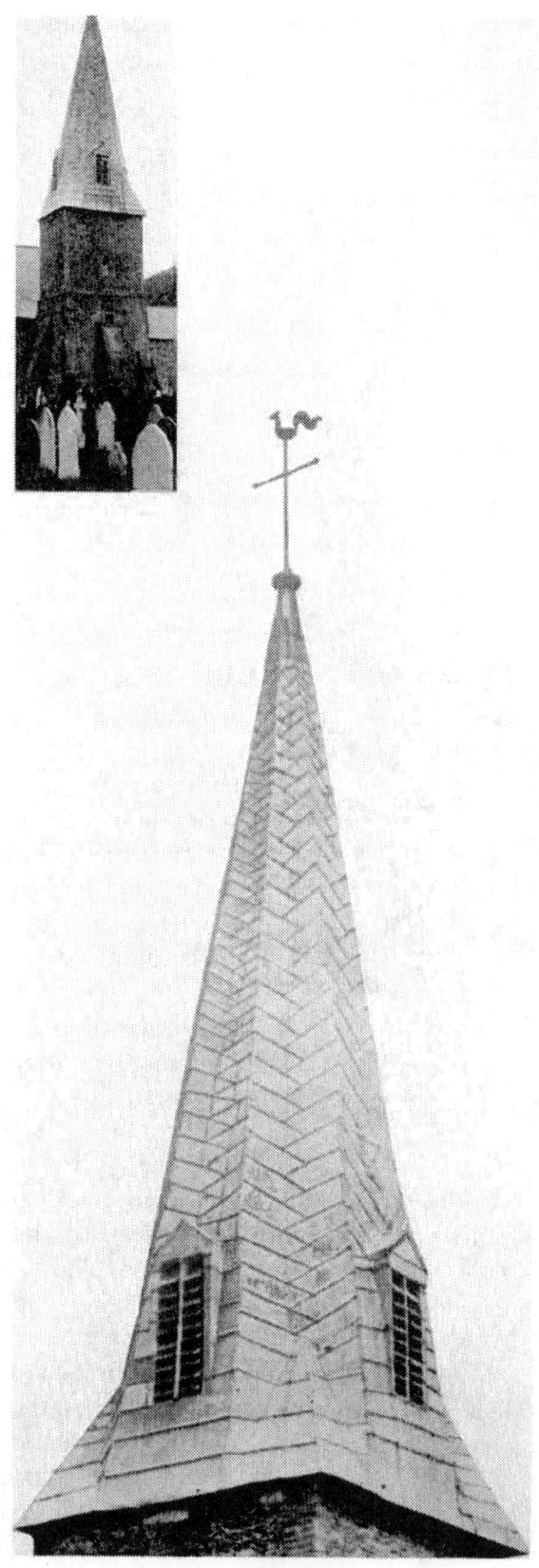

FIG. 165.—Hadleigh, Suffolk.

FIG. 166 —Braunton.

The essence of the broach is that the filling-in between the angles of the tower and the diagonal faces of the spire is of pyramidal form. Mr Bond says, when dealing with broach spires, "Just as the timber spire-form was copied in stone, so the stone broach was copied in wood, *e.g.*, at Braunton, Devon." He does not, however, point out that there are more broach-type than collar-type pathless leaded spires. Mr Prior, in his "History of Gothic Art in England," writes of "wooden lead-covered spires, first the models and then the copies of the stone." And again, "Almondsbury, Gloucestershire; Hemel Hempstead, Hertfordshire; and Braunton, which, being wood and lead productions of the Northamptonshire 'broach,' may be conjectured as originally due to its influence."

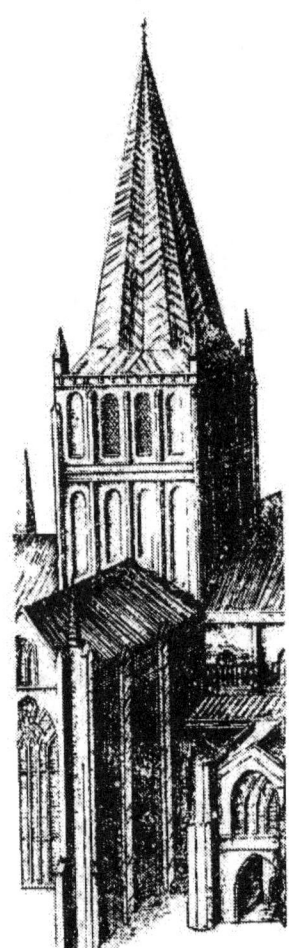

Fig. 167.
Hereford Cathedral.
From Dugdale's " Monasticon Anglicanum.")

So much may be admitted without suggesting that the leaded broach is a slavish or unintelligent copy of the stone broach. It is a question of carpentry. The construction of the collar-type is more congenial to wood than is the broach. The octagonal framing calls (but not very urgently) for strutting at the base. In the broach the main framing is strutted by single timbers running through the diagonal faces of the octagon; and this is not so satisfactory as the double strutting of the cardinal faces, which obtains in the collar-type.

The question should, perhaps, be considered rather from the point of view of weathering. The builder of leaded spires had a simple problem to face. He had to put an octagonal spire on a square tower, and to provide a weathering from the diagonal faces of the spire to the angles of the tower. In the case of shingled spires he elected to construct the collar-type; in the case of leaded spires he used both the collar-type and the broach-type, but the latter more commonly.

While it is true that in stone broach spires the pyramidal broach, borne on a squinch, buttresses the spire and has an important constructional function, it seems equally true that in

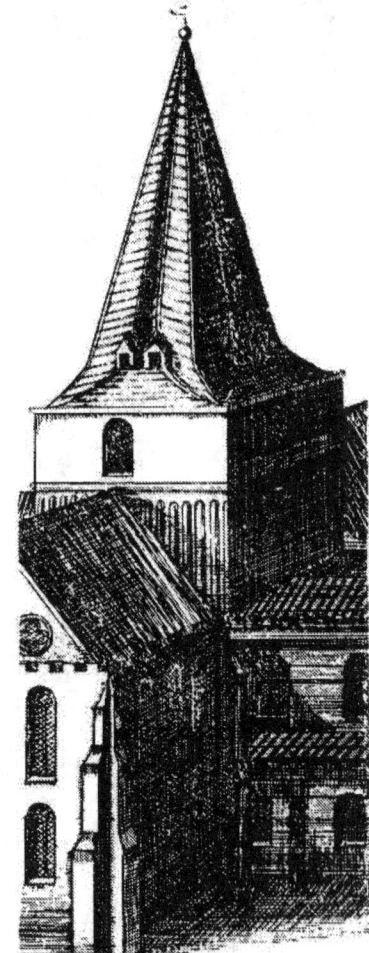

Fig. 168.—Rochester Cathedral.
(From Dugdale.)

timber spires the constructional significance of the broach or collar-type is less marked.

From the weathering point of view, the broach-type is as efficient as the collar-type, and the broach is far the more attractive.

Regarding the question of development, Mr Prior's view that the lead broach was inspired by the rise of the Northamptonshire stone broach is confirmed geographically. The leaded spires of broach-type in Devonshire, Gloucestershire, and Surrey are comparatively near Northamptonshire, while the farthest lead spires, viz., Ryton, Northumberland, and St John's, Perth, are of the collar-type.

The question as to the proportionate numbers of collar-type and broach-type respectively that existed in mediæval times is impossible of answer.

The grim comment on the English soldiers in the Crimea that "they showed a marked tendency to die," may fairly be applied to leaded spires. If the nation is happy which has no history, the national art of lead roofing must be unhappy indeed, for it has more history than being. This much is clear when we remember that not one of the cathedral leaded spires remains. Rude as are the sketches in Dugdale's " Monasticon Anglicanum," there are some indications of the various types, though it would be unwise to build a theory on the prints, which on such questions as these can do no more than fortify guesswork.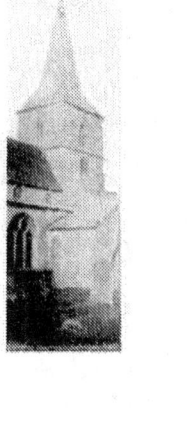

The central tower of Hereford Cathedral (Fig. 167) had a lead spire. It was apparently early and of collar-type.

The Chertsey Cartulary in the Record Office has a plan of the site of Chertsey Abbey, and a view of the Abbey Church shows a leaded spire.

At Rochester (Fig. 168) the central tower was also crowned with a spire which, perhaps, was of broach-type. The spire-lights are queer little features.

Among existing pathless collar-type spires that of Hadleigh, Suffolk (Fig. 165), calls for special remark. It properly belongs to the pathless class, although it now has a parapet. The latter is quite modern, and must, therefore, be disregarded for the purpose of classification. Before this addition of some thirty years ago, there was a wooden railing round the spire, which was called the cradle. This cradle was doubtless a piece of churchwarden carpentry, provided to make repairs easier. Originally, there is no doubt, the spire rose from the tower walls direct. The present parapet

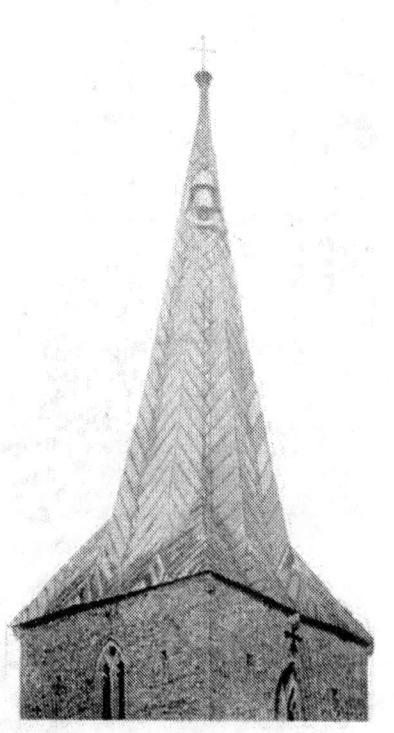

FIG. 169.—Ickleton, Cambs.

is a frank absurdity; it protects no footway round the spire, and is merely a frilling in stone.

Ickleton, Cambridgeshire, has a notable spire (Fig. 169). It is very low compared with the height of the tower, and has an odd treatment. The chief characteristic of the collar-type of shingled spire is that the sides do not run down

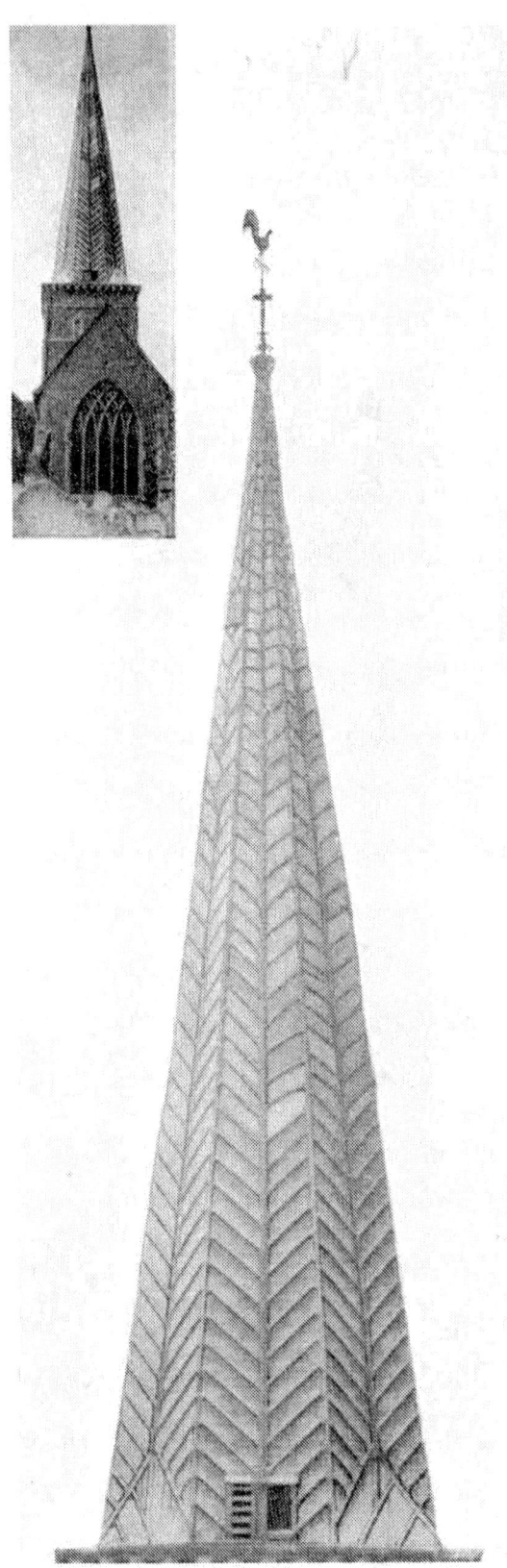

FIG. 170.—Godalming.

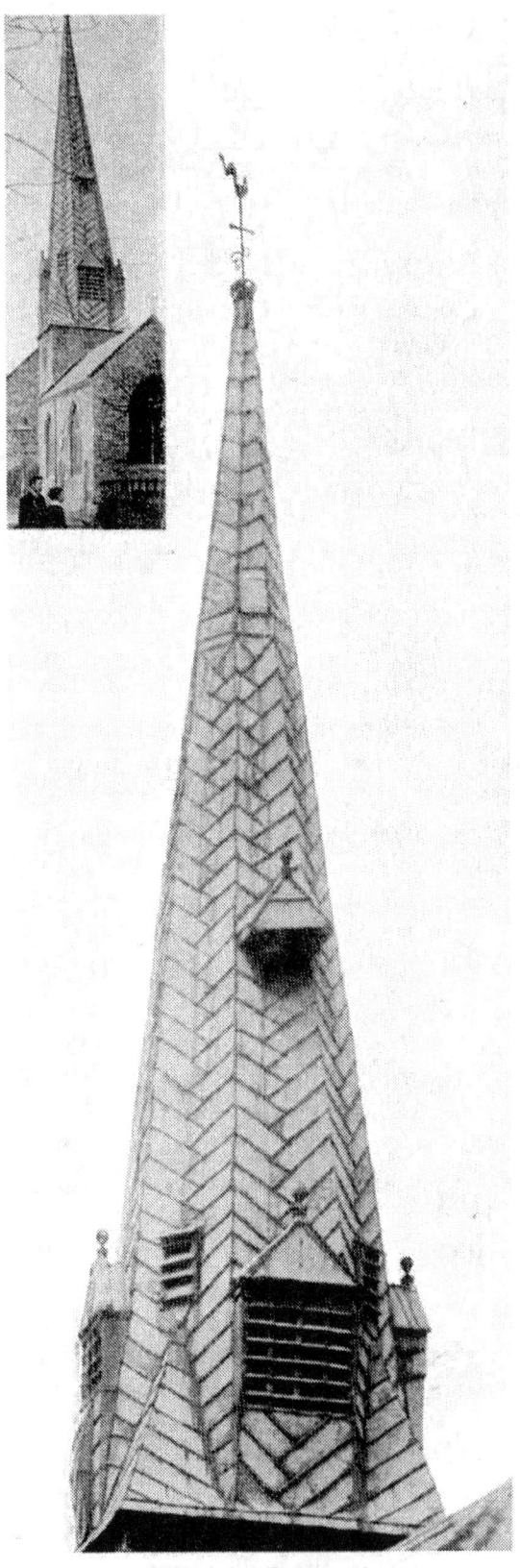

FIG. 171.—Barnstaple.

straight from the apex to the base, resting on the tower wall. At the collar the line both of the cardinal and of the diagonal sides breaks outwards. This is true of Merstham, Pembury, Plumpton, Tangmere, and Newhaven, all shingled. It is also true of St John's, Perth, leaded collar-type. It is, however, not the case with Hadleigh, Suffolk, and Ryton, Northumberland, both leaded collar-type.

The peculiarity of Ickleton is that, though it is broach-type, the sides break outward about half-way down the broach itself, and so give it a strong superficial resemblance to such shingled spires as Merstham. It is, in fact, a compromise between the broach and collar types, and supports the contention that the actual broach is as natural an angle finish for a timber as it is for a stone spire. Ickleton spire is of date 1351. The lead has taken on a delightful patina partly bluish and partly a brownish grey.

Of all lead spires Barnstaple is perhaps the most graceful and interesting (Fig. 171). It has stood for over five centuries. The alterations in the seventeenth century, when the spire-lights were opened, add considerably to its charm, as will be seen by a comparison with the neighbouring picture of Godalming,* which lacks the openings. It will also be noticed that the cardinal faces of Godalming spire stand a little within the wall of the tower, whereas at Barnstaple the lead sheeting overhangs. Very valuable is the sense of perfect roofing at Barnstaple which this overhanging gives. It gains over Godalming also by its much more strongly-marked broaches and the almost impertinent little opening with louvres at the point of the broach. The little twist is enough to give it interest, without inspiring nervousness as does the spire at Chesterfield. The arrangement of the rolls at Godalming (Fig. 170) is simpler and more regular than at Barnstaple. Of the two methods that of Barnstaple is the commoner and the more interesting. It takes the middle course between the severity of the Godalming rolls and the almost self-conscious irregularity that obtains at Hadleigh (Fig. 165).

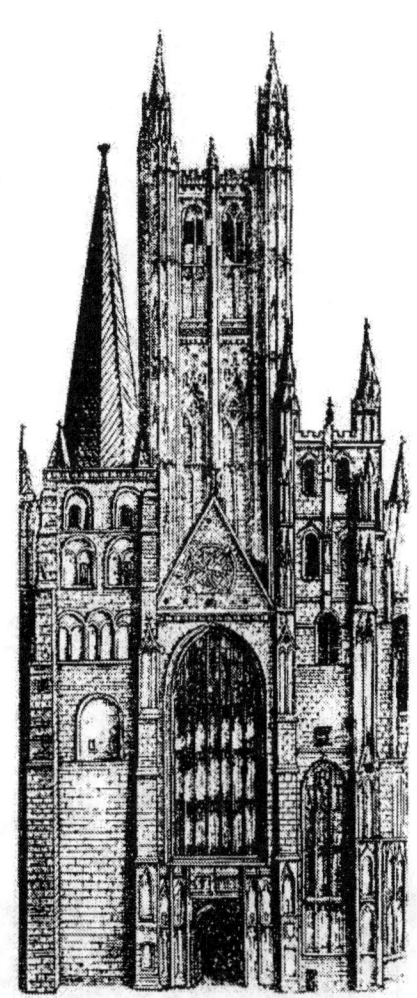

FIG. 172.—Canterbury Cathedral.
(From Dugdale.)

Almondsbury (Fig. 161) has, for its height, very small broaches; they strike the diagonal faces at a comparatively acute angle. With regard to the leading, the sheets are narrow, and the diagonal arrangement of the rolls is carried down to the base of the spire. There are no spire-lights, but very small openings for ventilation near the top. At Braunton, Devon (Fig. 166), however, there are gabled vertical spire-lights with luffer boards, and the rolls are gradually worked from a diagonal arrangement to the horizontal, half-way down the spire-lights, a treatment which adds much interest. At Swymbridge (like Braunton, near

* See Bibliography (Sundry), "History of Godalming."

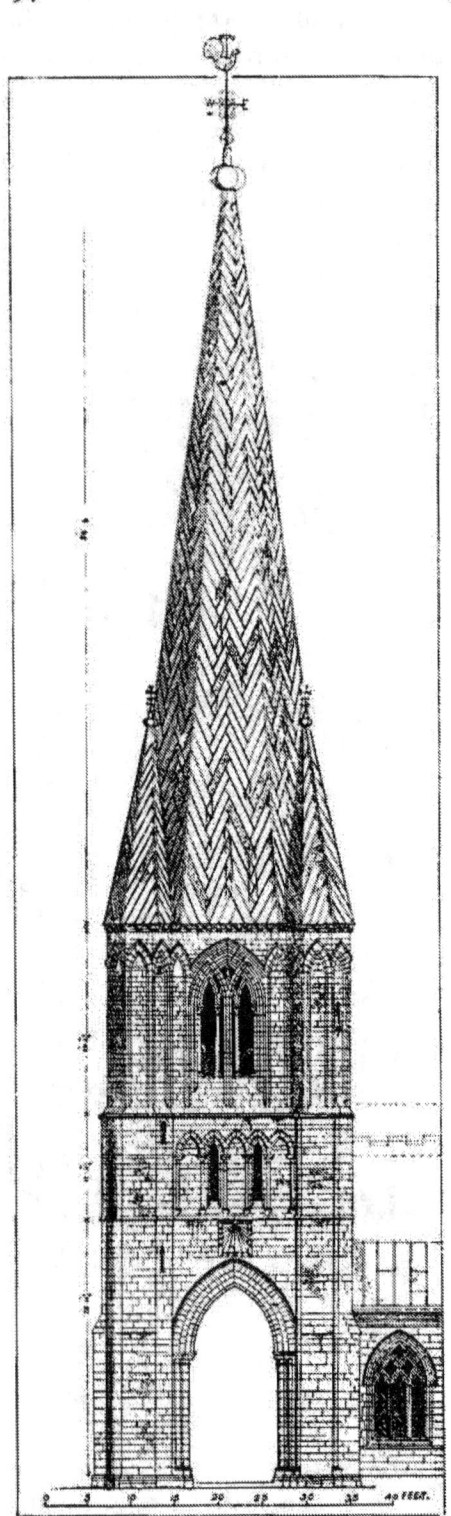

FIG. 173.—Long Sutton.

Barnstaple) the spire has gabled lights similar to Braunton, but the spire was restored a few years ago, and it may be that the existing spire is not an exact reproduction of the original.

Following the order of our classification we come to the pathless pinnacled type.

The west front of Canterbury is still probably the most interesting west front in England; but in losing the lead spire on the north-west tower of Lanfranc, it has lost half the charm of its irregular grouping. The drawing by Thomas Johnson, part of which is shown in Fig. 172, is one of the best in Dugdale. It shows the spire as being of more slender proportions than the view in Dart's "Canterbury." In this it agrees with the painting at Lambeth Palace. The spire was removed in 1705. The Dugdale drawing seems to show that the pinnacles engaged with the base of the spire in the same way as they do at Long Sutton. If this were the case Canterbury would be of the pathless pinnacled type.

The spire of Long Sutton (Figs. 173 and 174) is unique in England; it is certainly very beautiful. Professor E. A. Freeman, in his notes to Wickes's "Spires and Towers," is, however, very scornful about it. He says, "The examples of Witney and Oxford Cathedral show that pinnacles may be very well combined with a broach spire, either with or without turrets, at the corners of the tower. Sutton shows an unsuccessful attempt in the same direction . . . the effect is very bad, being neither that of pinnacles set on the squinches, nor that of turrets rising, as they generally do, higher than the tower."

Despite the eminence of the authority it will not be held generally that the effect is very bad. On the contrary, this spire and that of St Nicholas, Aberdeen (which was similar), seem quite extraordinarily successful, and, of the two, Long Sutton is the more cunningly designed. The plan at the joining of tower and spire is full of interest, whereas that of Aberdeen shows no particular invention. The achievement of the architect of Long Sutton is the more notable, in that we have all the grace and beauty that pinnacles add to a spire, without any surrender of the "roof" idea, which goes when the parapetted type of spire is adopted, as, for instance, at Norwich Cathedral and Kettering.

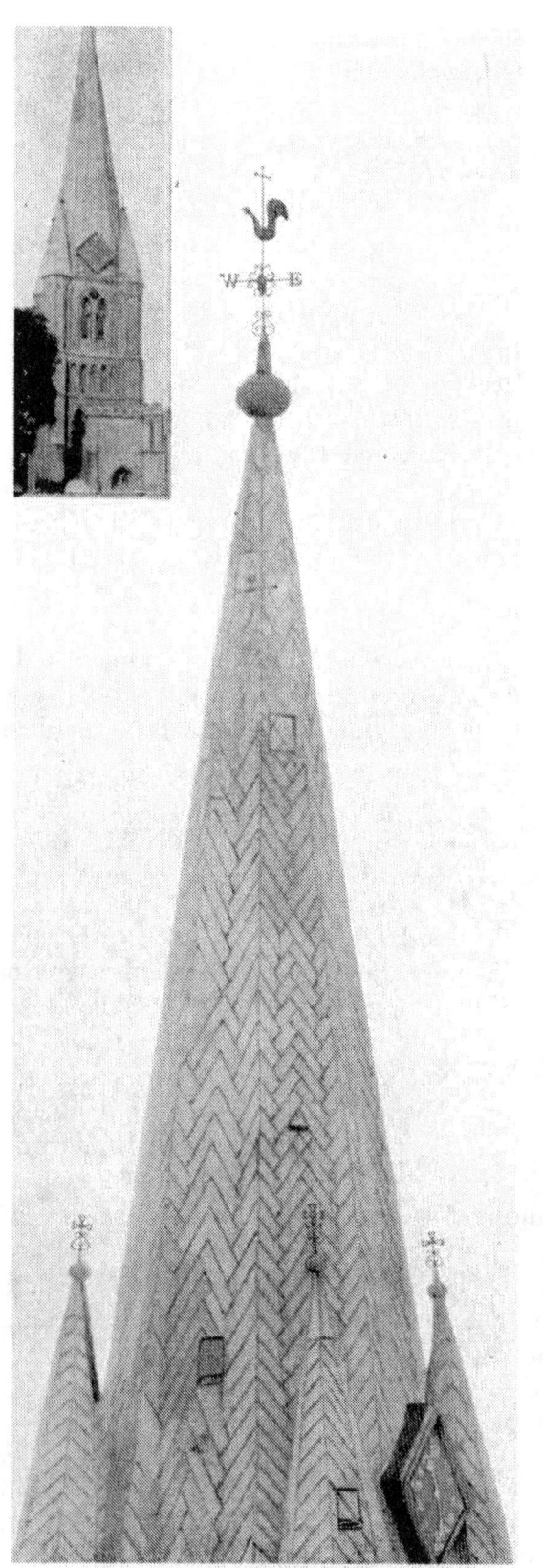

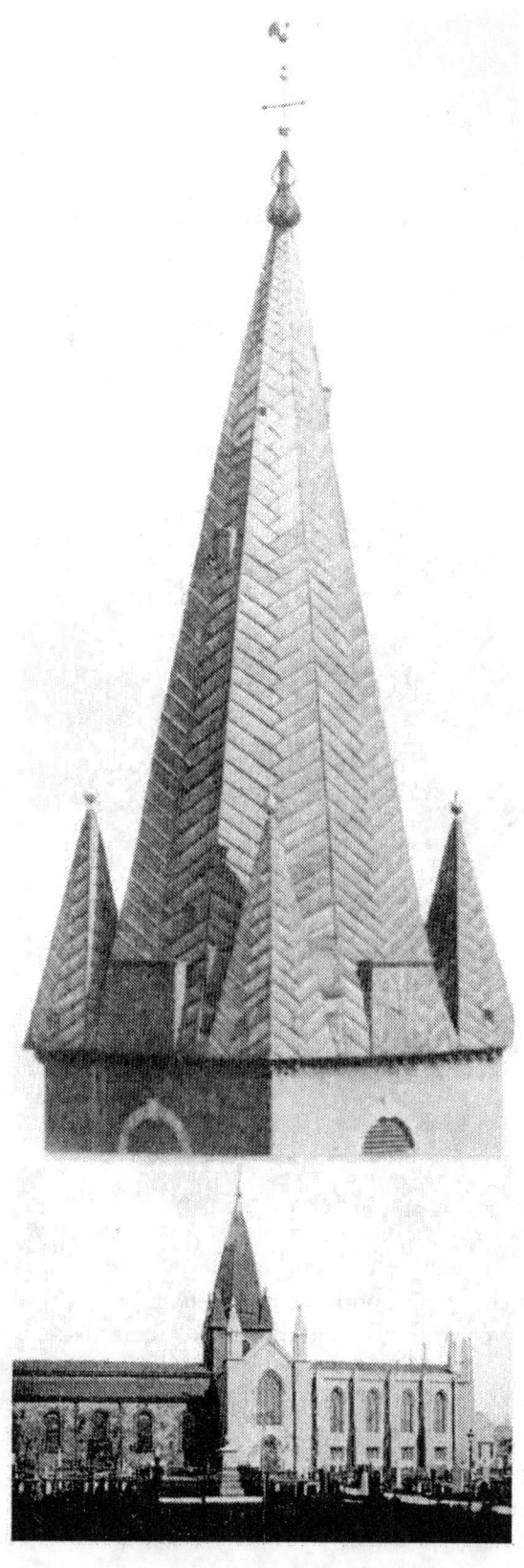

FIG. 174.—Long Sutton.

FIG. 175.—Aberdeen.

Mr Lethaby has pointed out the delightful effect which is gained at Long Sutton by the leaning inwards of the pinnacles, a refinement which Wickes apparently did not observe, for it is not brought out in his drawing. Probably Wickes had a poor idea of lead spires altogether, for the only other he shows is that of Wickham Market. Later students are less scornful. Measured drawings of St Mary's, Long Sutton, appear both in the "Spring Gardens Sketch Book" (vol. 5) and in the "Architectural Association Sketch Book" (vol. 1). A book on leadwork is not closely concerned with the insides of leaded spires, but these measured drawings are a liberal education in timber construction. The boarding to which the lead is fixed at Long Sutton is rough oak, 1 inch thick, and the height of the spire is 84 feet 6 inches.

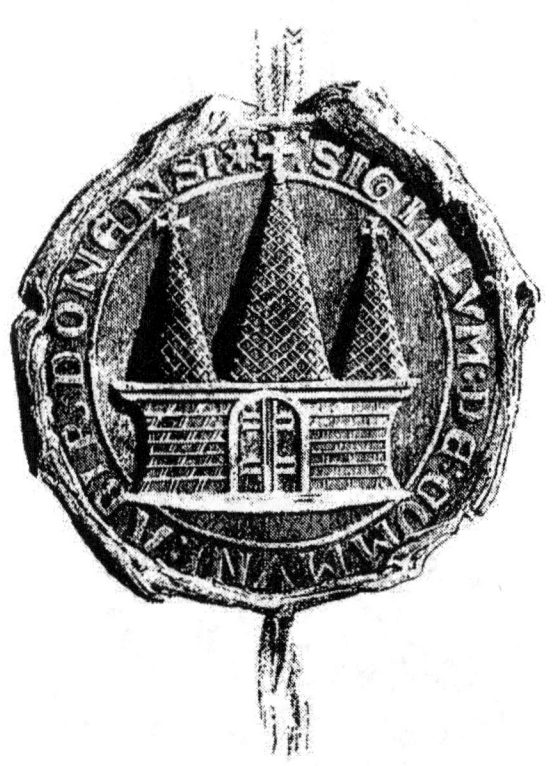

It is, of course, quite impossible to suggest a date for the earliest lead spires, but this much is clear, that they are much earlier than stone spires.

The towers drawn in the "Benedictional of Ethelwood" (tenth century) are covered with pyramidal roofs, but they can hardly be called spires; and though the drawing of these roofs suggests leadwork, one cannot build a theory on so uncertain a foundation. They may have been shingled. There is little doubt that Long Sutton is the earliest existing lead spire. Mr Francis Bond points out that it is "hardly clear of transitional detail," and Mr Prior also puts it as early as the latter part of the twelfth century.

Mr Bond in referring to the early spires and amongst them Long Sutton, says that they did not produce schools. While this is unquestionably and unfortunately true as to Long Sutton, it may be that the spire of St Nicholas, Aberdeen, may have been influenced by Long Sutton. There is no documentary evidence to bring in support, but it is a not

FIG. 176.—Early Burgh Seal, Aberdeen.

impossible theory. The lead for the spire and roof of St Nicholas was largely English, and why not the design? An English plumber, John Buruel, was employed to cover with lead the roofs of Aberdeen University in 1506, and the spire of St Nicholas was being built at this time. Buruel might not impossibly have seen Long Sutton spire, and advised his Aberdeen friends to follow so admirable an example.

Aberdeen seems to have taken to lead spires very early. The earliest of the burgh seals (Fig. 176) bears what was conjectured by Mr Astle ("Vetusta Monumenta," vol. iii., Plate 27) to picture a shrine of the patron saint. The three toy spires, which surmount the shrine, are represented as having reticulated coverings. The network probably indicates lead rolls. By way of comparison it is worthy of note that the existing spirelet of Sawbridgeworth, Herts, is leaded with a similar diamond pattern.

The many records that persist of the mediæval Scots plumbers give an agreeable vitality to the study of such of their work as remains. The most remote was one William of Tweeddale, a burgess of Andirstoun (St Andrews). The burning of the choir roof of Arbroath Abbey took him north, and he there contracted to "thek the mekil quer" and gutter it all about with lead. "Thek" is, of course, equivalent to thatch. The most notable clause in this mediæval contract provides that William shall, after the walls are parapetted, "dight" (or adorn) the work. Here are no specifications or bills of quantities, children of modern suspicion, but a large and free order to dight, and dight doubtless William did, though his handiwork has gone from our ken. His pay for the work was good, 25 marks (or £16. 13s. 4d.), but his honour greater, for he was to get a gown and hood, doubtless a token of his mastership. Nor were his comforts forgotten, for daily he received a penny for his "noynsankis," "noon-shenk," or noon-drink, *vulgo* beer-money. It is eminently characteristic that this great craftsman was not merely a master of other men, but master of his craft, for despite his hooded gown, he worked his lead with his own hand, and had but two labourers to help. The abbot found the lead, William found the brains to devise and the hands to work.

Towards the end of the fifteenth century the good burgesses of Aberdeen set themselves to build a new choir to their church of St Nicholas, and build they did for thirty-six years, with great scheming and stinting of themselves to find the wherewithal. Aberdeen was like London and Bristol in possessing a race of merchant princes. In 1474 David Menzies contracted with the Master of Kirkwork for "thre futhir of lead, ilke futhir contenand sex score of stanys, to be deliverit, God willand, gif wind and wethir will serve, betuix this and Pasch next to cum apon the key of Aberdeen."

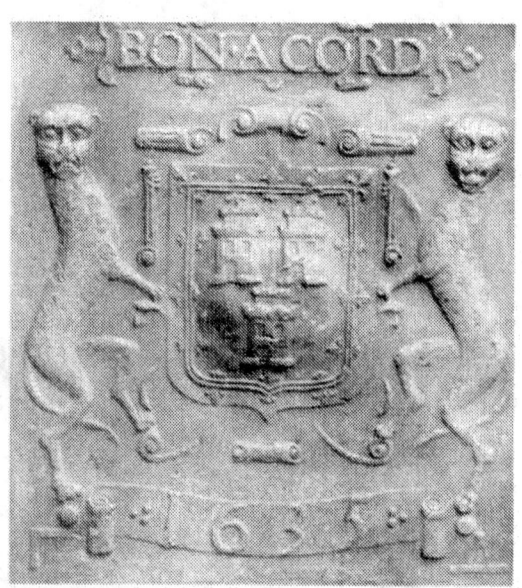

FIG. 177.—From St Nicholas, Aberdeen.

This David Menzies seems to have acted precisely the same part of general manager of the city's expenditure on their church, as did the famous William Canynge the younger at Bristol, when he "with the helpe of others of the worshipfulle towne of Bristol, kepte masons and workmenne to edifie, repayre, cover, and glaze the church of Redcliff," the St Mary Redcliff which is the chief glory of Bristol. This parallel from the south is given because it is good to emphasise what a great part the merchant adventurers played in the architectural energies of the Middle Ages. And, further, the works were almost contemporary—Aberdeen, 1474, Bristol, 1442. Canynge's work followed the fall of St Mary's spire, and Canynge's name, connected inseparably with Chatterton's forgeries, is a link with a tragedy of English literature.

To return to Menzies and his fellow-citizens at Aberdeen. From 1474 to 1510 the work at St Nicholas' spire went on, the lead being paid for largely by salmon, a staple export of the town. The carrying of the lead to Aberdeen was evidently no small matter,

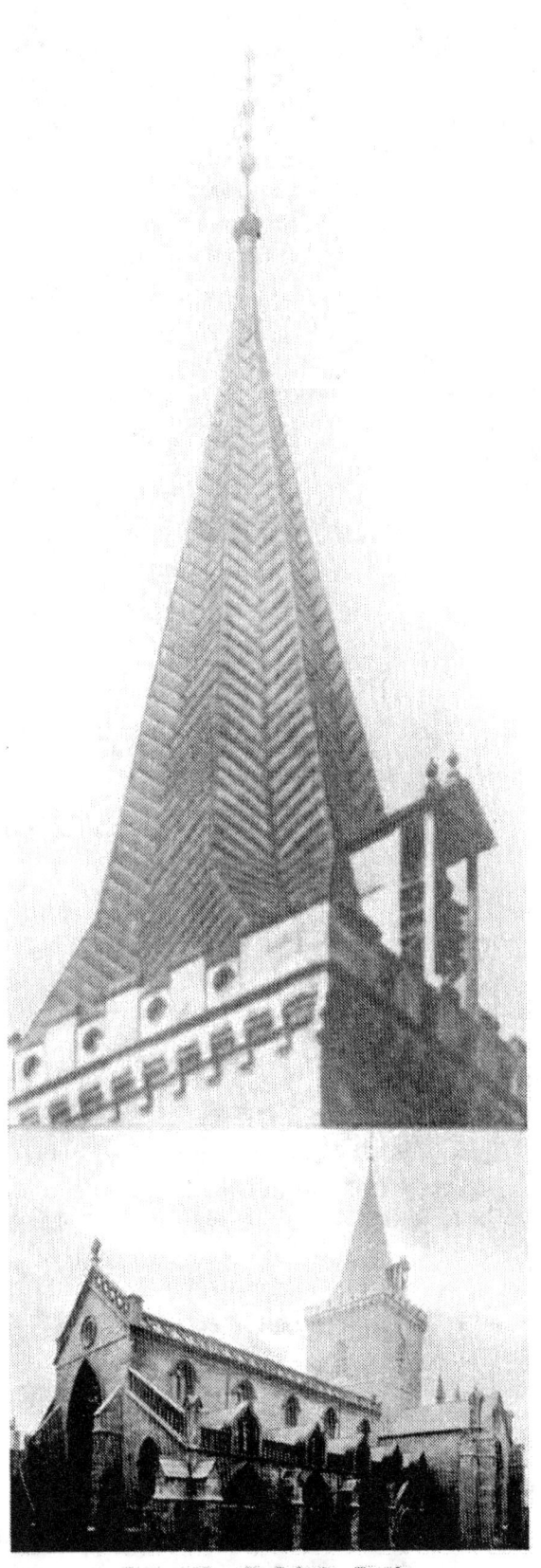

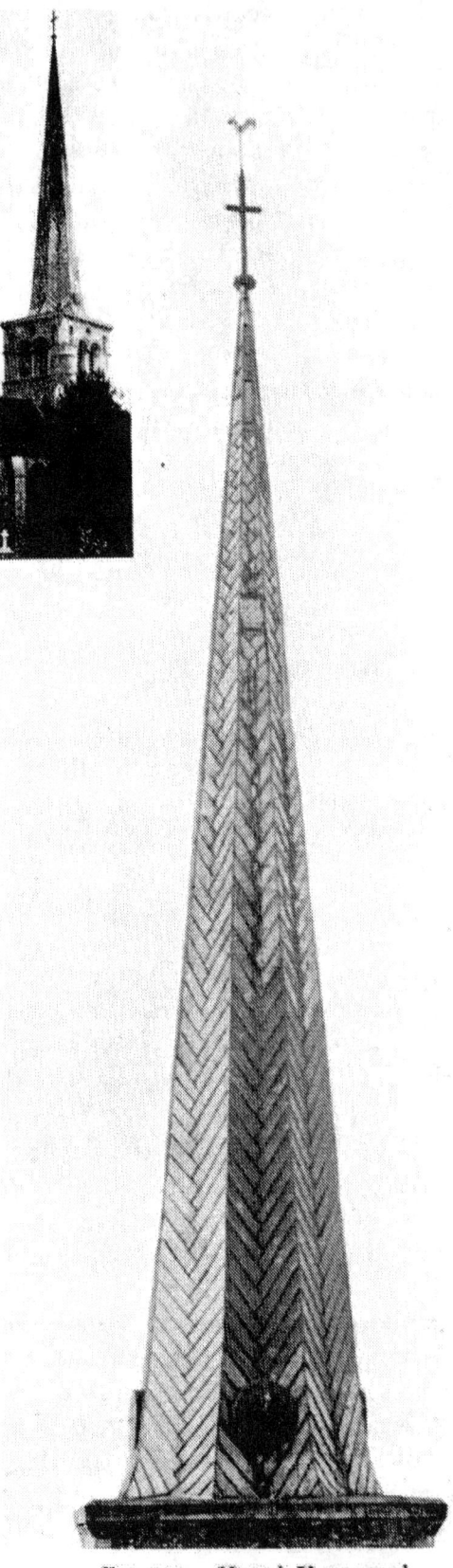

FIG. 178.—St John's, Perth.

FIG. 179.—Hemel Hempstead.

for in 1500 the Provost himself, Sir John Rutherford of Tarland, went as far south as Berwick to bring it home.

In the year of Flodden, 1513, their labours came to an end, for the records show that in November of that year Henry Reid "gifted" money for "up-putting of the weddercok," and John Cullan furnished the gold "for gilting of the weddercok." Fig. 175 shows the steeple as it stood from Flodden until 1874, when it was destroyed by fire. It is some consolation, and no little good fortune, that from such early photographic days the negative remained from which the illustration has been made. It would seem from the photograph that the Aberdeen pinnacles, like those at Long Sutton, bent inwards slightly. Aberdeen's records of the great spire do not end, however, with the story of its building. In 1546 the bailies ordained their Master of Kirkwork to send to St Andrews for a plumber "to reforme and mend the faltis of thair kirk." Again in 1559 "the lead thak" wanted repair, whether of the roof generally or of our spire is not recorded particularly. That further repairs to the leading were regarded as important works is clear from the admirable lead panel that came from the roof of St Nicholas, Aberdeen (Fig. 177). It bears the date 1635, the arms of the burgh, and its fine motto "Bonaccord." Another exists, made from the same pattern, but dated 1639, and is a rather sharper casting. The size of both is 1 foot 4¼ inches by 1 foot 6¾ inches. They serve no purpose save magniloquently to remind us of the pleasure of some Master of Kirkwork in his labours. The patterns were probably carved in wood (robust and masculine work it is), pressed into the casting sand, and cast by the plumber on one of his roofing sheets. With the timber work of the great spire we are not so concerned as with its lead covering, but the name of the "wright" who probably framed it remains, John Fendour. In those days there were no nice distinctions as to-day, between carpenter, joiner, and carver. Fendour was a "wright," a worker in wood, and a master at his work. All woodwork, massive or intricate, came from his hand. In 1495 he was building the roofs of St Nicholas, and in 1507-08 he made and carved the choir stalls and screen.

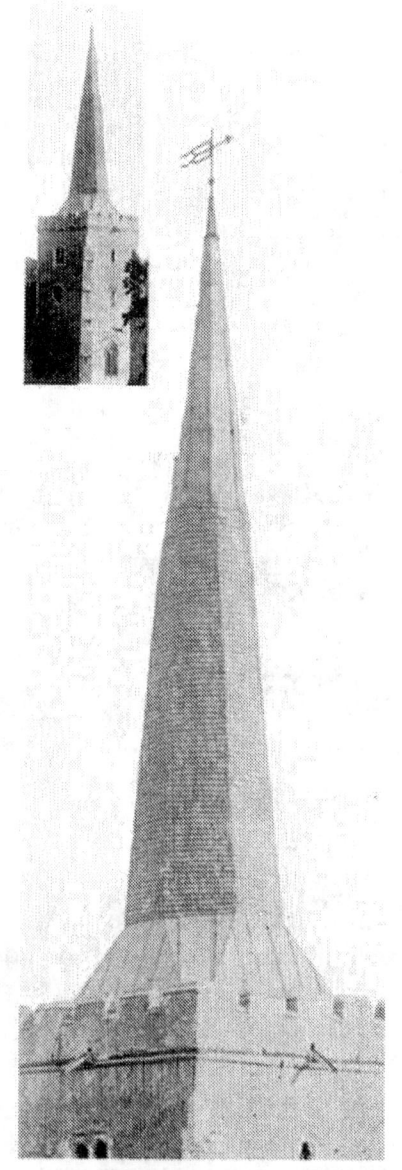

FIG. 180.—Danbury, Essex.

Passing now from the pathless spires we come to the parapetted examples, and Class I., the collar-type. It is unusual for collar-type spires to stand within a parapet, but there are at least two examples, and one, i.e., St John's, Perth, is important (Fig. 178). The parapet is heavily corbelled out, and in proportion to the tower the spire is very low and squat.

In connection with St Nicholas', Aberdeen, we have already met Fendour, the carpenter. In 1510 he agreed with the great Bishop William Elphinstone (an heroic figure in mediæval Aberdeen, an episcopal Mæcenas) to build the great central leaded spire of St Machar's Cathedral, Old Aberdeen. Build it he accordingly did, but no trace remains, save the written contract. It was to be after a form and pattern given by the bishop to Fendour, to be substantially hewn and joined "as the steeple and prik (spire) of the kirk of Saint Johnstoun is." Here we come into contact with the existing. This likeness of the cathedral spire to that of St John's, Perth, must, however, have been rather in the method of timber construction than in the actual shape and proportion. This seems to be proved by the freestone spires of the cathedral built by Elphinstone's like-minded successor, Bishop Gavin Dunbar, for he ordered them to match his predecessor's work. So closely, even slavishly, were his lordship's orders followed, that there appear in the stone spires sham dormers. Now dormers are proper enough to a timber spire needing ventilation, but not needful in a stone spire. The cathedral did not long

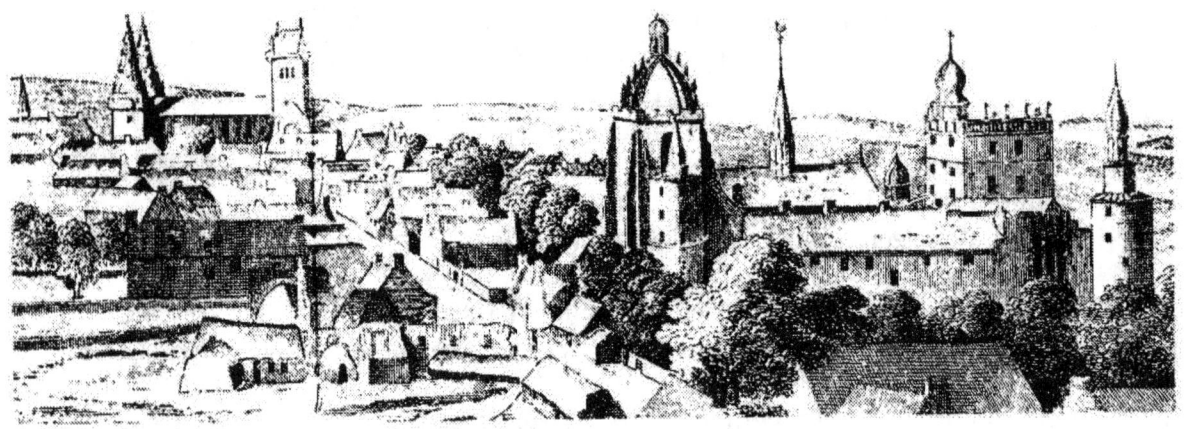

The Cathedral.   King's College.

FIG. 181.—A Reproduction of Part of the Prospect of Old Aberdeen in Slezer's "Theatrum Scotiæ," 1693.

enjoy its leaded spire. After having been despoiled of its lead and its bells, in 1560, it fell into ruin. Unhappily, not even an old drawing remains, such as Van den Wyngaerde's "View of London," dated 1543, showing the spire of Old St Paul's. Slezer's "Theatrum Scotiæ" (Fig. 181) shows Dunbar's spires, but the great tower is covered with a low roof. The contract is, however, of peculiar interest as showing the great importance attached to the St John's spire. The outside bellcote is obviously a late addition.

At Danbury, Essex (Fig. 180), there is an interesting if somewhat cross-bred collar-type spire. It is in fact an epitome of various methods of covering a timber spire. The lowest part from the collar downwards is covered with copper. The top part is leaded, and the middle is shingled. It is stated that the structure of the spire dates from 1402; but in 1749, when it was struck by lightning, the apex was burned. Perhaps the amount now leaded indicates the extent of the damage and of the restoration.

The parapetted broach spire of Hemel Hempstead (Fig. 179) is probably of the fourteenth century, and is one of the loftiest remaining. On the east face of the spire, shown in the illustration, will be seen an oblong lead plate about 12 feet from the top. This plate covers a hole which was probably left for purposes of repair. At Chesterfield there is a similar opening. Among broach spires Hemel Hempstead is not a very convincing example, since the parapet covers all but the top of the broach, and the spire looks straight-sided.

At Durham (Fig. 183) and Ely (Fig. 182) Cathedrals the western towers appear to have been crowned with broach spires which came within the parapets. At Ely the spire was very slender. In 1174 Bishop Geoffrey Ridal built the west end and steeple. In 1454 Bishop William Grey "bestow'd great sums of money on building the steeple and west end of his church." It is quite likely that the broach spirelet was Grey's work of 1454. It could not have been a copy of Ridal's

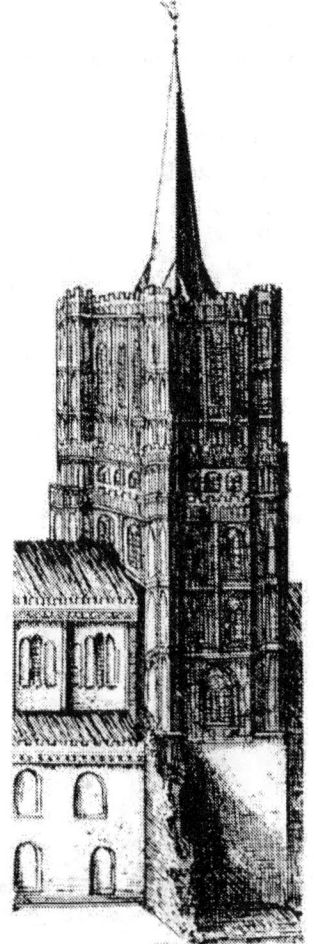

Fig. 182.—Ely Cathedral.
(From Dugdale.)

steeple of 1174. Ridal's work was probably on the lines of the pyramidal roofs (they can hardly be called spires) of Southwell Minster, which are illustrated in Fig. 163.

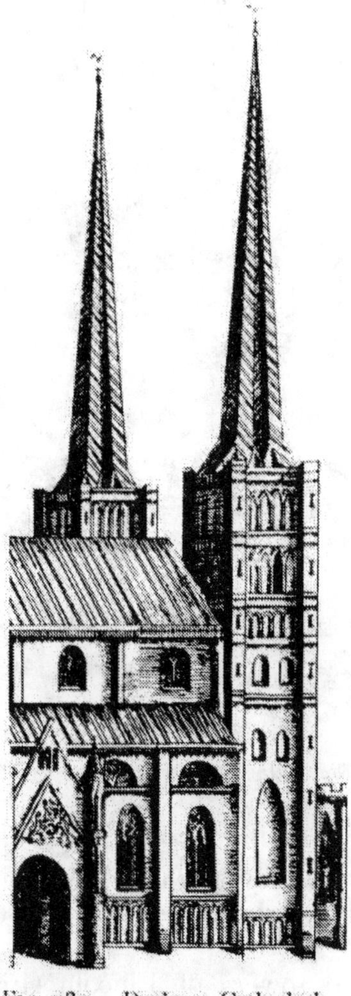

Among parapetted spires and indeed among all leaded cathedral spires the place of honour must be given to Old St Paul's. In Fig. 184 is reproduced a rare engraving which shows the spire. Apart from its intrinsic charm it emphasises the proud way in which St Paul's dominated London. The print cannot be claimed as in any sense contemporary, for the spire was destroyed in 1561. It is undated, but is said by those who are connoisseurs in these things to be of not very early in the seventeenth century. A great merit of the engraving is its (comparative) wealth of detail, which is absent from Braun and Hogenberg's view, drawn by Joris Hoefnagel, and also from Wyngaerde's. The latter was published about 1545, but is very sketchy. The important features of this spire, in its relation to those that remain, are its pinnacles. These "assert (to use Mr Prior's phrase) the English principle of angle accentuation." If the engraving is to be trusted so far in detail, the pinnacles themselves were of two stories and stood within the parapet. The Cowdray engraving shows the tower and spire of St Paul's. It suggests that the pinnacles, of which there were eight, engaged with the spire itself, and were separated by a pathway from the parapet. If this was in

Fig. 183.—Durham Cathedral.
(From Dugdale.)

fact the case, the spire occupied a position midway between the pinnacled type, *e.g.*, Long Sutton, and the parapetted type, *e.g.*, Fig. 192, Minster. Dugdale's *St Paul's* gives the height of the spire as 274 feet and of the tower and spire together as 520 feet. Stow's figures are 260 and 260, and the engraving (of Fig. 184) says, "This spere wch was of tīber coverd with lead was in height 260 foot." The first steeple built in 1221 had become weak in 1315, and was thoroughly repaired "and a new cross with a pommel well gilt set on the top thereof." This pommel was large enough to contain ten bushels of corn. In 1561 lightning and the ensuing flames

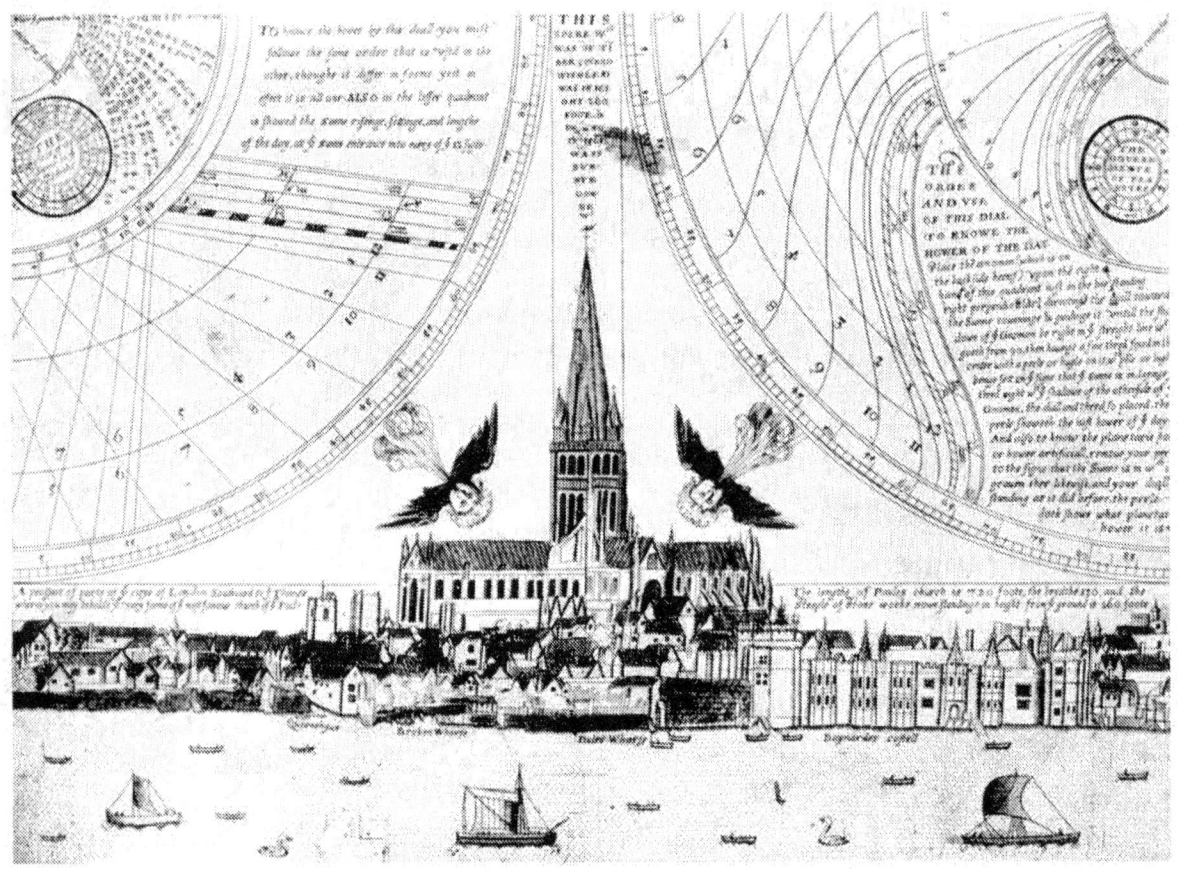

Fig. 184.—Old St Paul's.

(Reproduced by permission from a print in the possession of the Society of Antiquaries.)

destroyed in four hours the proudest English spire. There seems to have been an idea of rebuilding it in 1639. On 29th October the Chamber of London received £150 "towards the work of the steeple." Perhaps, however, "steeple" is here used loosely, and refers only to the tower.

Quite different were the spires on the west and central towers of Lincoln (Fig. 185). They were obviously of the parapetted type, and stood well within the walls, leaving a path between the spire and the parapet. This path cuts off the spire from the pinnacles. Though the leaded pinnacles remain on the three towers of Lincoln, they cannot be regarded as organic parts of the spire, as are those at

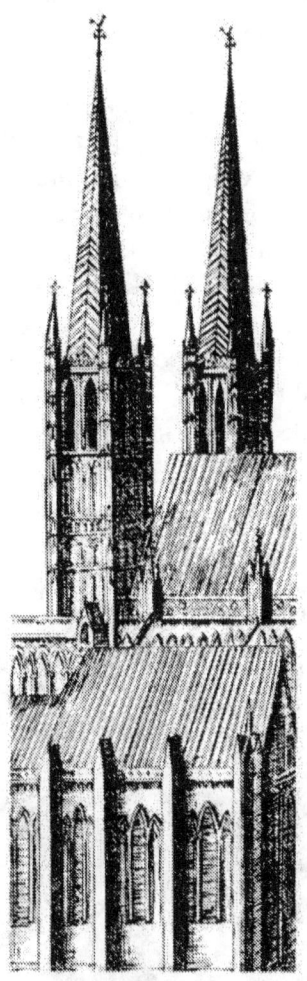

FIG. 185.—Lincoln Cathedral.

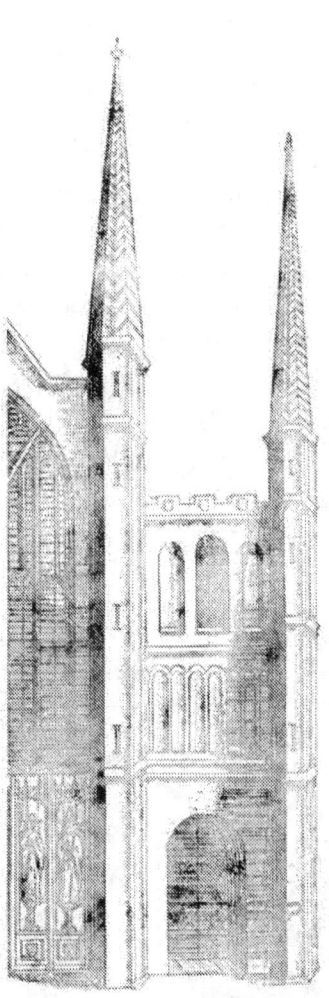

FIG. 186.—Norwich Cathedral.

(From Dugdale.)

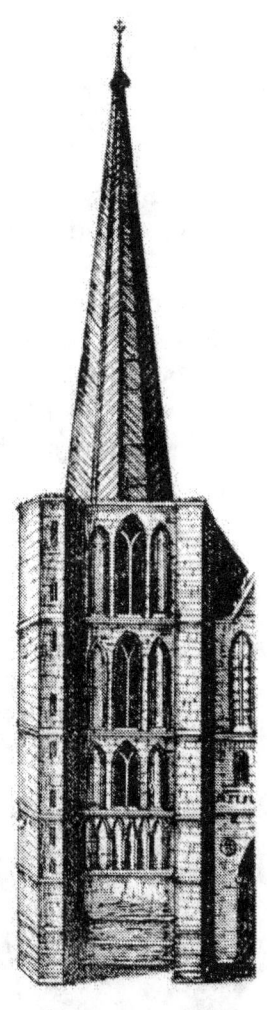

FIG. 187.—Ripon Cathedral.

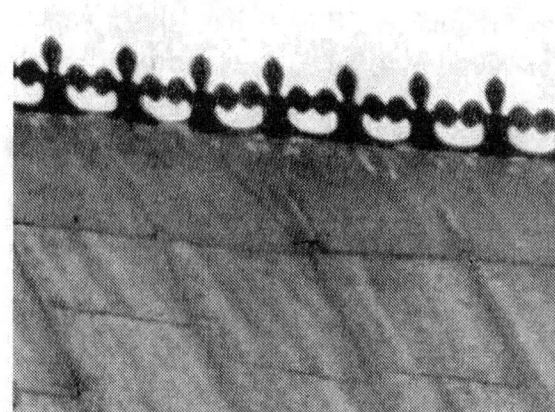

FIG. 188.—Lead Cresting, Exeter Cathedral.

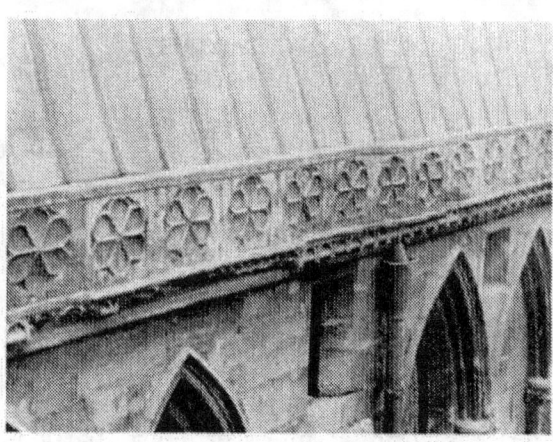

FIG. 189.—Parapet Gutter, Lincoln Cathedral.

Long Sutton. In Fig. 190 is illustrated the top of the central tower with its leaded pinnacles, melancholy reminders of what has gone. The pinnacles were probably restored by Essex in 1775, when the flimsy stone battlements were put up.

The top of the central spire of Lincoln is said to have been 524 feet from the ground. This figure sounds suspiciously like a local attempt to say 4 feet better than Old St Paul's, but as the spire was destroyed in 1548 by a tempest, the question remains unsettled. Whatever the height, the effect of the three spires must have been unique. Every one who does no more than pass Lincoln in a train must be impressed by the dominance of the cathedral towers. When the height was doubled by spires, the effect must have been amazingly increased.

Other notable details at Lincoln are the lead-covered wood parapets (Fig. 191) and gutter (Fig. 189). The former from the ground looks like stone. It is on the west side of the south-east transept, and exactly copies the bulk of the stone parapets.

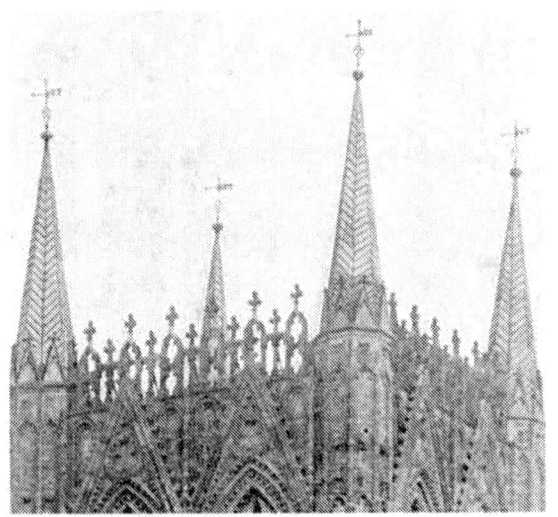

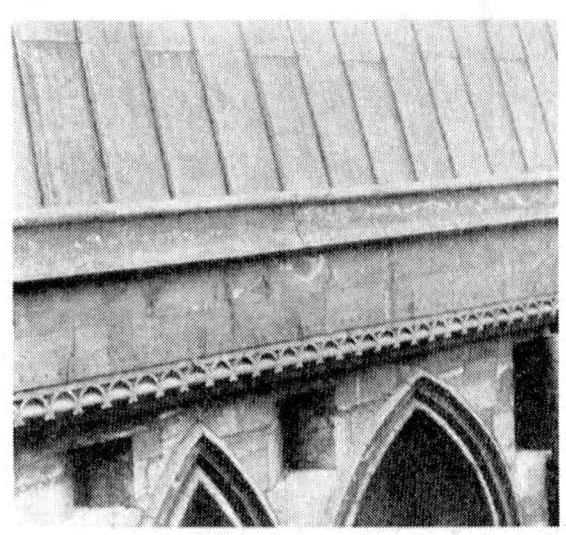

FIG. 190.—Lincoln Cathedral.   FIG. 191.—Leaded Parapet, Lincoln Cathedral.

The latter has sunk tracery panels spaced not too regularly. These have been copied at Canterbury Cathedral. Here also may be illustrated the lead cresting from Exeter Cathedral (Fig. 188).

One half of the west front of Norwich Cathedral is shown (Fig. 186) for the sake of the very lofty pinnacles, which were as large as the spire of a parish church.

At Ripon the two west towers (one of which is illustrated in Fig. 187) and the central tower had lead spires, all apparently of the straight-sided type without broaches.

Few spires show the delightful whiteness, to which lead will weather with age, so well as does Minster. In the corner photograph of Fig. 192 it will be noticed that the spire shows even whiter than the sky. Of this type of spire Professor Freeman, in his notes on Wickes's book, is so sweeping as to say that "when the spire rises within a mere ordinary battlement without any connection with the tower, the effect is always unpleasing." If this severe standard were approved, the parapetted

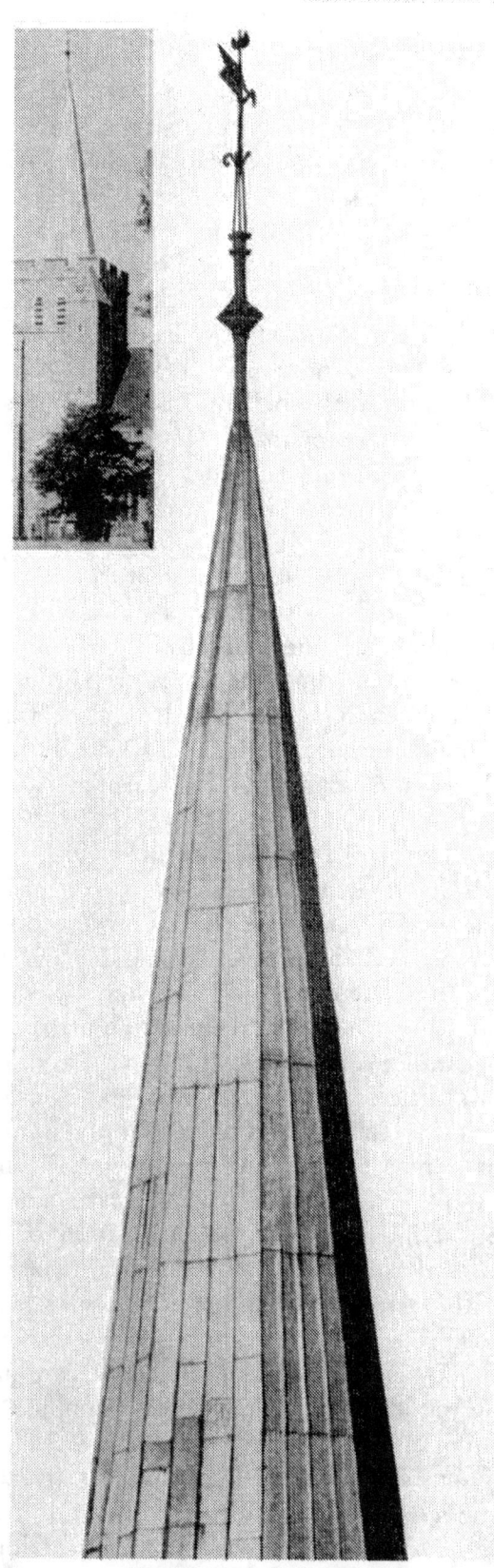

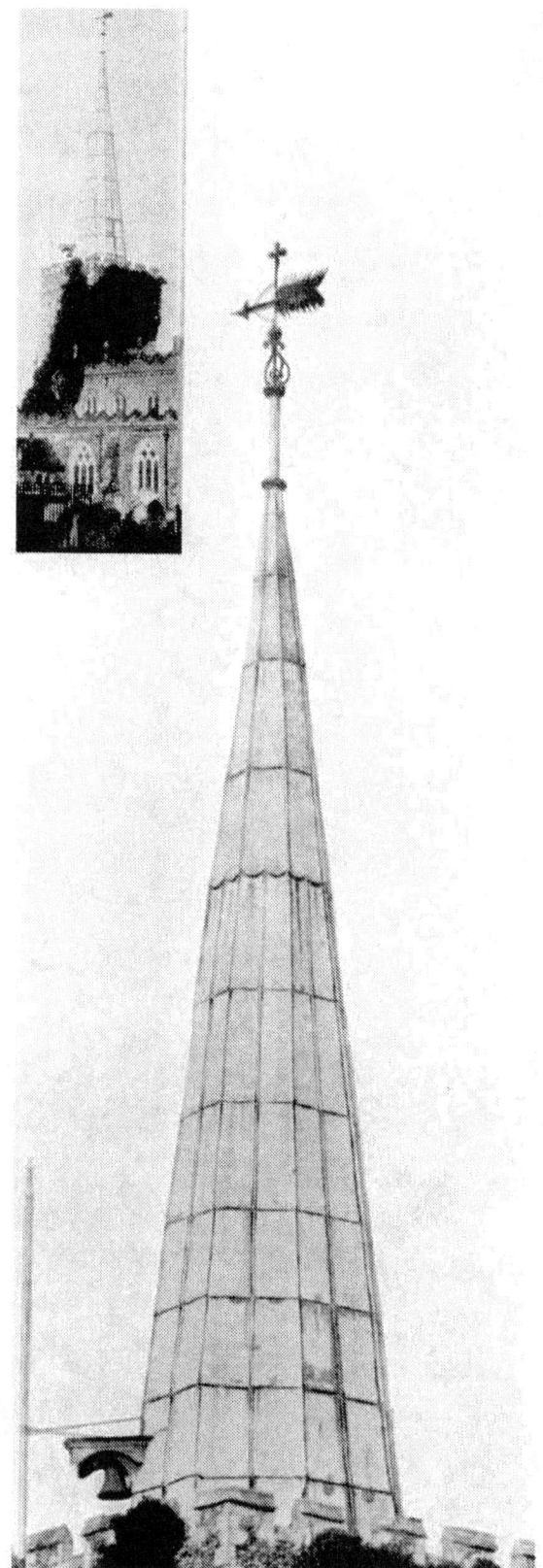

Fig. 192.—Minster.

Fig. 193.—Great Baddow.

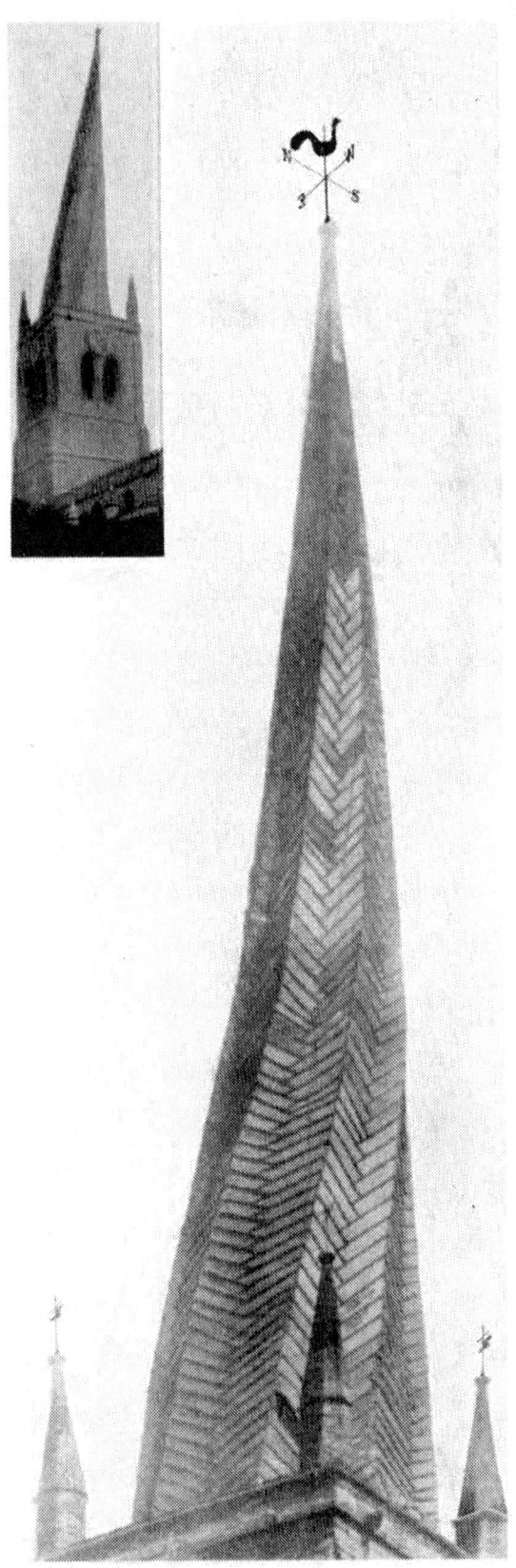

FIG. 194.—Chesterfield.

straight-sided spires and the spirelets would be ruled out. Only the pathless spires would pass the test, for there are no lead spires resembling the later stone spires which were connected with the parapet by pinnacles and flying buttresses.

The rolls at Minster are vertical only, as are those at Great Baddow, Essex (Fig. 193), where on each face there is only one roll between the angle rolls, and this ceases at the fourth horizontal division from the top. The little bellcote is an interesting addition, but apparently recent.

Harrow, on the other hand, is prodigal of rolls, there being three on each face between the angle rolls (Fig. 162). The spire is of the fifteenth century. On the lead near the base of the spire are writ large the names of the churchwardens of 1823, under whom the spire was repaired, and curiously enough, also the legend " Hannah Patman, plumber, 1823." This leadworking lady was carrying on the business of her deceased husband.

The spire of Chesterfield (Fig. 194), with its amazing twist, is a cause of such controversy that one needs, when dealing with it, to behave even as Agag, and walk delicately. John Henry Parker, by writing that "the lead is so disposed as to give the appearance of the spire being twisted" was not a little misleading. Some have gathered from this that the spire has an apparent but not a real twist. Happily a good photographic lens is not so subject as the retina to optical illusion, and the illustration is quite emphatic as to the reality of the twist. As to the cause of the twist it is generally thought that the warping of the main timbers is responsible. Equally careful investigators, however, have examined the timbers, and have declared with equal emphasis, indeed with equal heat (*venenum archæologicum* is not far behind *odium theologicum* in fervour), that the timbers

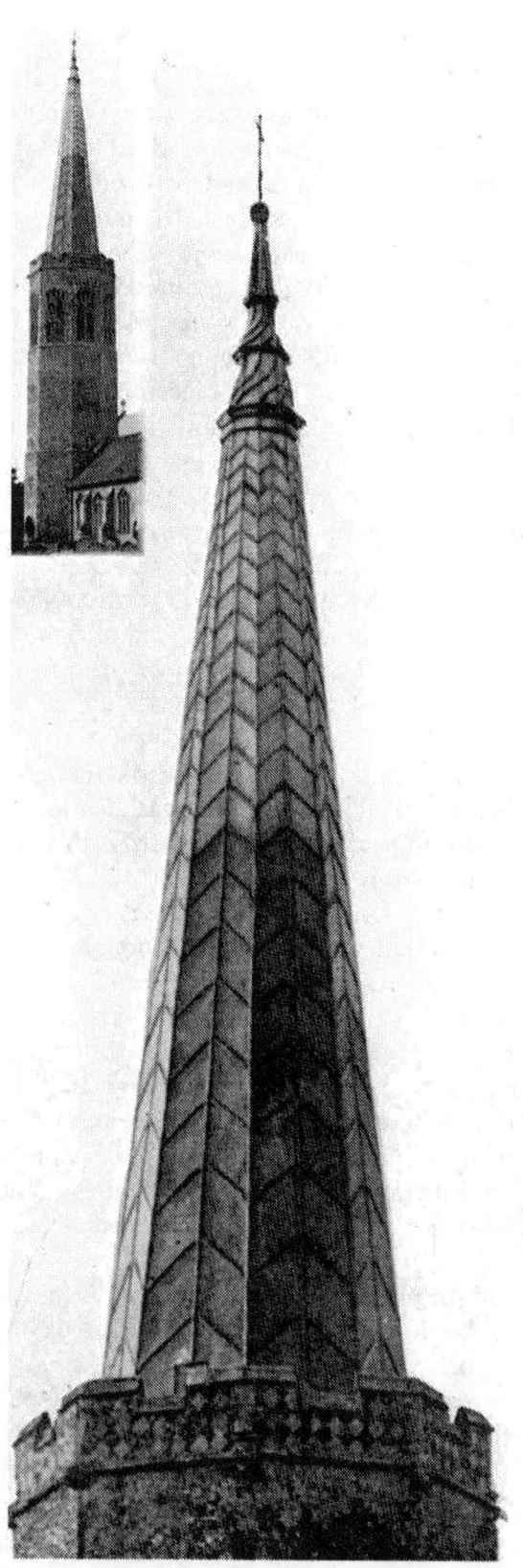

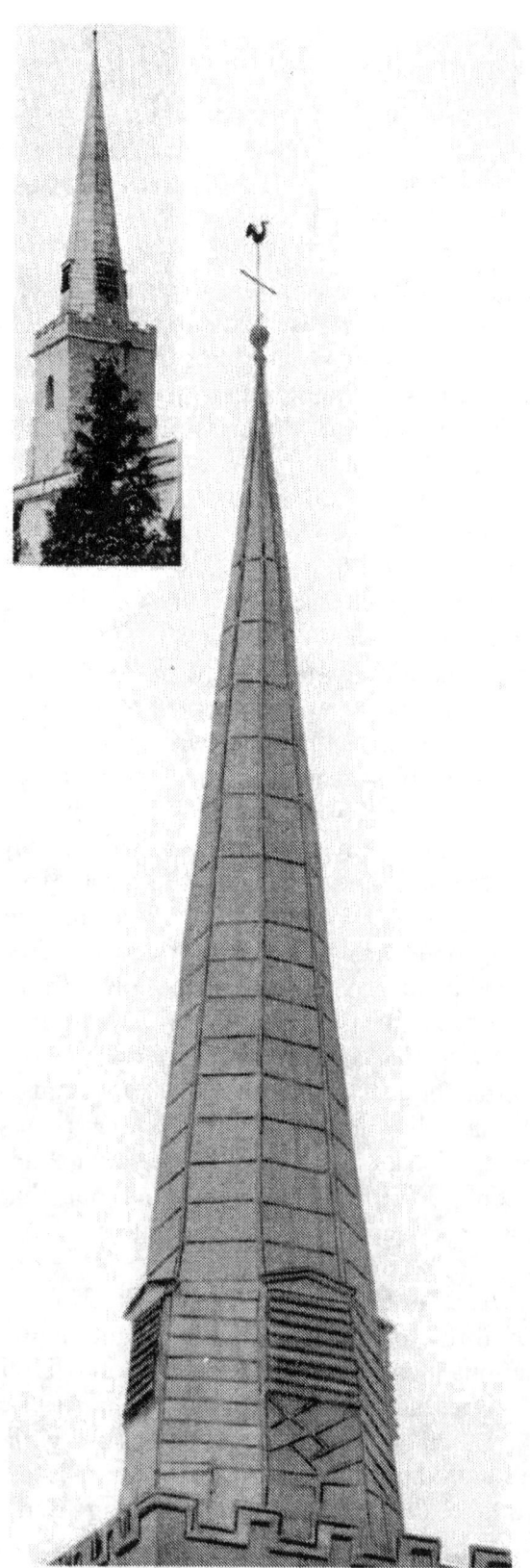

FIG. 195.—Wickham Market, Suffolk.

FIG. 196.—Much Wenlock, Salop.

show *every sign*, and that they show *no sign* of having warped and sprung at the joints.

One is a little suspicious when "spirals" are imported into architectural discussions. Some people want to read spirals into everything. Assuming, however, that we may properly look for a purpose in the twist of Chesterfield, the spiral theory seems just tenable. About 1370 practically the whole structure of Chesterfield parish church was rebuilt. The nave and tower are good ordinary work of the period, and we are asked to assume that the architect determined on a spire which should give extraordinary distinction to an otherwise ordinary church. The whole structure of the spire rests on four massive beams which are built into the top of the tower, crosswise, forming on plan nine small squares. The corner squares are intersected diagonally by cross pieces which take the diagonal faces of the octagon. From each corner of the middle square rise the great stanchions which form the real core of the work. The spire is built in sections from 18 to 20 feet in height, and it is affirmed that each succeeding section is intentionally twisted at a regular degree above the one beneath. Obviously such a construction leads to all manner of difficulties in the direction of keeping the spire at all plumb. The theorist goes on to affirm that when the steeple rose to about two-thirds of its height the builder got alarmed at the amount it was out of plumb, abandoned the system of twist, and made for the summit by the straight route. This theory is set out for what it may be worth. It is not vastly impressive, but experts in the mysteries of carpentry must be left to settle the point. That the twist is due to the great weight of the lead, and the warping of imperfectly seasoned timber seems a simpler explanation. It should be remembered that Chesterfield is not alone in possessing an erratic shape. The lead spire at Walsingham, Norfolk, though not so large, is considerably bent at a point about one-third from the top. The shingled timber spire of Cleobury Mortimer is also badly twisted.*

One other point with regard to the Chesterfield spire deserves mention. The herring-bone arrangement of the rolls produces an optical illusion which, though more noticeable to the eye when looking at the actual spire, is also to be observed in the photograph (Fig. 194). It might be thought that the plan of the spire, instead of being a plain octagon, is an octagon of which the eight faces recede in V fashion inwards, or (to put it another way) that the plan is a sixteen-sided star, and that an imaginary line connecting the outer points of the star would form an octagon. This is not, of course, the case; the suggestion of a star-shaped plan is purely an optical illusion. It may also be pointed out that the rolls are of herring-bone arrangement, as is more common with pathless spires, while vertical rolls are more usual with parapetted examples.

At the Church of Ottery St Mary is a delightful octagonal spire standing well within the parapet, and so low and squat as to be almost of the proportions of the octagonal leaded roof of the Chapter House of York Minster.

Wickham Market (Fig. 195) has especial interest in that it has an octagonal spire on an octagonal tower. A pleasant variation from the ordinary apex is afforded by the mouldings which encircle it, the rolls on the two little stages so made being arranged spirally. One cannot help wishing that some builder of lead spires had built an

---

* The "twist" theory, shortly described above, is set out in a long article in the *Derbyshire Courier* of 14th November 1903.

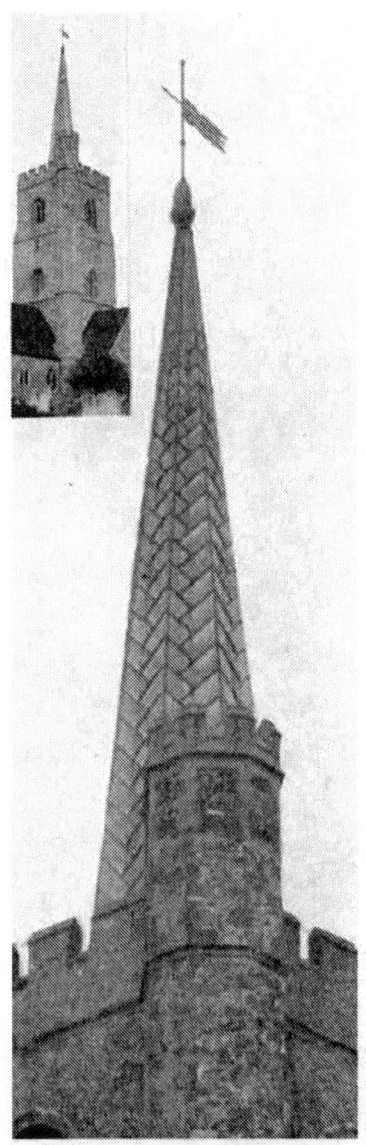

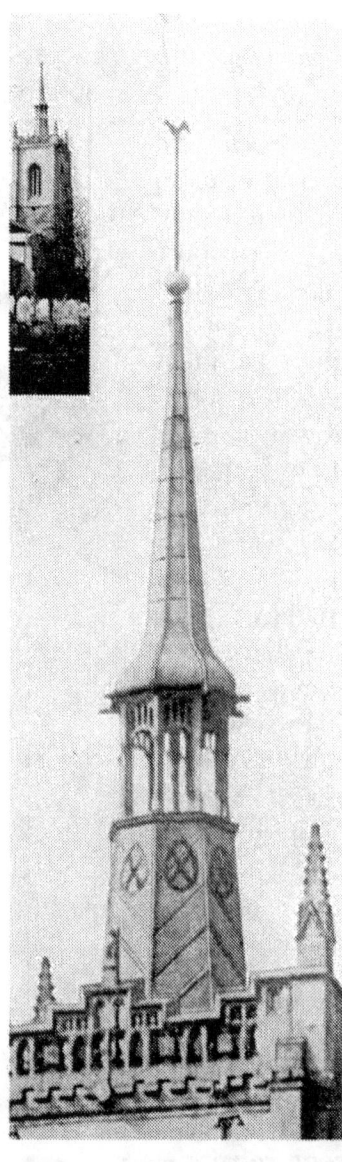

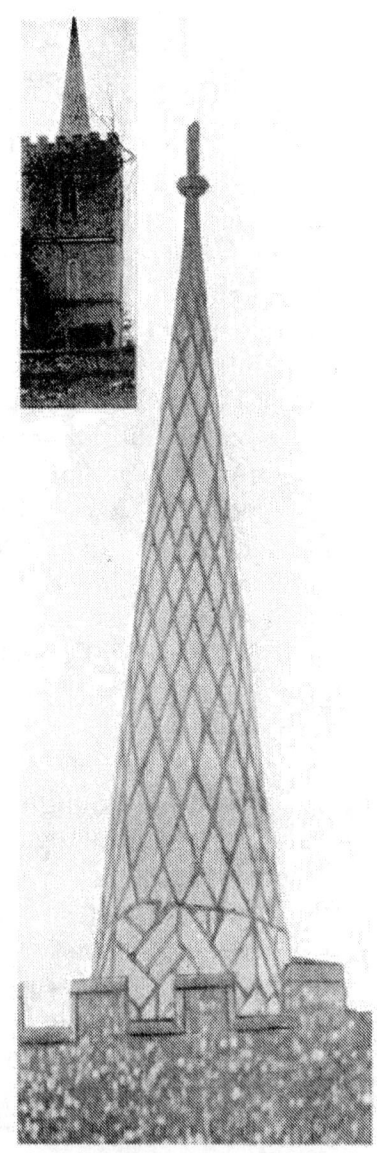

FIG. 197.—Ash, Kent.   FIG. 198.—Swaffham, Norfolk.   FIG. 199.—Sawbridgeworth.

THREE TYPICAL LEADED SPIRELETS.

octagonal or, better, sixteen-sided spire, and arranged the main rolls in strongly marked spirals from the base up. The result would be unrestful, but as it is presumably the business of a spire to aspire, it would have been an interesting experiment, and certainly amusing.

Much Wenlock, Salop (Fig. 196), has no vertical rolls between the angle rolls, and consequently the horizontal sheets are very narrow. There are openings with meagre luffer boards, and below them some rolls arranged in network fashion, which gives variety. This spire was erected in 1726, but the tower is of the thirteenth century, so probably the present spire took the place of an earlier one.

St Margaret's, Lowestoft, has a lead spire of the straight-sided type standing well within the parapet, and calls for no special remark.

For the highly Gothic person, the parapetted spirelets, such as those at Hitchin and East Harling, can have no justification, except a purely decorative one. To people who want to justify everything, a broach spire is a roof, and bells can be hung in it. For the large plain spire standing within a parapet there is less excuse, and for spirelets none at all. They are, however, very delightful things, and should be jealously preserved. A few years ago a good lead spirelet at Brandon, Norfolk, was taken down without any faculty being obtained. The criminal does not appear to have been dealt with in any suitable Gilbertian way, such as with melted lead, an omission one cannot sufficiently regret. There was a similar spirelet on St Alban's Abbey. Perhaps it was grimthorped. At St Alban's nothing is astonishing, but the spirelet has gone.

Sawbridgeworth, Herts (Fig. 199), has a charming spirelet. The diamond shaped arrangement of the rolls on the upper part is unusual, and of happy effect. The larger diamonds coming above smaller give a pleasant irregularity. The haphazard arrangement on the lower part is possibly the result of comparatively recent repairs.

Ash, Kent (Fig. 197).—Of this there is little to say save that the little spire groups oddly with the corner turret.

Bramford, Suffolk, has a plain spirelet of considerable merit.

The most notable spirelet is that of East Harling, Norfolk (Fig. 200), which dates from 1450. It is not only the most ambitious in England from the leadworker's point of view, but the most beautiful. The spirelet proper stands on an octagonal drum with vertical sides, also leaded. This con-

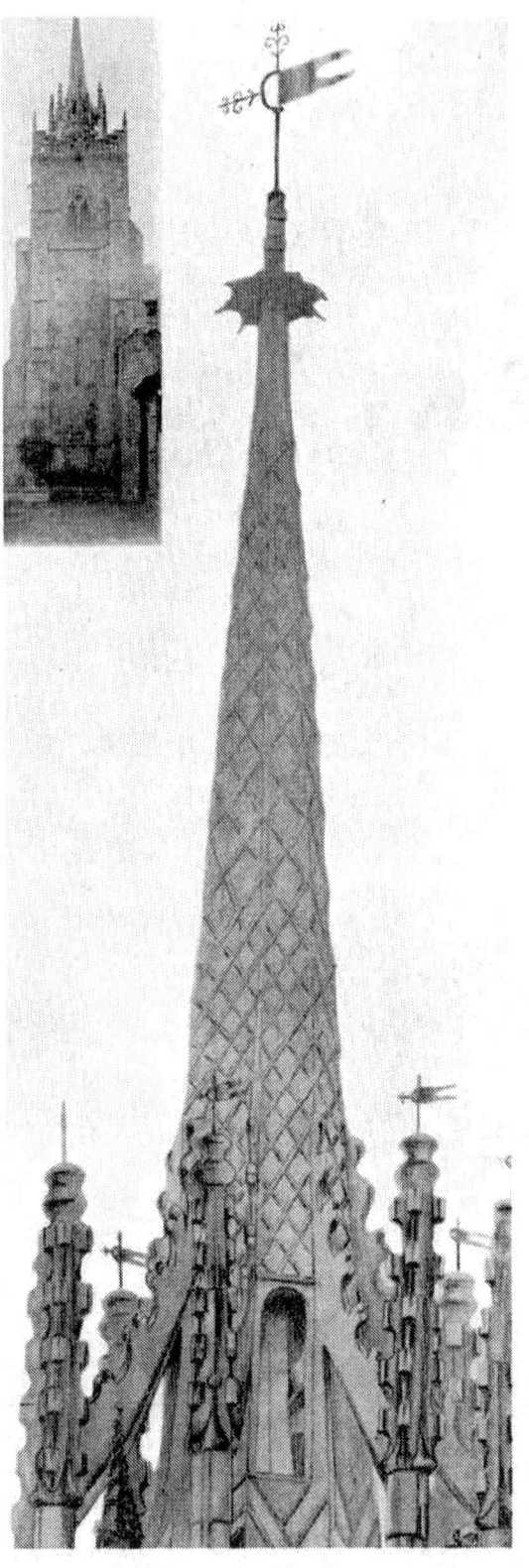

FIG. 200.—East Harling.

junction of spire and drum is an imitation *in petto* of the octagonal intermediate stage between tower and spire that we find in stone at Wilby and Exton. There is in Dugdale a drawing (Fig. 202) of a very notable feature of Hulm Abbey, Norfolk, which is of cognate character. The lower stage of the spire was apparently circular and altogether leaded, and seems to have been in a general way the ancestor of the East Harling treatment. At each point of the East Harling drum there rises a leaded pinnacle, and from each pinnacle two flying buttresses are thrown to the spire. The upper tier of buttresses is crocketed with seven crockets to a buttress. Mr Leonard Stokes's sectional drawing (Fig. 201) in the "Architectural Association Sketch Book" (vol. i., Plate 18) shows the woodwork only down to the roof of the tower, but the beams run down to and rest on the sills of the window in the belfry story. The timber work is of oak throughout. As to the leading, the metal is dressed over each face of the pinnacles and lapped on the edges. The rolls on the spire are solid (without wood core) and they form reticulated patterns which vary not only on different faces, but between the top and bottom of the same face. The leading on the lower parts of the main pinnacles has been restored in

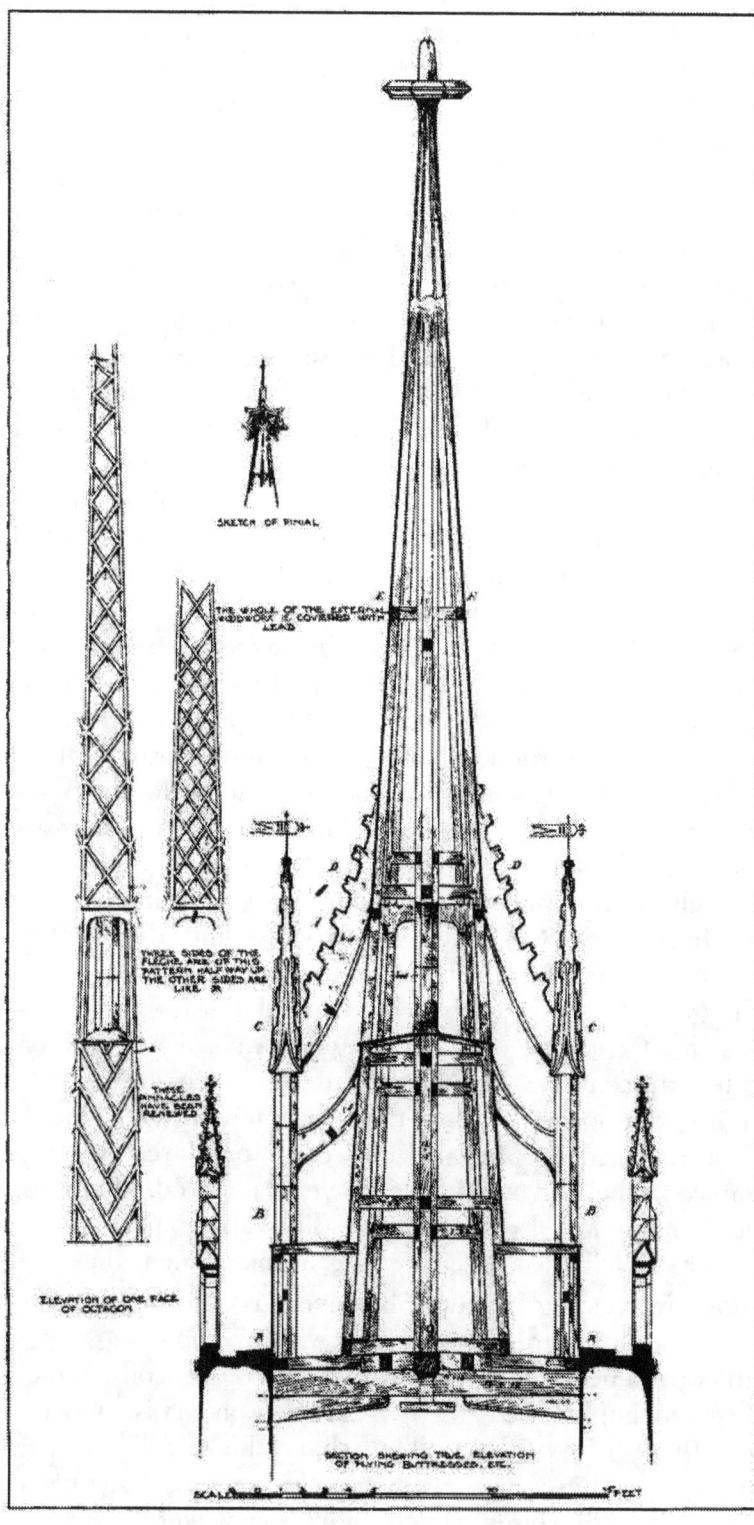

FIG. 201.—East Harling.

recent years, as also the leading of the drum, but the spire proper and the tops of the pinnacles, if not the original work, are obviously of a most respectable antiquity. The finial at the apex of the spire is of umbrella form, not unlike that on the lead flèche at the Law Courts. The total height of the spire is 52 feet 6 inches.

The churches of St John and St Peter, Duxford, have little lead spirelets, one being leaded in diamonds and with the "umbrella" top as at East Harling. The spirelet of Swaffham is very interesting, if late (Fig. 198). It was restored in 1896 but so piously as to rob the word "restoration" of its sting.

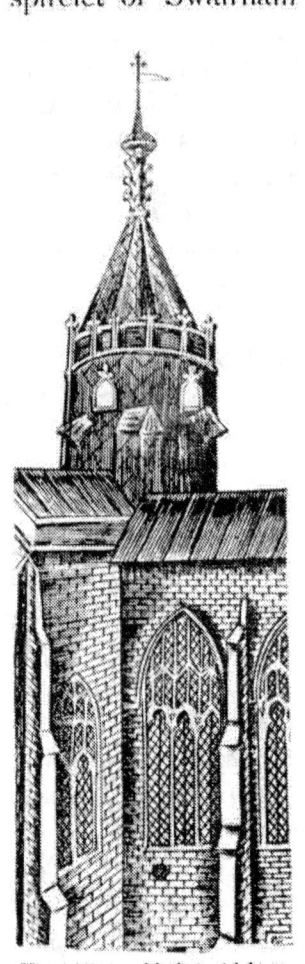

The history of the spire is so interesting as to deserve extended mention. The tower is of 1507-1510. It is not known whether a spire was built then, but probably not. It is likely that the first spire was built about 1600. In 1777 the spire was taken down because, as the vestry minutes state, it was observed to be out of perpendicular. Upon this one of the churchwardens and the vicar employed Mr W. Ivory, an eminent architect of Norwich, and Mr Robert Treegard of London, a retired builder, to take a survey of the spire. After survey they reported that the spire was dangerous and must be taken down. A vestry meeting then made order that Mr Frost, carpenter, "do forthwith repair the spire at an expense of £80." Apparently the joint wisdom of the eminent architect and the retired builder was flouted, and the spire only ordered to be repaired. The strenuous Mr Frost, however, "without further application to the wardens, proceeded to take the spire down entirely and to rebuild another." In 1778 the wardens are presented with a bill for £437. 0s. 5½d., the 5½d. doubtless for moral and intellectual damage consequent on the original contract only having been for £80. After much wrangling they settled for £387. 0s. 5½d. One feels that Mr Frost's honour was secure. He gave away £50, but he triumphs with 5½d., altogether a charming picture of the engaging ways of contractors in the eighteenth century. To return to the spire itself. The drum was not taken down in 1896, though some of the decayed timbers were replaced by new. The open oak arcading was entirely renewed, the old work being very debased, doubtless some of our friend Mr Frost's work. The upper part of the spire has been rebuilt to precisely the same dimensions and

FIG. 202.—Hulm Abbey, Norfolk.
(*From Dugdale's "Monasticon Anglicanum."*)

details as before. By far the most interesting feature, however, is the ornamentation of the drum. Cross keys and swords are surrounded with a moulding, egg-shape in outline, and 1½ inch thick. These doubtless came from the spire which Mr Frost pulled down, as they were simply fixed by two large iron nails, assisted by two hooks at the top to hang them in position. They have been refixed with every care. Probably such ornaments as these were common features of mediæval lead spires, and have disappeared as the spires which now exist were repaired and releaded. At Shipdham, Norfolk (Fig. 203), there is a debased Gothic steeple which has even more parts than a Wren composition. Between the domical roof, which is its lowest element, and the ogee spirelet which crowns it, there are two lanterns, separated by an ogee roof trimmed with ridiculous pinnacles. It is altogether a wild exercise in timber and lead.

The hand of the destroyer has been unhappily active in doing away with the leaded spires of parish churches as well as of cathedrals. St Nicholas, Great Yarmouth, until 1803 had a lead spire. The old spire was 186 feet in height, rather loftier than the present one. It had been struck by lightning in 1683, and, whether from that cause or through shrinkage of the framework, was crooked. In 1807 the tower was repaired and the spire altogether rebuilt.

The spire of Shakespeare's church at Stratford-on-Avon is of stone and 83 feet high. The tower, however, was originally crowned by a timber spire covered with lead, and about 42 feet in height. This was taken down in 1763, and the present spire of Warwick hewn stone built in the following year.

At Thorpe-le-Soken, near Frinton-on-Sea, there is a spirelet in a curious middle state of dissolution. The lead has gone, but the open timber framework remains. The district was an important military area in the Great Rebellion, and local tradition credits Oliver Cromwell with stripping many roofs and steeples to provide his men with bullets. This may be true, for we find the Lord-General writing to his cornet: "We shall want some lead—the steeples have plenty." It is fair to Oliver's memory, however, to point out that many Cromwell legends

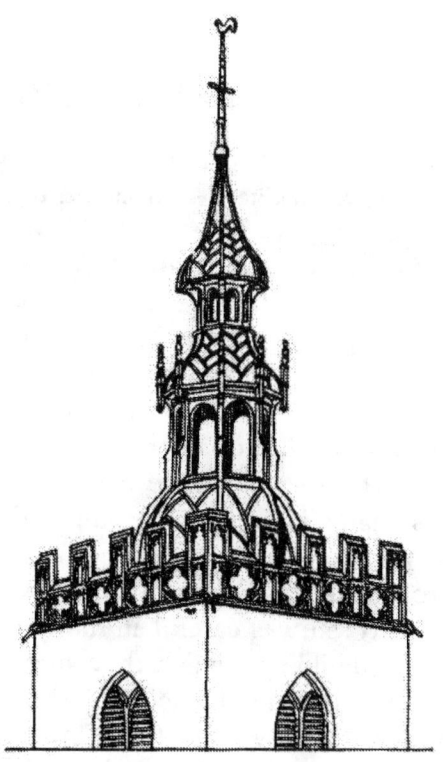

FIG. 203.—Shipdham.

when critically examined prove to be attributable to Thomas Cromwell (or, better, Crumwell), the complaisant Vicar-General, who understudied Henry VIII. in his ruffianism.

It admits of little dispute that much "Cromwell" defacement of England's buildings should properly be laid at the door of Thomas and not of Oliver. Moreover, Oliver destroyed either from military necessity or from religious conviction, unhappy in its operation, but sincere; Thomas, from sheer rapacity, the less pleasant from being covered by an ecclesiastical posture.

CHAPTER VI.

LEADED STEEPLES OF THE RENAISSANCE.

Wren's Steeples and the Sky-line of London—A Classification—Class (*a*), The Two True Spires—Class (*b*), The Spire-form Steeples—Some Destroyed Steeples—Scottish Examples—The Character of Wren's Work.

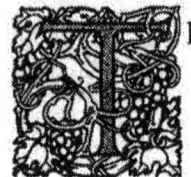 HE lead steeples and domes of the Renaissance period fill an important niche in architectural history; but they do more. They have an eminent place in any survey of the art of Sir Christopher Wren, and they are largely accountable for the sky-line of the city of London. If Wren's achievements in this direction were cut out, very little would be left either of the sky-line or of this phase of the history of leadwork in England.

If we could have accompanied the late Mr Samuel Pepys, M.A., F.R.S., on one of his many jaunts in his galley down the Thames to Greenwich before 1666, we should have observed a sky-line, which, save for the dome of St Paul's, was not greatly different from that which Canaletto drew in 1767 (Fig. 207).

Wren was careful in many of his new churches to preserve the outstanding features of the buildings which they succeeded, and by the leaded dome of St Paul's he re-established the dominance of the cathedral, which was to some extent lost with the destruction by fire of the great leaded spire of Old St Paul's in 1561. Splendid as are the steeples of Wren's parish churches, Canaletto's view (Fig. 207) (taken from the gardens which are now the site of Somerset House) shows how entirely St Paul's governed the sky-line of London. To-day it is different. St Paul's is still the supreme feature of the City (as Turner said, " The dome of St Paul's *makes* London "); but commerce is crowding out the parish churches. Mr Pepys' galley being unavailable, a journey on a steamboat from Temple Pier to Cherry Gardens Pier* makes melancholy travelling.

Seen from the Temple, Cannon Street station is a hideous incubus on the City sky-line. It blots out all the Monument except from the gallery upwards (not a great loss perhaps), and every spire, save the tip of St Magnus, while the bridges at Blackfriars cut out the foreground. The City of London School on the left, with its lead lantern of unsatisfactory outline, almost wholly hides St Paul's. The miserable spikes on the corners of Cannon Street station add insult to injury, for they are, in outline, vulgar caricatures of the steeple of St Magnus. They serve only to remind us of what

* The "Diary," 13th June 1664: "Thence having a galley down to Greenwich, and there saw the king's works, which are great a-doing there, and so to the Cherry Garden, and so carried some cherries home, and after supper to bed, my wife," &c.

a wealth of steeples the station blots from sight. Maybe they are a mark of the engineer's feeble compunction. Once past Blackfriars Bridge, the ten-storied warehouses of Thames Street make a wall impenetrable save for glimpses of St Benet's, Paul's Wharf, and St Nicholas', Cole Abbey. St Margaret Pattens, and of course St Magnus, complete the list of what commercial London has left to be seen from the river. It is only from a lofty vantage ground like St Paul's or the Monument that one can now get any general grasp of the grouping as Wren left it. The two photographs of Figs. 204 and 205, taken from the top of the Monument, show how little the church towers and spires count now that the office buildings are so high. They do, however, emphasise the contrast between the blackened lead spires and the white towers; in Fig. 205, the lantern of St Edmund's, Lombard Street, against the Royal Exchange, and St Peter's, Gracechurch Street (on the extreme right), against the mass of St Michael's, Cornhill.

To attempt any classification of the domes, lanterns, and steeples of Wren's London is a difficult task, for in nothing did Sir Christopher Wren show the almost wanton luxuriance of his art more markedly. For the twenty-eight towers that are crowned with either spire or lantern, Wren employed stone for only nine, and leaded timber for nineteen. Lead may, therefore, claim the first place in his affections as a spire material. These nineteen we may divide into three classes.

> *a.* True spires.
> *b.* Spire-form steeples.
> *c.* Lanterns.

This is a loose and arbitrary classification, but Wren's masterful way of playing with architectural elements and combining them in astonishing ways makes havoc of any orderly description. He created within the square mile of the City more forms of steeples than all the architects of the Middle Ages, and if, as was inevitable, some pay the penalty of rash experiment, others make an assured success.

The attempt to set out the lines on which Wren proceeded is hampered at every turn by lack of evidence. We have little clue as to some of his more curious designs, but these were probably less arbitrary in their creation than may appear to us in the absence of such indications.

That Wren was a close student of his predecessors in the art of building is easily proved, but his debt to mediæval sources is not generally realised. Imperfections of detail ought not to obscure an appreciation of the fact that his grasp of Gothic principles is rarely at fault. There is much in Wren's work otherwise inexplicable which may be traced to the wide catholicity of his mind. It is not only difficult but impossible to point to another architect of his epoch, who, with anything approaching his success, seemed so nearly to have reconciled the opposing ideals of classicism and romanticism. To the union which he thus achieved must be ascribed the marvellous picturesqueness which, united with imposing mass, makes St Paul's the unique masterpiece amongst Renaissance churches.

In connection with his large use of leaded timber spires it must be remembered that Wren was an architectural economist, and the results he achieved are the more notable, when considered in relation to the very limited means which were generally at his disposal. This is especially the case with the parish churches of the City. The use of leaded spires enabled him to give distinction and character to churches, where limitation of cost put stone spires out of the question. His followers, however, in many cases departed from

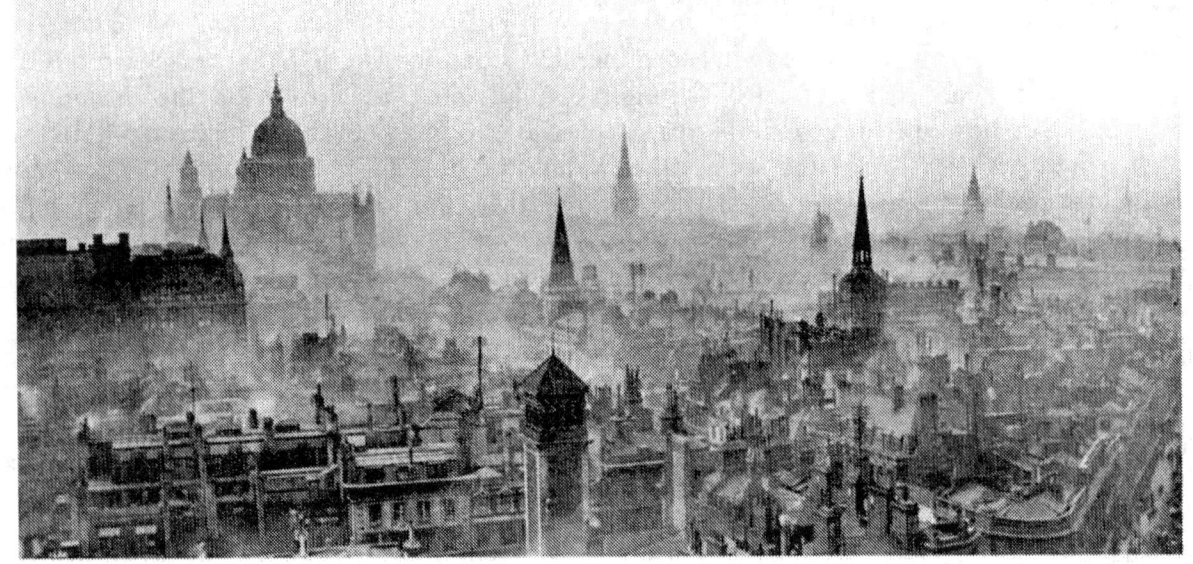

St Swithin's.   St Mary Abchurch.   St Lawrence, Jewry.

FIG. 204.—The City from the Top of the Monument.
(King William Street on the right.)

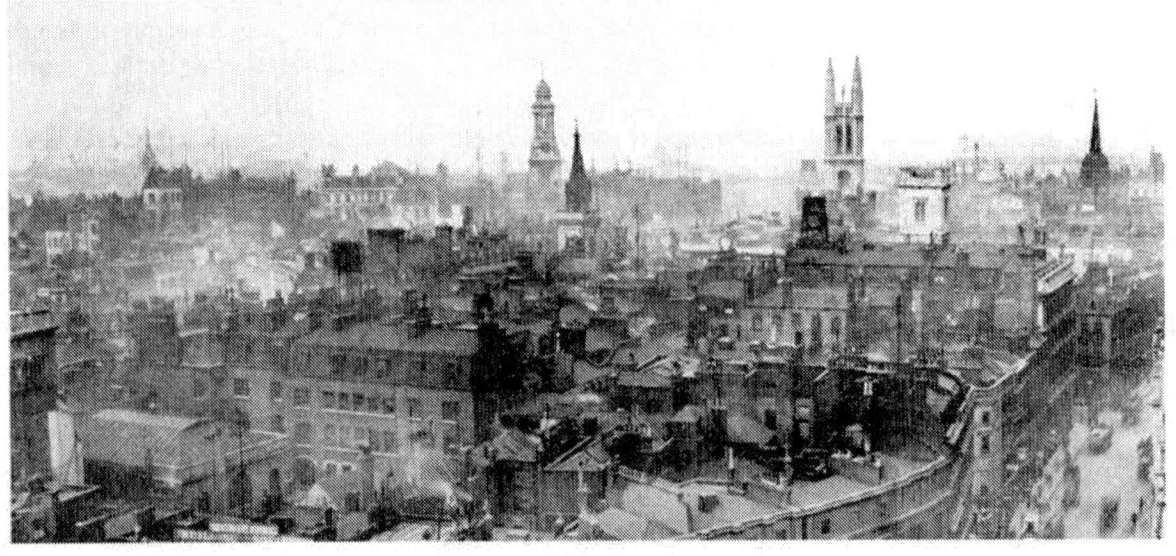

St Margaret's, Lothbury.   St Edmund's, Lombard Street.   St Peter's, Gracechurch.

FIG. 205.—The City from the Top of the Monument.
(Gracechurch Street on the right.)

the excellent precedent which he had set. Some of the later classic spires would have been quite reasonable in leadwork, which allows of a certain quaintness of design, whereas a great masonry obelisk, such as we see in South-east London, is merely an architectural oddity.

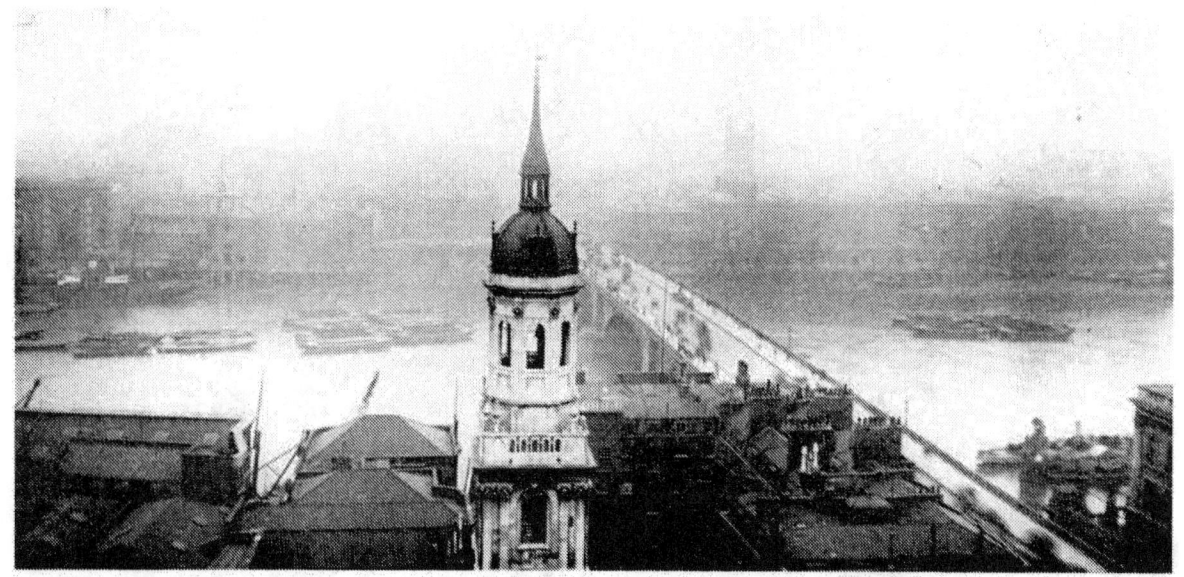

FIG. 206.—St Magnus from the Top of the Monument.
(Looking across London Bridge.)

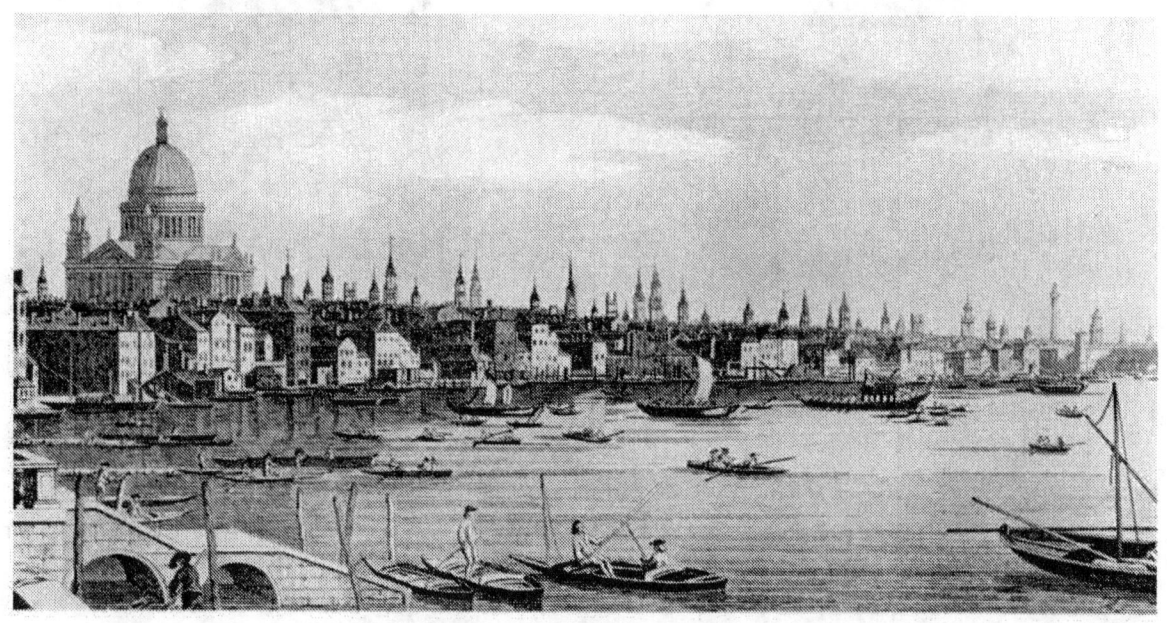

FIG. 207.—Canaletto's View of London (Part of), 1767.

Among the nineteen leaded steeples there are only two which can be described as true spires, St Swithin's, London Stone, and St Margaret Pattens, Rood Lane.

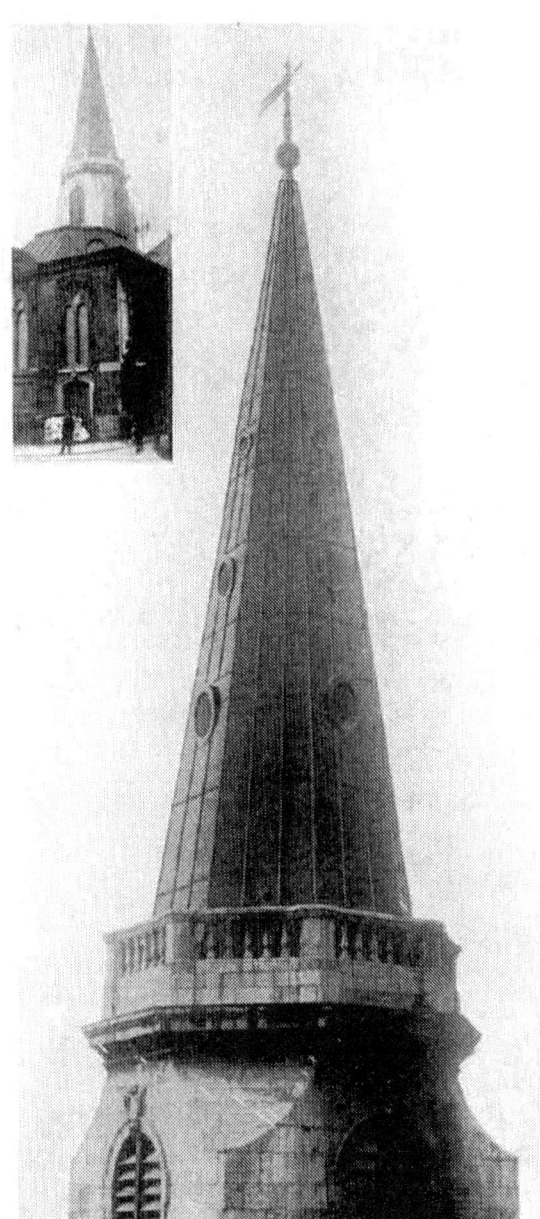

FIG. 208.—St Swithin's, Cannon Street.

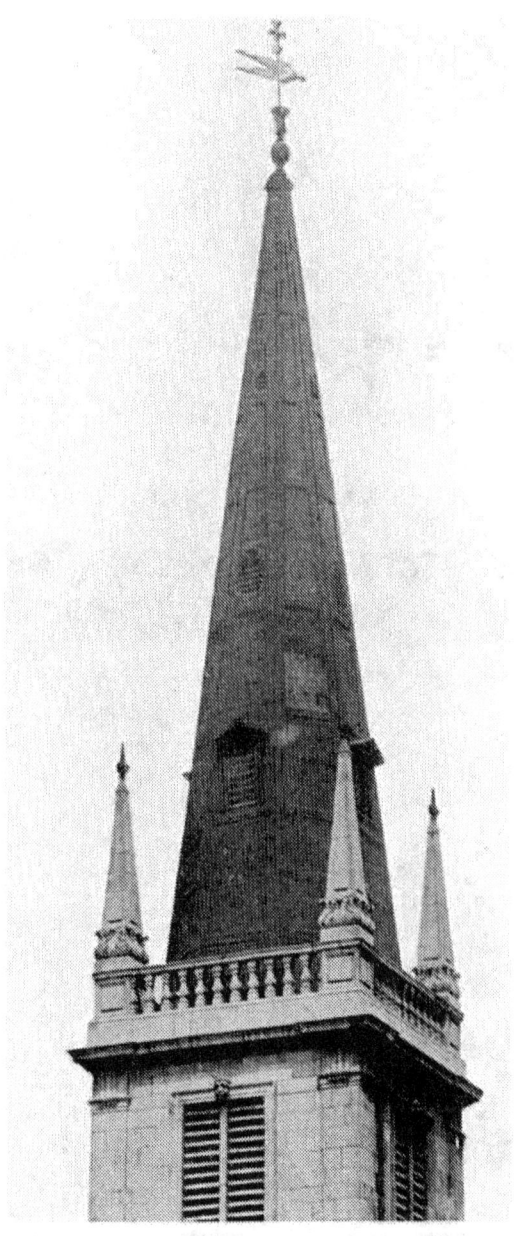

FIG. 209.—St Margaret Pattens, Rood Lane.

Their peculiar interest lies in the fact that in them Wren is in debt to his predecessors.
They are, in their essential lines, Gothic.   With St Swithin's this is especially the case.

Mr Andrew T. Taylor in his admirable book,* suggests that the towers which have
no steeples would stand them, and that those with steeples could do without them.
While this is true of the majority, it is not wisely said in respect of St Swithin's.   The
top of the tower was obviously designed purely in relation to the spire which surmounts

* "Towers and Steeples designed by Sir C. Wren," published 1881.

it. Without the spire the scooped-out splays at the top angles would be meaningless and even absurd. Wren's problem was both simple and old, how to step from the square of the tower to the octagon of the spire. He attacked it with his usual queer mixture of boldness and compromise. The mediæval architect did not tamper with his stone tower. It began square and finished square. The change to the octagonal was effected in the timber work, and in two main ways: by framing a collar (*e.g.*, Ryton), or by constructing broaches (*e.g.*, Godalming). Both of these methods involved diagonal bearers across the corners of the tower. At St Swithin's, Wren took a characteristic short cut. By trimming the tower angles to a splay he secured solid masonry to take both the cardinal and diagonal sides of his spire, and so simplified its timber construction. There is, moreover, another element of compromise. The method of recognising the step from the square to the octagonal by obvious construction had hitherto been used only on towers without parapets. Wren, however, emphasises the break with a cornice topped by a balustraded parapet, and so gets the best of both worlds. The leading of the spire itself is purely Gothic in feeling. The oval shape of the spire-lights alone betrays its seventeenth-century origin. Mr A. T. Taylor thinks the scooped-out splays of the tower not very happy, on the ground that the diagonal view brings them into painful obtrusiveness. If this be the case, the photograph of Fig. 208 shows the splays at their worst, but the worst does not seem very bad. Though the splays may fairly be said to obtrude, obtrusiveness is one of Wren's strong points, and even then the delicate frilling of the balustrade tones down not only the incidental coarseness of the splays, but also the inevitable baldness of the progression from tower to spire.

St Swithin's may be taken as Wren's exercise in lead spires in the earlier Gothic manner, which regarded a spire primarily as a roof, and, secondarily, as an architectural feature. St Margaret Pattens (Fig. 209) is of the later type of parapetted spire (*e.g.*, Chesterfield), which, standing well within the lines of the tower walls, abandons the idea of a roof altogether. More significant, however, of the abandonment of the Gothic spirit while retaining the Gothic form is the treatment of the leading. The vertical rolls of St Swithin's are replaced at St Margaret's by a series of sunk panels, which cannot be regarded as so suitable a treatment for lead.

This change may be attributed to Wren's desire to emphasise horizontal lines that would counteract the verticality of the spire proper. Sir Charles Barry in his last work, the Halifax Town Hall, proceeded on the same lines in the bold and vigorous spire that dominates his building and raises it out of its sunken valley site.

These examples may be placed as Renaissance translations of a Gothic original, and be regarded as an example of the power of tradition in English building, even with (or perhaps especially with) such giants as Wren and Barry.

The splendid stone spire of St Antholin's, which was wickedly and quite needlessly destroyed in 1875, was panelled in a similar way to that of St Margaret Pattens. St Antholin's was finished by Wren in 1682 and St Margaret Pattens in 1685, and it is not unreasonable to suppose that the great success of this treatment in stone tempted Wren to essay the same in lead. The likeness of the two spires is carried out even in the character of the spire-lights, which have similar pediments, but the towers are quite unlike. At St Antholin's an intermediate octagonal stage with semicircular buttresses on the diagonal faces marked the progression from the square of the tower to the octagon of the spire.   In the case of both these churches, Wren was careful to reproduce in general form the pre-Fire churches, both of which had lofty spires.

Mr Reginald Blomfield groups the steeples of St Mary-le-Bow, St Bride's, and St Margaret Pattens as "of their kind the most perfect specimens of Renaissance architecture in England."

While it may be presumption to criticise anything that Mr Blomfield may say about Renaissance architecture, there seems room for the view that the steeple of St Margaret Pattens is partly in intention and wholly in outline a Gothic spire.

Though it has admittedly all the simple beauty which Mr Blomfield claims for it, it can hardly be claimed as being in Wren's habitual manner.   Mr Blomfield suggests that Wren's Gothic efforts such as St Mary Aldermary may have been "academical exercises for the entertainment of his (Wren's) friends."   The lead spire of St Swithin's, though Gothic in feeling, has a character at once natural and convincing, and does not need to be explained as an architectural humour.   It and St Margaret Pattens are not in the same category as the seventeen other lead steeples, which owe little to the Gothic spirit and are *sui generis.*

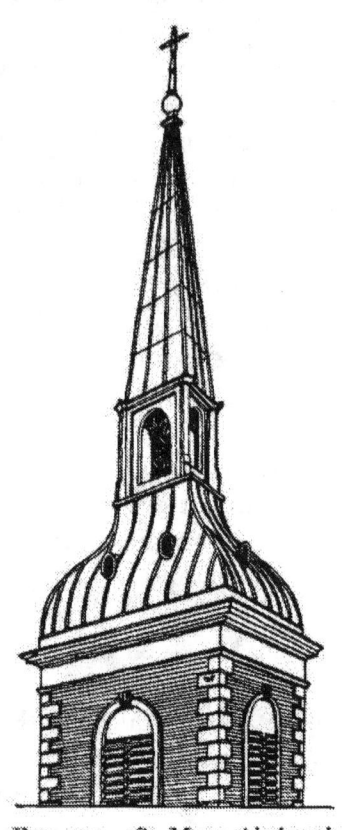

FIG. 210.—St Mary Abchurch.

We next come to Class (*b*), the spire-form steeples. It is a lame description, but may serve roughly to group the eight existing steeples which are neither true spires like St Swithin's, nor simply lanterns like St Edmund's, Lombard Street.   They are essentially hybrids, cunning compositions sometimes brilliantly successful, *e.g.*, St Martin's, Ludgate; sometimes more curious than beautiful, *e.g.*, St Mary Abchurch.   They can be classified roughly by separating those whose terminal is an octagonal spirelet (St Peter's, Gracechurch; St Martin's, Ludgate; St Augustine's, Watling Street; St Lawrence, Jewry; and St Magnus, London Bridge) from the three which have a terminal square on plan (St Mary Abchurch; St Margaret, Lothbury; and St Mildred, Bread Street).   Of these the two latter have abandoned the last flavour of Gothic feeling, for the topmost member is a frank obelisk.

While it is undoubtedly the fact that the amazing variety of Wren's steeples, both of stone and of leaded timber, is to be attributed to the luxuriance of his genius, some root in the past is to be found.   The outstanding difference between the spire-form steeples and the true spires of the mediæval builder is in the complex composition of the former as compared with the simplicity of the latter.   A general glance at the illustrations will show that each spire-form Wren steeple has three main divisions, which are usually—

(*a*.)   A domical or ogee roof;

(*b*.)   A lantern (either with open lights, as at St Martin's, or fitted with luffer boards, as at St Mildred's); and,

(*c*.)   A spirelet or obelisk.

In early mediæval work there seem to have been few important compositions of this kind. The steeple at Hulm Abbey, Norfolk, of two stories, consisting of a

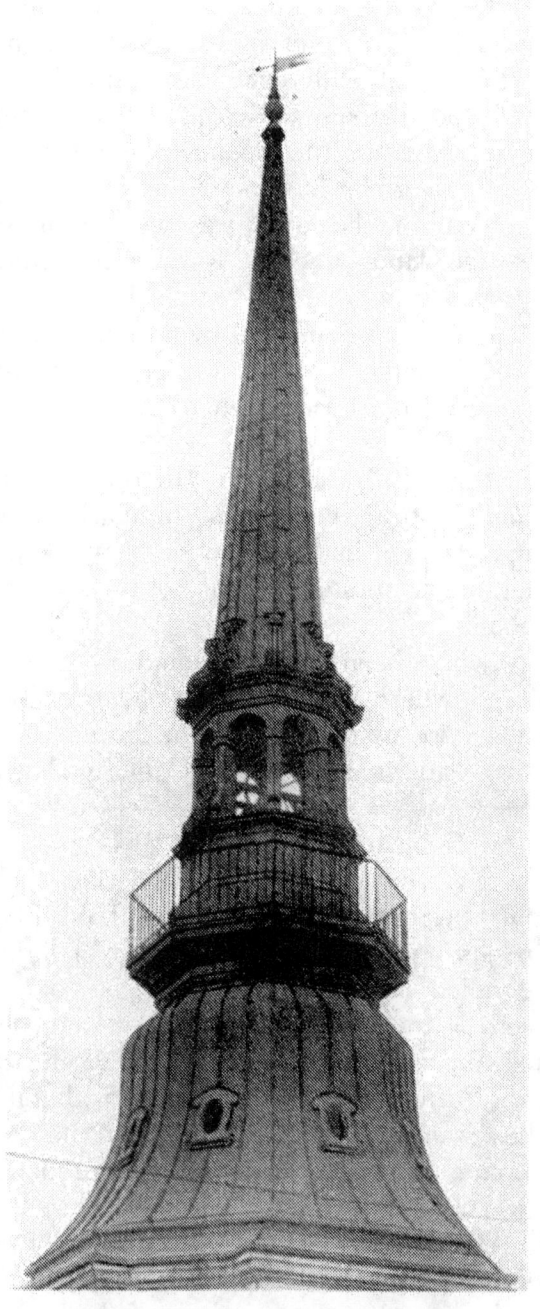

FIG. 211.—St Martin's, Ludgate.

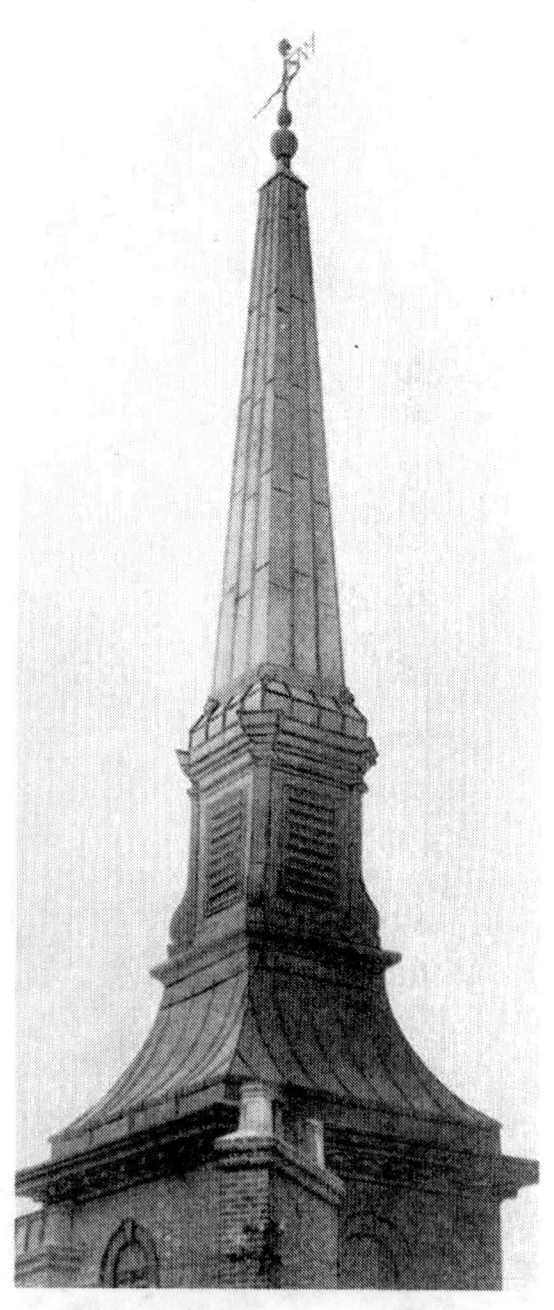

FIG. 212.—St Mildred's, Bread Street.

circular lantern and a short spire, was the nearest approach (Fig. 202), and there were doubtless many more spirelet structures of timber covered with either shingles or lead which may be taken as the groundwork from which later varieties have developed. Fire has, however, left but few.

When we come to the late Gothic spirelets of the fifteenth century, *e.g.*, East Harling (Fig. 200), we are on more solid ground, and the later forms of Swaffham and Chelmsford point in Wren's direction.

The vital difference between Wren's spire-form steeples and the great Gothic lead spires is in the open-arcaded lantern, which the former have and the latter have not. The mediæval spires were glorified roofs, the later steeples were architectural features.

So much may be said by way of examining the general features of Class (*b*) before proceeding to a description of the examples so grouped.

St Martin's, Ludgate Hill (Fig. 211), is doubly attractive. It is singularly interesting *per se ;* its slenderness is a miracle of judgment in its relation to the bulk of St Paul's.

It has been already pointed out that Wren nowhere grapples with the transition from square to octagon in the lowest story of his lead steeples, as did the mediæval people.

At St Martin's (as at St Swithin's) the change is effected at the top of the tower. From the tower walls, octagonal on plan, there springs an ogee roof with oval lights. The railed balcony is a bold device, but its success is the more apparent when one compares the steeple of St Mary Abchurch. In the latter church the lantern with open arches stands direct on the top of the ogee roof, and the effect is meagre and unhappy (Fig. 210).

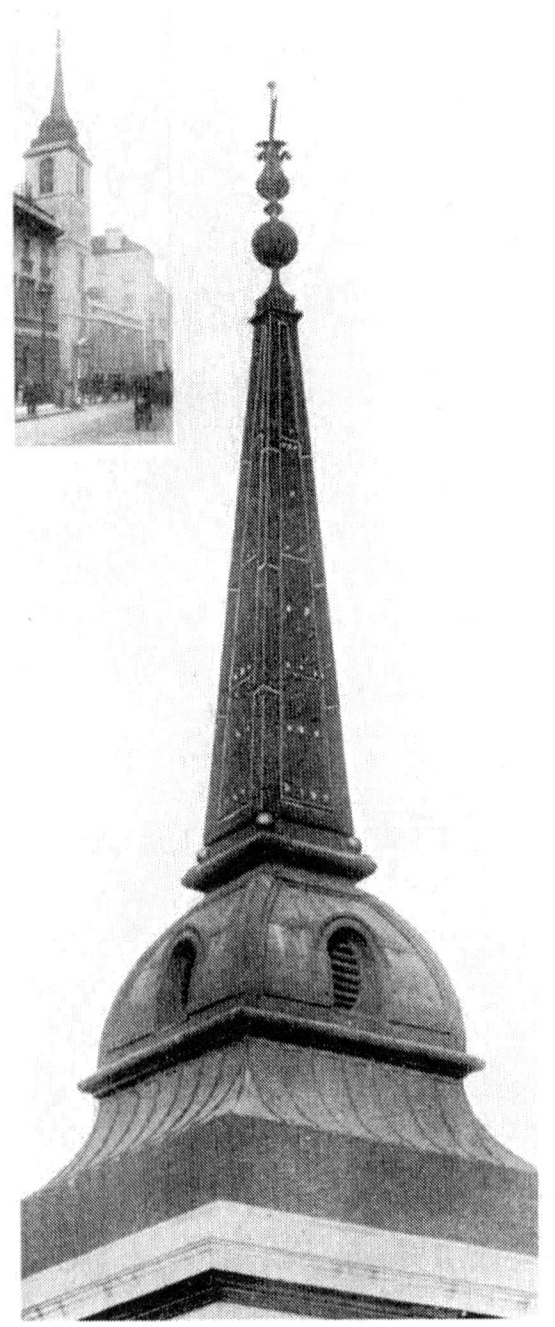

FIG. 213.—St Margaret's, Lothbury.

At St Martin's the sharper pitch of the ogee roof, the cornice supporting the balcony, and the fact that the openings of the lantern are only in its upper half, lead the eye gently from the tower to the top of the graceful spirelet. The angle trusses at the base of the spirelet add a touch of delicate scholarship which is far removed from Wren's sometimes brutal plainness.

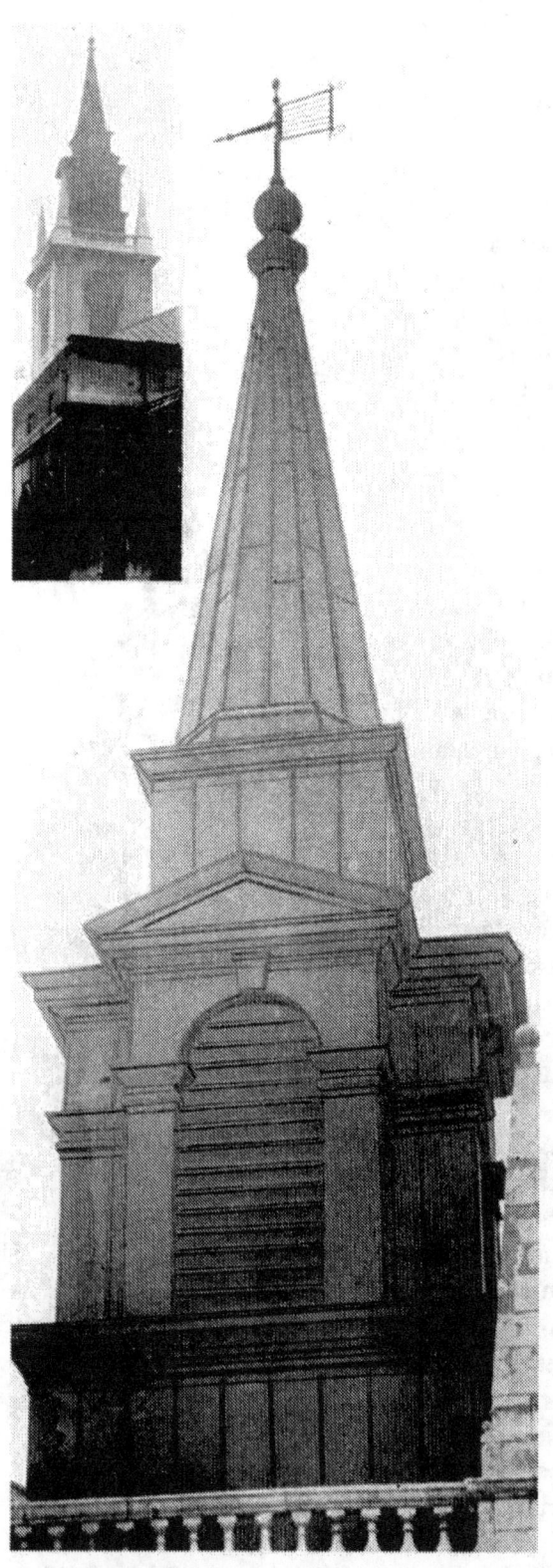

FIG. 214.—St Lawrence, Jewry.

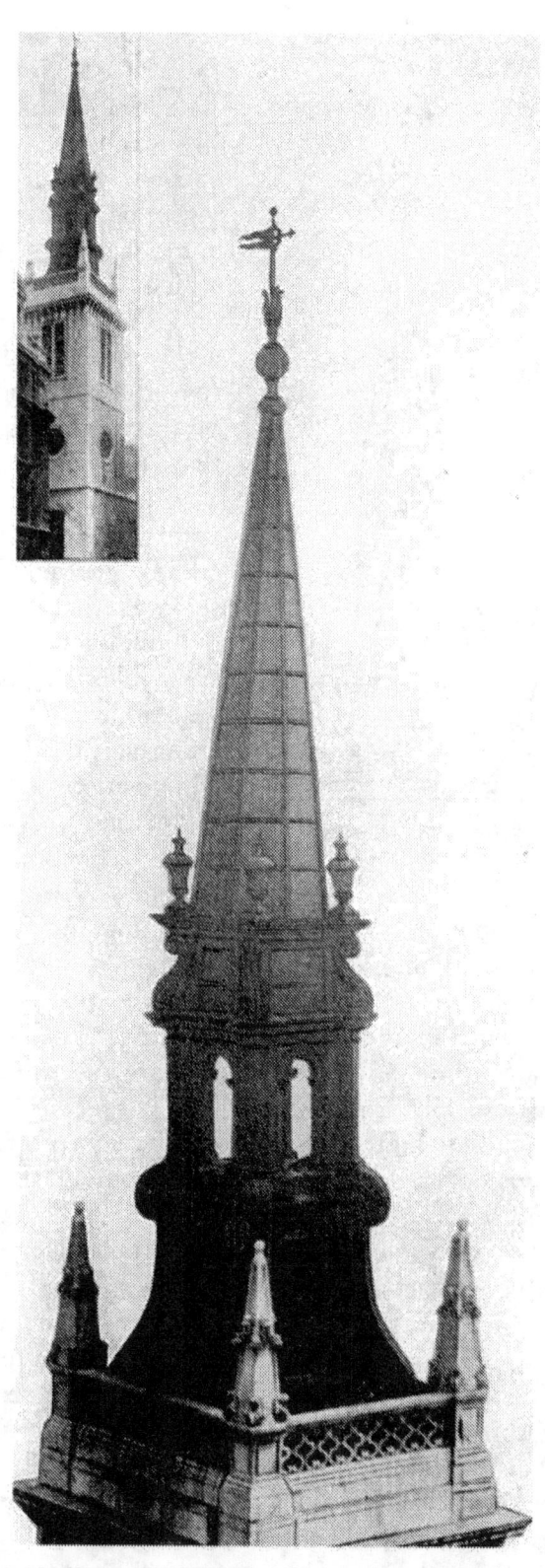

FIG. 215.—St Augustine's, Watling Street.

St Mildred's, Bread Street (Fig. 212), is a good deal less inspired; indeed, it verges on the dull. The concave, pyramidal roof supports a square lantern which has rather feeble louvres, and the lantern is crowned with an obelisk. The steeple of St Lawrence, Jewry (Fig. 214), shows Wren in his strongest mood. The sharp breaks between the three square stages of the lantern, which are accentuated by the vigorous cornices and the solid proportions of the octagonal spire, combine to give an effect which is certainly coarse. It only just escapes being oppressively heavy. It is worth noting that the gridiron vane is symbolic of the patron saint. It is likely that this is a post-Wren detail. Wren was essentially a man of large view. In detail he constantly failed. Indeed, when one remembers the enormous number of buildings for which he was responsible, it is astonishing that the details are so good. In strong contrast to this very masculine composition is the steeple of St Augustine's, Watling Street (Fig. 215). The outline seems almost trivial. We have here a notable

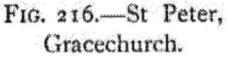

FIG. 216.—St Peter, Gracechurch.

example of Wren's practice of making his tower very plain and lavishing detail on his steeple. St Augustine's tower up to the cornice is plain to the point of baldness. The piercing of the parapet and the pinnacles are very gay, and the outline of the steeple is as free as the vases make it spotty.

The lantern is not in happy proportion. Its three divisions below the octagonal spirelet seem rather an effort, and it is too lofty for its bulk. In effect it looks attenuated. It is very elegant and clever, but Homer seems rather to have nodded. Here again, as with St Martin's, Ludgate, the idea was doubtless to effect a contrast with the mass of the cathedral, but it will readily be admitted that St Augustine's comes far behind St Martin's in result. The two are within a year of each other in date. It is an unhappy thing that the commercial buildings of the City are so insistent to put barriers between Wren's various churches, and in particular to make it so difficult to realise their relationship to St Paul's.

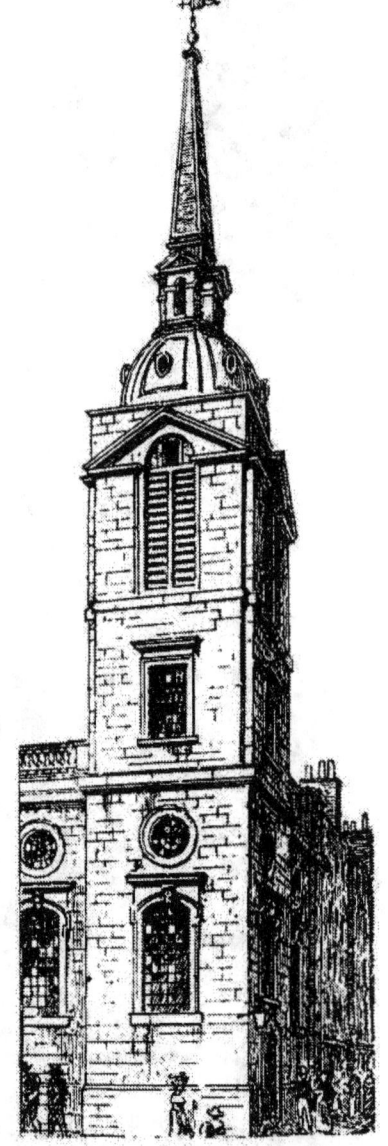

FIG. 217.—St Benet, Gracechurch (destroyed).

It has been well said that St Paul's bereft of the surrounding steeples would be like a mother bereft of her children.

Some authorities on Wren's work are rather scornful about the steeple of St Margaret, Lothbury (Fig. 213), but for what sound reason it is difficult to understand. It is the direct antithesis to such work as St Augustine's, Watling Street. The bold curves of the concave pyramidal roof and of the square cupola which comes above it, the simple massive mouldings of the cupola, the deep reveals of the lights, and the obelisk standing on gilt balls at the angles, all go to make up a "solid masculine and unaffected" steeple. Were such a crime permitted as the destruction of St Margaret's (and the destroyer, as Voltaire said of Habakkuk, is *capable du tout*), we should lose a piece of Wren's work, which, if it is not startling, is eminently sound and characteristic. Without being hysterical, it is perhaps allowable to add that the steeple rising above the Bank and Throgmorton Street is a witness to the unseen which we can hardly afford to lose without more than the loss of a Wren church.

The leaded steeple of St Peter, Gracechurch (Fig. 216), is simple. The plain dome with four small round lights is surmounted by an octagonal lantern and spirelet. It is, I believe, the only spire-form steeple by Wren which has a dome base circular on plan. The exquisite lantern of St Benet, Paul's Wharf, is also circular at its base.

In Fig. 206 appears St Magnus, London Bridge. Finished in 1705, the tall, square tower changes into a stone octagonal lantern, which is covered with a lead cupola. On this there stands a lead lantern, and above that a diminutive spirelet. Here we have the spire element treated with scant courtesy, in fact, as little more than a finial to the lantern and cupola.

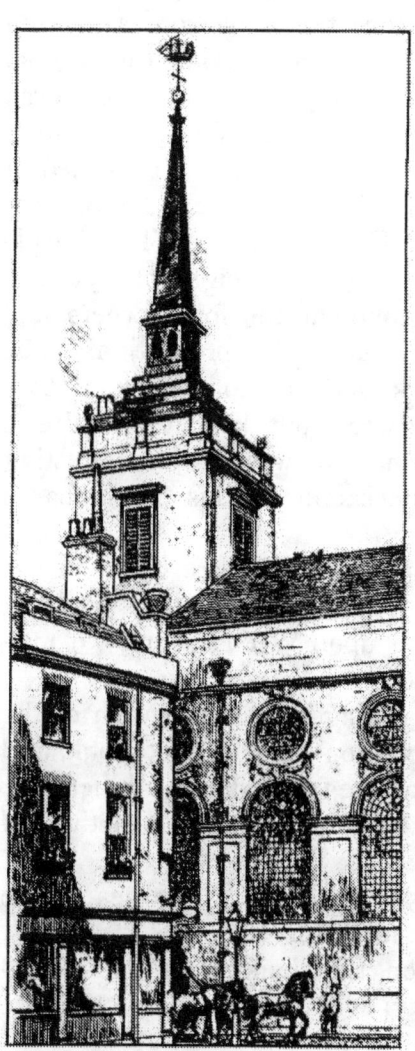

FIG. 219.—St Michael, Queenhithe (destroyed).

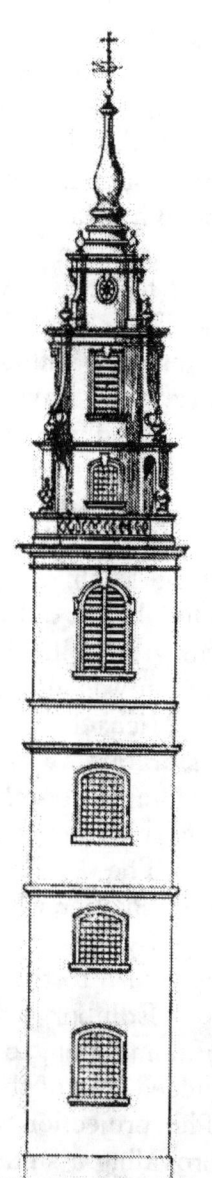

FIG. 218.—St Michael, Crooked Lane (destroyed).

The destroyed steeple of St Benet, Gracechurch (Fig. 217), rose to the height of 149 feet. Wren here, as in other churches, maintained the main feature of the pre-Fire church, which, as Visscher's view shows, had a lofty spire. Wren finished his building in 1685, and it fell to the destroyer in 1867, to the discredit of all concerned.   While no two spires of Wren's designing are alike, the general outline of St Benet, Gracechurch, and its composition of dome, lantern, and obelisk, furnishes the nearest approach to a favourite type.

Of Michael one may fairly complain that he is a saint of ill omen in the matter of lead spires.   The churches dedicated to him in Crooked Lane and Queenhithe have perished.   The former had a lead spire for its most notable feature.   The tower stood at the west end, and was united to the church by its eastern wall only.   Mr W. Niven, F.S.A., found a measured drawing, with plans, section, and elevation, in the British Museum, and the elevation is reproduced in Fig. 218.   As St Michael's was demolished as early as 1831 to form the approach to the present London Bridge, it is almost forgotten.   The pre-Fire church had a steeple, and, as Stow records that in 1621 the whole roof was "with strong and sufficient timber rebuilded, and with lead new cast covered again," the original spire may have been leaded.   The Fire made entire rebuilding necessary, and Wren completed the tower and spire in 1678.   The steeple was of unusual form.   It rose in three stages, circular on plan, and tricked out with buttresses and vases.   It finished at the apex in an extraordinary spike, suggestive of the product of a gigantic lathe, altogether a very roguish composition, and reminiscent of some of the Dutch steeples.   The steeple of St Michael, Queenhithe (Fig. 219), was very small, rising to a height of 135 feet. The obelisk did not rise squarely on its pedestal, but on globes at the four corners, and the great gilt ship in full sail which served as the vane was big in proportion.   The church was altogether an admirable example of Wren's work, and was done away in 1876. St Michael's, Wood Street, had a timber spire, but it was built later than Wren's restoration, was covered with copper, and of little charm.   It was an uninteresting building altogether, and as some city churches have to be sacrificed, this St Michael's was suitable for handing over to the destroyer.

The details of the actual leadwork of some of the foregoing steeples are given in the next chapter where also will be found descriptions of Class (c) of Wren's leaded steeples.

By way, however, of throwing the light of comparison upon Wren's work, we may here turn to the consideration of some Scotch leaded spires.

Edinburgh has one lead spire (Fig. 221) on St Mary Magdalen, the church of the Hammermen, to which guild the plumbers belong.   Its ogee top gives it a late look, and indeed it is of the seventeenth century, but there is no departure from traditional methods. The projection at the base like a sentry-box seems a somewhat cumbrous method of providing a suitable door to the roof of the tower.

The building of the spire occupied from 1620 to 1625, and in the latter year there appears in the accounts of the Edinburgh Hammermen the following item :—

"Thomas Weir his compt of the leid imployit upon the theiking of the steipill extending to ij͡c iij͡xx v stane viij lib. (265 stones 8 lbs.) at xxvj . viij the stane is iij͡c Liiij lib. (£354 Scots)."

Examination of the records of the Edinburgh building trades, and particularly of the Hammermen, fails to reveal either the word plumber or any reference to plumbing as a separate craft during the sixteenth and seventeenth centuries.   Leadwork seems to have been left to the wrights (carpenters) and masons.

At a brewery in Leith, which was St Ninian's Church, there remains a lead lantern with the edges decorated with a spotty cresting similar to those at Aberdeen. St Ninian's (Fig. 220) was built about 1670, and while Wren did nothing just of this shape, it is of the same family as the London lanterns.

The Bishop Elphinstone of Aberdeen, to whom reference has already been made, did not confine his architectural enthusiasms to church building. He was the founder of the University, built a great deal of it, and roofed his building with lead. The bishop was obviously bent on getting the best men he could for his work. In 1506 we find him employing no less a person than the plumber to the King of England, one John Buruel. Unhappily, we cannot judge of Buruel's work, for none remains. About a hundred and fifty years later the plumber was again abroad at King's College Chapel. Fig. 222 shows the very beautiful flèche, as to the date of which there is room for much doubt. Some facts can, however, be set down. In June 1638, a report was made by the Dean of Guild that it was "neidful that . . . the litle stipill be bothe theikit with leid and repairit in the timber wark."

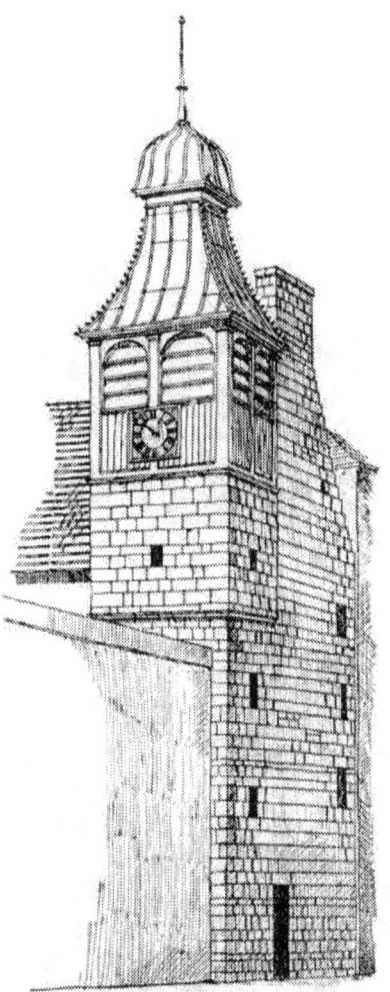

FIG. 220.—St Ninian's, Leith (now a brewery).

If the steeple was old enough in 1638 to need repairs, it was probably sixteenth-century work, maybe as early as 1506, when the chapel was roofed with lead. In Gordon's "View of Aberdeen," done in 1660, the flèche appears, as also in Slezer's view of 1693 (Fig. 181). The initials C. R. on the spire make difficulty by their "husky" character. They can hardly be so early as the repairs, which, presumably, were done after the report of 1638. We may perhaps conclude that the general form of the spire was the same all through the seventeenth century, and that whatever repairs were done in 1638, it was again thoroughly re-leaded about 1680, when the C. R. initials and other ornaments were added.

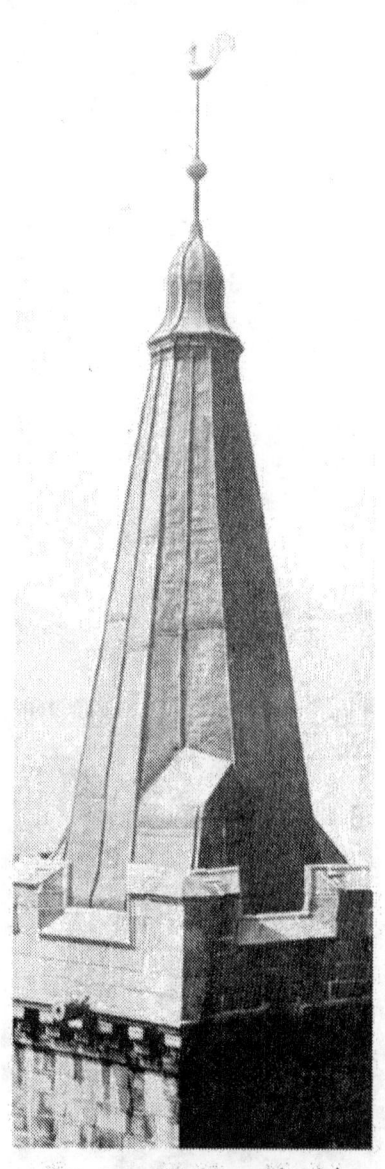

FIG. 221.—St Mary Magdalen, Edinburgh.

The notable features of the spire are in its hexagonal instead of, as usual, octagonal plan, and in the wealth of surface ornament. In the panels are crowns, thistles, fleurs-de-lys, and stars. In the most elaborate of the English leaded spires, East Harling, richness of effect is secured by the pinnacles and flying buttresses. The spire itself relies for interest on the reticulation of the lead rolls which pleasantly diaper the surface. The decoration of the King's College flèche was approached in a very different spirit. The surface was left plain and free from rolls, so that scope might be given for the invention of a formal design. It is altogether a work of scholarship rather than of fancy, an affair frankly of decoration rather than of construction, but very successful. In cleverness of invention it is comparable with Wren's London spires, but the small surface decoration is quite unlike Wren.

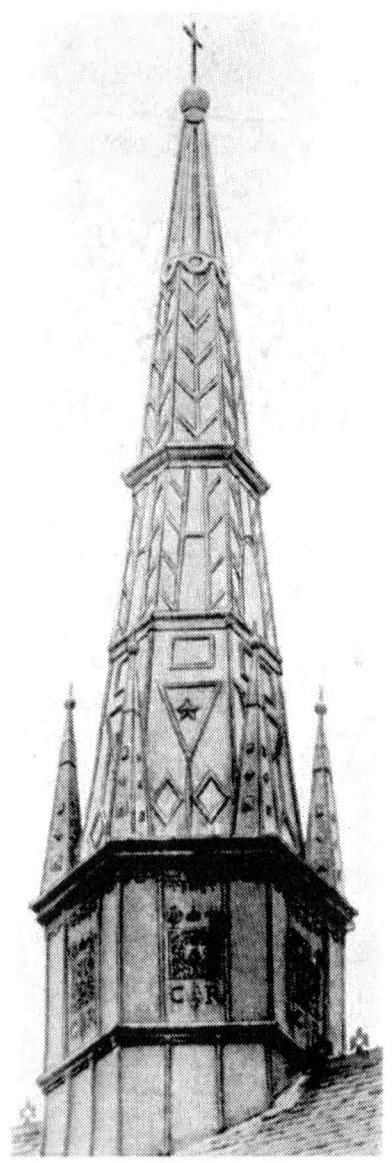

FIG. 222.—King's College Chapel, Aberdeen.

King's College, Aberdeen, had other lead spirelets. Reference to Slezer's view will show four besides the chapel flèche. Gordon says: "The southe syde hes upon everie corner two halff round towers with leaden spires." That on the right is curiously bulbous, if it is a fair representation of the original, which is doubtful.

The spire of Robert Gordon's College (Fig. 223) brings us into touch with a famous name. The architect of the building was the father of the brothers Adam, and practised in Edinburgh. His connection with the lead spire, and indeed with the whole building, is somewhat slender. The actual work is provincial in character, and represents, doubtless, the view of the Aberdeen mason and plumber as to what Adam ought to have designed.

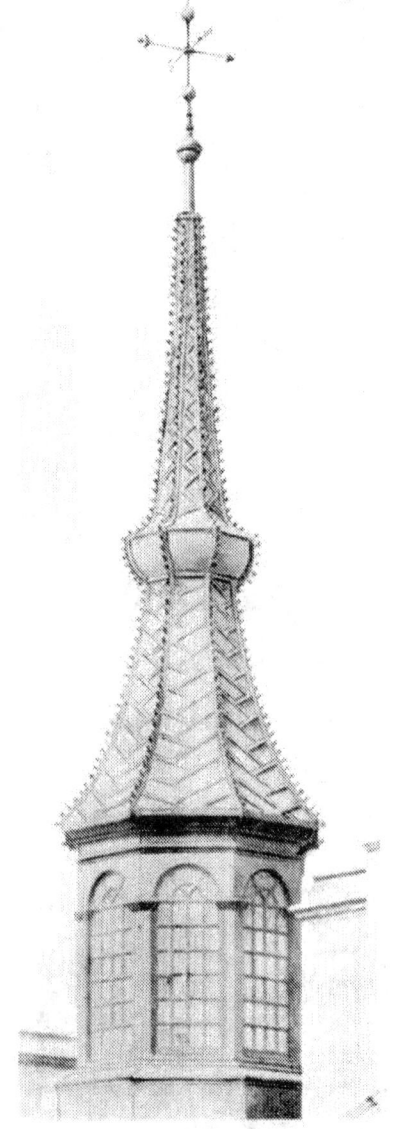

FIG. 223.—Robert Gordon's College, Aberdeen.

It lacks the refinement one would expect, and is probably a free translation of Adam's plans. The house was finished about 1744, but was not occupied at once by the boys of the foundation. It served, therefore, as a convenient barracks for Cumberland's men in the '45.

The rolls on the spire are merely decorative, bossed over wooden batons, and not honest seam rolls. They were a short cut to texture, and helped the belated Gothic feeling which the fleur-de-lys edging stimulated. The fat, moulded collar, half-way up, is a clever feature. We find this repeated on the Tolbooth spire in a modified form (Fig. 224).

Of the latter Gordon wrote in 1661, "builded it wes anno 1191, and not long since enlarged and adorned with a towre and high spire covered with lead, wher they have ther commone bell and prissone." It was rebuilt by John Smith, architect, about seventy years ago. He made extremely careful sketches and measurements of the original work, a piety for which we may be grateful. The steeple as it stands represents the original work very well. The point, of some value to establish, however, is comparative rather than historical. If the Gothic trimmings of these Aberdeen steeples be for a moment disregarded, they might be, both in their elements (of ogee roof, lantern, and concave spire) and in their outline, Wren steeples. Wren cannot, therefore, be regarded as the inventor of the type of Renaissance steeple which in varied forms is seen in so many City churches. He was probably influenced by the steeples of the Netherlands and Spain. He could hardly have seen many during his French tour. Even if he did, he was then more occupied with the works proceeding at the Louvre and other examples of the grand manner.

In his treatment of the lead itself Wren,

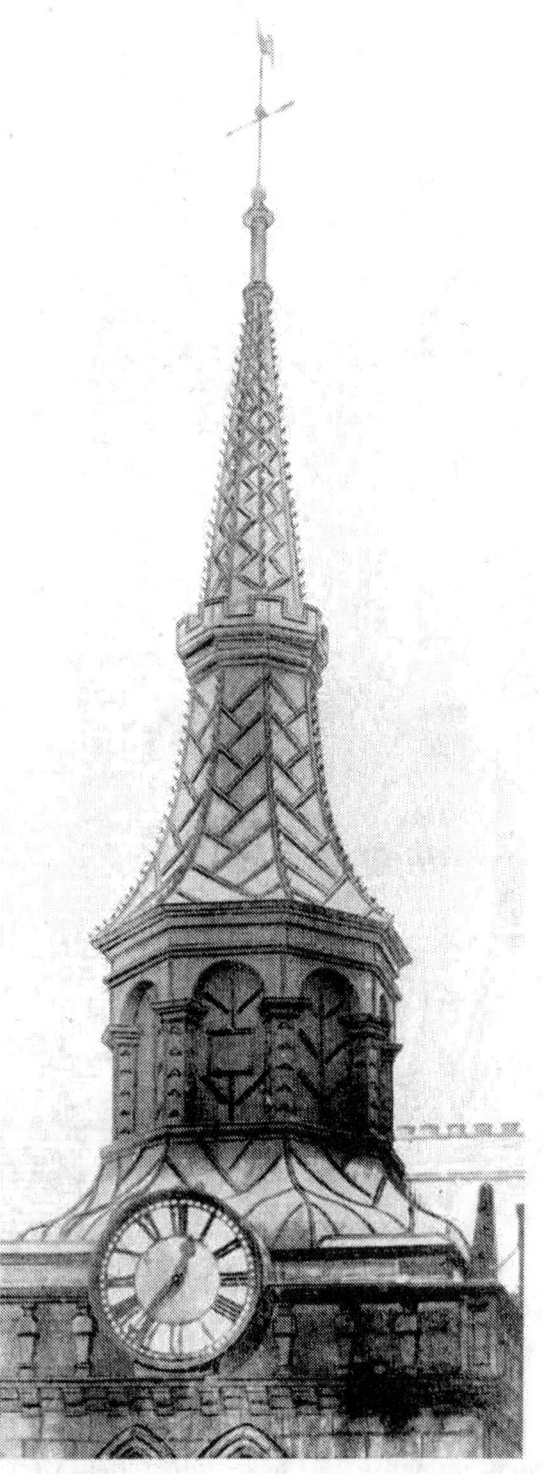

FIG. 224.—The Tolbooth, Aberdeen.

in practically every case, discarded the mediæval character which is so insistent at Aberdeen. In no case does he make a pattern on a steeple with the rolls, still less does he employ such rollicking ornament as a fleur-de-lys edging to the ribs of a spire and a battlemented collar. It is amusing, if not very profitable, to speculate as to what Wren would have done by way of an academical exercise in Gothic leadwork if he had attempted something on the same lines as his other Gothic details. One may, perhaps, be permitted to regret that he rejected any such temptation if it came to him. That he liked lead as a material is abundantly clear from the great extent to which he used it. It is equally obvious that he neither realised its decorative possibilities nor thought of it otherwise than as the most efficient roofing material, and as giving a broad colour contrast when used to crown a white tower. Wren thought and designed on broad lines. The quality of mystery in architecture and the sense of craftsmanship, which developed in the Middle Ages on parallel lines, were no stumbling-blocks to him. Had he conceived of the former as a necessary equipment for the architect he would certainly have dismissed it as foolishness. It is obvious from the details of St Paul's Cathedral that he took a keen delight in good craftsmanship, and the bad detail in many of his parish churches, *e.g.*, the plasterwork of the dome of St Stephen's, Walbrook, was doubtless a source of irritation. He was, however, a victim of the times he lived in. The Civil War had shattered the trades, and the difficulties in obtaining an adequate number of skilled workmen must have

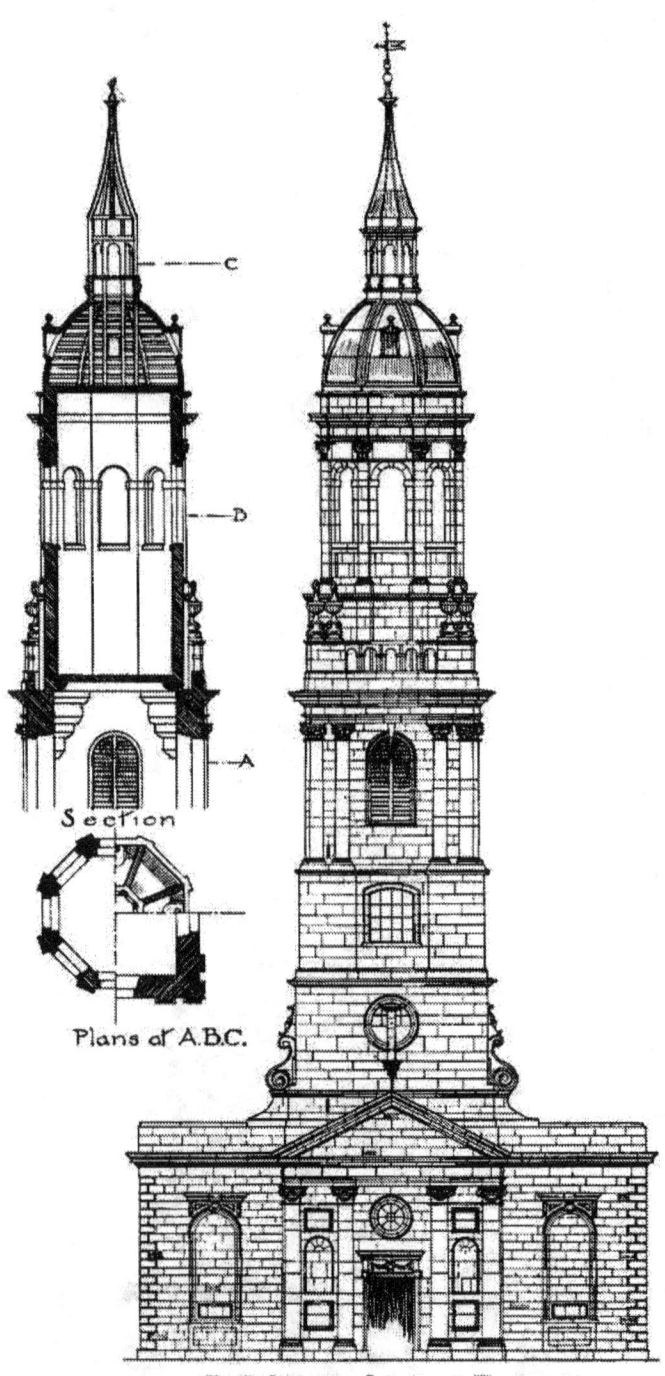

Section

Plans of A.B.C.

ST MAGNUS. LONDON BRIDGE.
FIG. 225.

been immense. These difficulties were accentuated by the Great Fire, which threw the building trades into the confusion that follows infinite overwork.

For every reason, therefore, it is idle to look in the mass of Wren's buildings for the tenderness and fancy in detail and for the beauty of execution which marked the leadwork of Gothic times and of the early Renaissance. Their place is taken, however, by a vigour of invention and a sanity of treatment which are characteristic of the man and of the idea behind his work.

CHAPTER VII.

LEADED DOMES, LANTERNS, AND WALLS— A LOST FOUNTAIN.

Curves in Roof-lines, a Slow Development—The Use of Lanterns—Wren's Treatment of Domes and Lanterns— Class (*c*) Constructive Details of their Leadwork—Archer's Work—The National Gallery—Nonsuch and Cheapside—The Great Fountain of Windsor Castle.

THE leaded domes and lanterns of Wren's London churches are not only of great intrinsic interest, but have an important place in the development of the roof idea as applied to towers. The dome of simple curve is a frankly foreign element in English architecture, and became acclimatised only by slow stages. With the cupola of ogee curve it was different. The genius of native building accepted with enthusiasm the ungeometrical and flowing line when it arrived by way of the ogee in the first half of the fourteenth century. For a time it was supreme and rioted freely, and sometimes absurdly, but still mostly in such decorative positions as were afforded by niches and tombs. Hopelessly bad structurally, the ogee arch was rarely powerful enough in its attractiveness to take other than a decorative place. In English mediæval architecture, at least, it never affected external roof-lines until Perpendicular times, and then only in rather trivial ways. At King's College Chapel, Cambridge, which was building from 1446 to 1540, the corner turrets finish with ogee finials, and these, and others like them, were the forerunners of the numerous ogee-roofed turrets of the early Renaissance, such as those at Hampton Court and at Abbot's Hospital, Guildford. Even in the case of the example at King's College, however, there is obviously no intention seriously to employ curves in roof work. Such finials are decorative trivialities employed to finish rather unimportant elements such as corner turrets. We have still no evidence of a desire to introduce curves into the crown of a tower. Where a tower was to be topped with a notable feature, a spire composed of straight lines in one combination or another was still the only treatment. (Such towers as St Giles's, Edinburgh, and the Cathedral, Newcastle, are excepted, where curved flying buttresses uphold a spirelet, but these from their rarity can scarcely be regarded as traditional.)

The development of Perpendicular tower building tended greatly to the elimination of the spire, as in the Somersetshire churches, where the wealth of pierced parapet and pinnacle took the spire's place.

Had the provision of a stage above the tower proper remained an organic essential of the treatment of church towers, perhaps something in the nature of a great domed lantern would have been evolved in late Perpendicular times on the lines of the lead cupolas on the turrets of Hampton Court.

As it is, we have to wait for the full tide of the Renaissance before the dome comes into its own, and to look to Sir Christopher Wren in particular for its noblest expression.

The description "lantern," applied to such steeples as St Benet, Paul's Wharf, deserves attention. The original purpose of a lantern is obviously to give light, and the notable lead lantern of Horham Hall, near Thaxted, Essex (Fig. 226), is the best possible example of this use. It is, in fact, a beautiful architectural expression of the same need

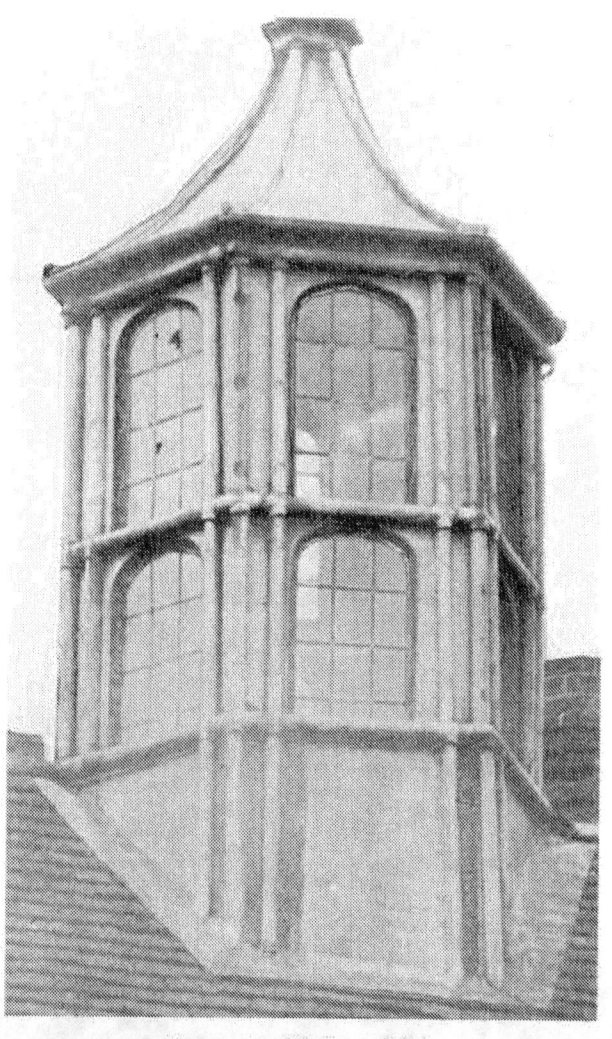

FIG. 226.—Horham Hall.

as is served by the range of vertical roof lights in a modern billiard room. At Horham Hall the provision of light is the first consideration, and the craft of the plumber is spent on emphasising the window openings by vigorous vertical and cross lines rather than on beautifying the roof. Horham Hall was built at the beginning of the sixteenth century, and there is nothing in the design of the lantern to contradict so early a date.

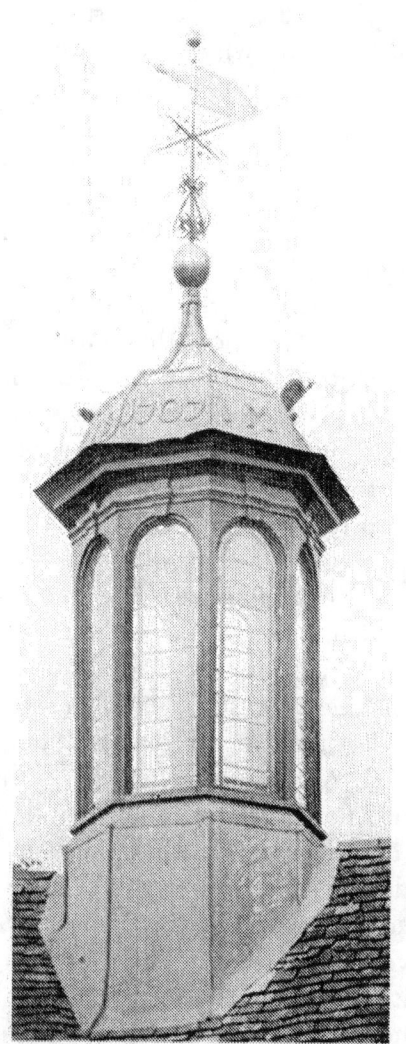

FIG. 227.—Christ's Hospital, Abingdon.

At Christ's Hospital, Abingdon, Berks (Fig. 227), the lights of the lantern were untouched by the plumber, who spent his energies on the ogee roof, with no little help from the smith on the vane. The hospital was founded in 1553, so the lantern dated 1707 marks a period of renewed activity. A pleasant feature of this Abingdon lantern is the placing of lead ornaments on the roof itself. About half-way up, gilded crowns stand out and break the ogee outline, and are doubtless examples of many like decorative gaieties which have gone from other roofs with the passage of time and thoughtless repair. Abingdon is rich in lanterns, for the exquisite market-house (attributed to Christopher Kempster, who worked under Wren at St Paul's) has a lantern of great delicacy of detail.

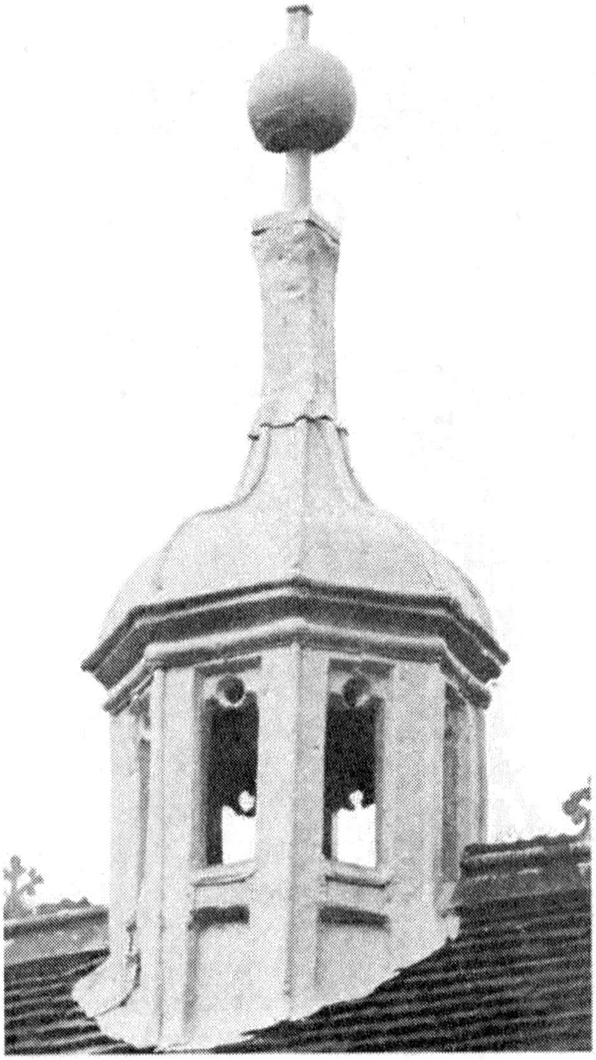

The leaded lantern of Barnard's Inn Hall, now the Mercers' School (Fig. 228), is probably as perfect an example as can anywhere be found of the right adjustment of the elements of light opening and roof. The point where the tip of the ogee joins the finial has been very clumsily repaired, but even with this blemish the composition is altogether delightful. It is complete plumber's work. There is no shirking of the technical difficulties involved in sheeting with lead the mullions of the lights (as at Abingdon where the wood is left unprotected), and the proportion between the cusped openings and the sturdy mullions could not be bettered.

This lantern, however, is purely an architectural feature. It does not light the hall, and may be regarded, therefore, as of the type of roof flèche (as, for example, that of King's College Chapel, Aberdeen, Fig. 222). The ceiling of the hall is comparatively modern, and it may be that there was in the original ceiling an opening below the lantern, which would in that case have served to ventilate. The "lantern" idea is altogether absent from the handsome lead turret roofs of Hampton Court (Fig. 229). The richness of treatment there, the wealth of crocket and pinnacle and the great applied roses, make the roofs worthy successors of the most decorative of English lead spires, that of East Harling, Norfolk.

The composition is simple and natural. The lower octagonal stage takes up the lines of the brick turret, and is surmounted by an ogee cupola.

Like the Barnard's Inn lantern, the feeling is wholly Gothic, though the rather non-

FIG. 228.—Barnard's Inn Hall, London (now the Mercers' School).

descript shape of the eight little finials gives an uncertain touch and indicates the arrival of new motives. The neglect by Wren of the decorative possibilities of frankly ornamental leadwork cannot be more acutely recognised than by comparing the wealth of detail in the Hampton Court turrets with the sobriety of, say, the lantern of St Benet, Paul's Wharf.

Fine detail there is at St Benet's, but it is in the wooden cornice mouldings. The leadwork is subsidiary and protective. In Wren's most ornamented steeple, St Edmund's, Lombard Street, the decorative urns were apart from the structure. At Hampton Court the ornament is organic, and has relation to the lines of the roof.

With Wren the ogee form developed into the bell-like outline of the lead roofs of the western towers of St Paul's. The form is more severe but still picturesque.

Turning now to Wren's use of the dome in connection with the lanterns surmounting church towers, we take up again the classification begun in the last chapter and deal with Class (c). Possibly Wren's finest lantern is at St Benet, Paul's Wharf (Fig. 230).

There is a peculiar interest attaching to this church, as Wren's great predecessor, Inigo Jones, was buried in the pre-Fire church in 1651. Unhappily his monument was destroyed when the church fell to the flames. The church was rebuilt by Wren in 1685.

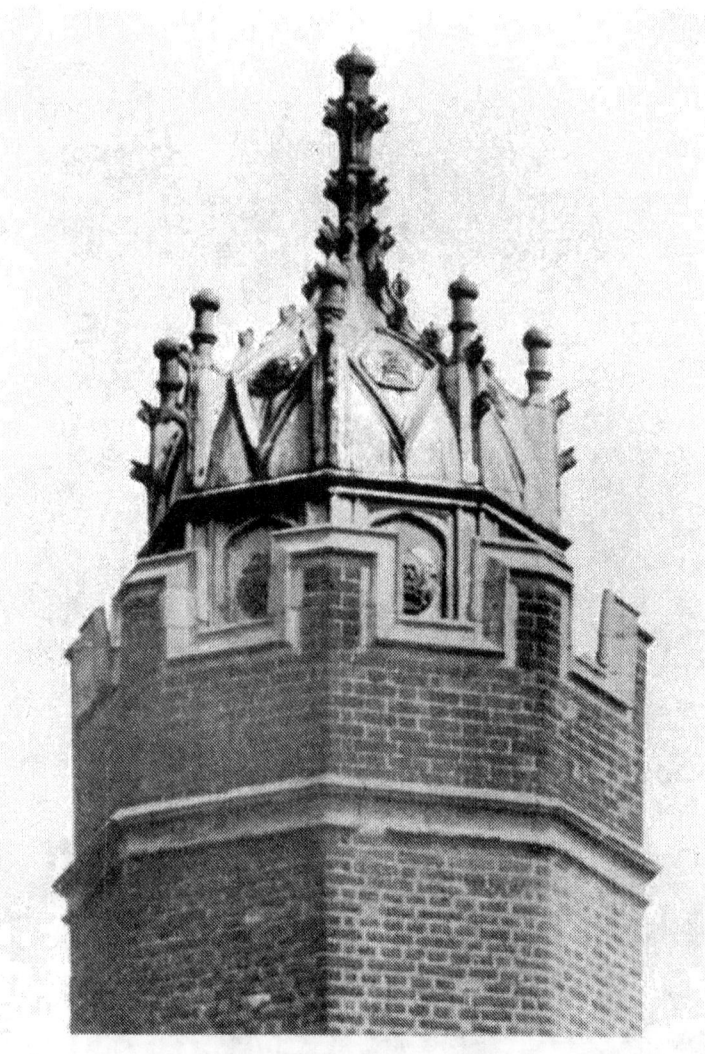

FIG. 229.—Hampton Court.

and not only the exquisite lead lantern but the whole building is a miracle of sane and simple art. The photograph of Fig. 230 is of happy effect in showing the little lantern of St Benet against the bulk of St Paul's.

It is impossible, within the compass of this book, to do more than touch on St Paul's, the greatest of all English leaded domes. It is not, moreover, in the same category as the lanterns of the City churches, which all meet the same architectural need, viz., that of furnishing a suitable crown to a square tower. At St Paul's the plan below the dome is circular, and the treatment is altogether *sui generis*.

In earlier chapters stress has been laid on the texture value in lead roofing of the rolls, which make the junction between adjoining sheets of lead.

At St Paul's, Wren has emphasised this surface treatment by having the lead dressed over great moulded ribs, a feature which has been carried much further in Italy. In San Michele's great dome at Montefiascone the dome surface is constructed with reversed arches giving a moulded contour of ribs and hollows all covered with lead.

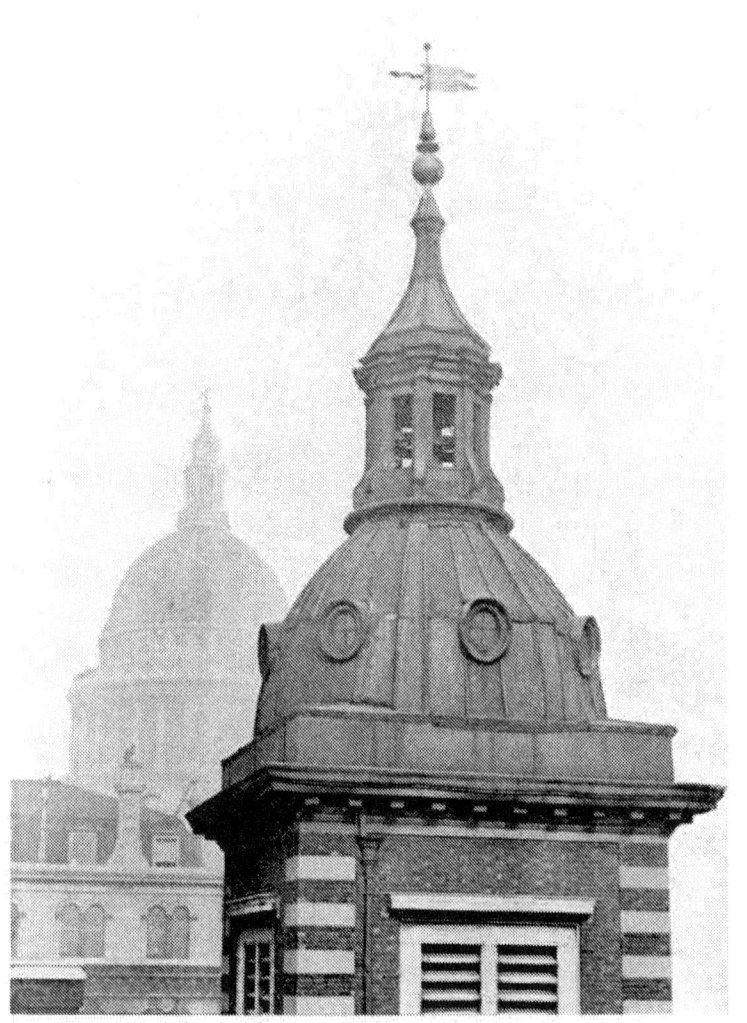

FIG. 230.—St Benet, Paul's Wharf.

In Rome are several domes with highly developed ribbing. In general effect of outline the leaded dome of the Brompton Oratory follows this later type, and gives an idea of their character. In the dome of the Salute Church at Venice we have the supreme example of a plain ribbing which hardly interferes more with its surface than the simplest of welts could do, so that if lead sheets be used at all its characteristic joint lines could scarcely be less emphasised.

A passing reference must also be made to the great domes of Santa Sophia at Constantinople by way of comparing the characters of Byzantine and Renaissance domes. Perhaps the out-standing features of Wren's more conscious art are the elaborate lanterns surmounting the domes proper, and the fact that where the dome is seen also from the inside, as at St Paul's, the inner and outer lines do not agree. In the case of lanterned domes surmounting towers, as at St Benet's, this discrepancy does not arise, as the inside of the dome is not visible. It goes, however, to show that Wren's chief idea in St Paul's dome was to create an architectural feature dominating London, and to establish a relationship between the cathedral and the steeples of the parish churches.

Returning to the smaller domes and lanterns covering towers, that of the destroyed church of St Benet Fink bore a marked general likeness to those of St Benet, Paul's Wharf, but with one notable difference.

At St Benet Fink (Fig. 231) the cupola was square on plan, at Paul's Wharf we have a true dome, circular on plan. Wren here goes about his work in a straightforward way. There is no attempt to mask the change from square to round by corner vases or any like device which might have tempted a lesser man, and the steeple is by so much the gainer in breadth and simplicity. We may note a similar directness in the domes flanking the tower of St Clement Danes.

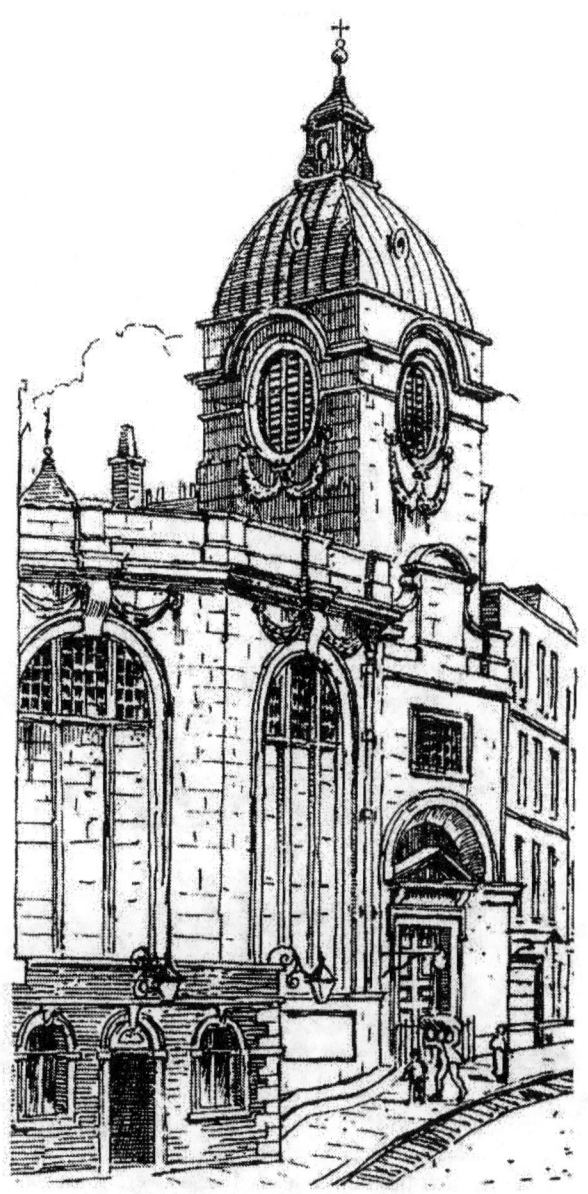

St Benet Fink was rebuilt by Wren in 1673 and demolished in 1844. It stood on the south side of Threadneedle Street, where the late Mr Peabody now sits in bronze. The cupola with lantern was a fine feature of one of Wren's most ingeniously planned churches. The site forbade a rectangular plan, so Wren turned it into a decagon and attached the tower to its western face. It will be noted that this lantern, though similar in design to that of St Benet, Paul's Wharf, is smaller in proportion to the cupola, and the cupola lights are less important. The illustration of Fig. 231 shows what London has lost in losing St Benet Fink.

The two Wren lanterns, which defy classification perhaps more vigorously than any other of his church steeples, those of St Nicholas, Cole Abbey, and St Edmund, Lombard Street, may perhaps be grouped together on the ground of a likeness in curious outline. The former was rebuilt in 1677, and the latter in 1690. Both are characteristic work, examples of Wren's wealth of invention. The lantern of St Nicholas (Fig. 233) has been a good deal abused, and not altogether without reason. Wren's use of a railed balcony at St Martin, Ludgate, was a bold stroke, which is justified in the result. Hardly so much can be said for the like feature at St Nicholas, Cole Abbey, and above it Wren seems to have lost himself in a kind of architectural marine store.

FIG. 231.—St Benet Fink.

At St Edmund's, Lombard Street (Fig. 232), the lantern is coherent and of admirable proportion. The lantern with its louvred lights forms a satisfactory stage between the tower and the little concave spire surmounting it, but perhaps in none of his steeples did Wren break away more violently from traditional treatment. It is unfortunate that St Edmund is so little visible. It is only from St Clement's Lane that it can be seen at all satisfactorily. From Lombard Street the steeple is hardly within sight, so narrow is the street and so lofty the tower. During the latter part of 1907 the lantern needed re-leading, and the opportunity was taken to remove the twelve flaming vases which, as the illustration shows, formed so notable a feature. They were of wood covered with lead; the wood had rotted; restoration was certainly needful. The failure to replace them is, however, serious.

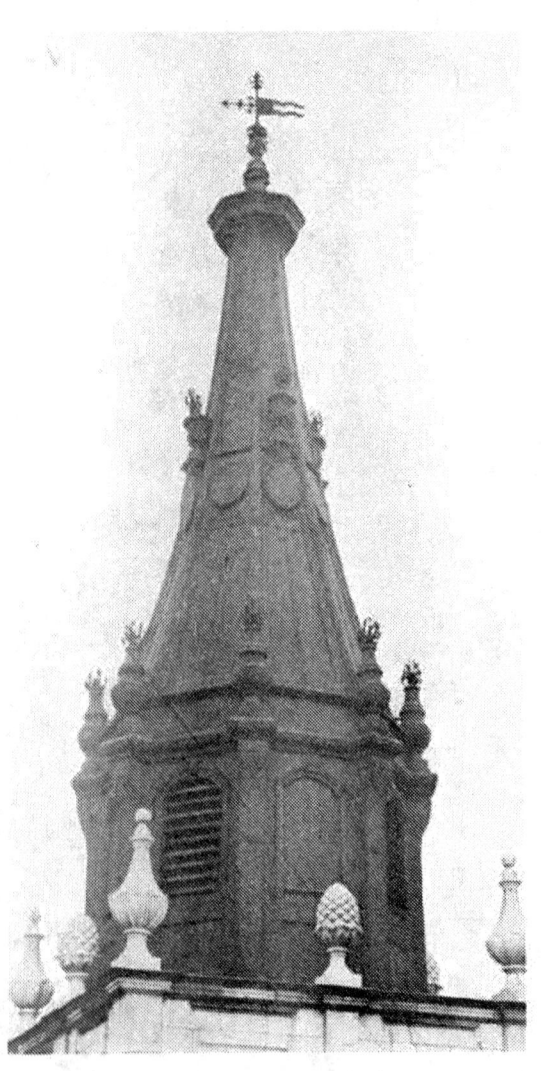

They were a characteristic feature of Wren's design, and the plea of lack of money for the work sounds absurd in Lombard Street.

A few notes may be added here as to the workmanship of the leadwork on some of Wren's steeples, described in this and the last chapters.

In the case of St Swithin's (Fig. 208), the top of the spire is a rough tree post sitting on a stiffening floor. The spire is boarded with 6-inch battens 2 inches apart on a framing like a stud partition, braced by 8 inches by 5 inches angle rafters, and has uprights 5 inches by $2\frac{1}{2}$ inches. The main ribs at the angles of the octagon, at the base of the spire, are 12 inches by 8 inches and have a bracing 7 inches by 5 inches in shape of St Andrew's Cross, halved together and held by axle pins, with wedges. There are many rough iron straps.

The lead sheeting has vertical welts which are $1\frac{1}{2}$ inches wide and project $1\frac{3}{4}$ inches. In the top sheet of each face of the spire there is no welt, in the next two lower sheets there is one middle welt. The next sheet has a spire-light. The six next sheets are in three widths, divided by welts. Each sheet is 5 feet $4\frac{1}{2}$ inches deep, and there are ten in all. Each sheet has two clips. The welts at the angles do not differ from those on the faces. The oval lights touch the spire faces at the bottom and stand out perpendicular. Their lead

Fig. 232.—St Edmund's, Lombard Street.
(Photographed before the leaded vases were removed.)

covering is in two sheets; the division comes at the middle horizontally. The lights have at the back an oval cup for weathering purposes, which reaches to half their height. They have been made in ship's carpenter fashion with curved ribs and open battening arranged like the boarding of a boat.

At **St Augustine's**, Watling Street (Fig. 215), the plumbing is of a much more elaborate kind. At the base of the big consoles the face sheet on each side is turned over to form a welt on the back of the console on both edges so that the effect of fluting is given.

The base of the spire has angle pilasters, the edges of which are formed with welts in the same way, and the lead-covered cornice is returned to form caps for these pilasters. The welt is 1⅛ inch. No clips are used for the sheets, but they are fastened with lead-headed nails. There are no soldered dots.

The louvres are not covered with lead. The impost of the arch is a solid block of wood covered with lead, and the shield at the top of the arch is a casting. Without ladders it is impossible to reach the vases, but they are almost certainly castings.

The mouldings generally are of some complexity, and the lead has been well dressed over them and nailed freely.

At both St Swithin's and St Augustine's the leadwork seems to be that originally fixed.

At St Nicholas, Cole Abbey (Fig. 233), it has been renewed altogether, as has also the iron railing. The panelling on each face is 12 inches by 4 feet 6½ inches with 5-inch by 2½-inch mouldings, and the cornice is 12 inches.

The loss of interest caused by the re-leading of the steeple is very marked. It is certainly a point to be insisted upon, that in any restoration repairs only should be permitted so that the original plumbing method is scrupulously followed. The lead should always be recast in the sand, as is the practice at Westminster Abbey, and no modern milled lead and wooden rolls, &c., should be used.

St Margaret Pattens (Fig. 209), is notable for the great size of the lead sheets, which are cast, and a full eighth of an inch thick. At the base of the spire they are nearly 8 feet wide and about 6 feet deep.

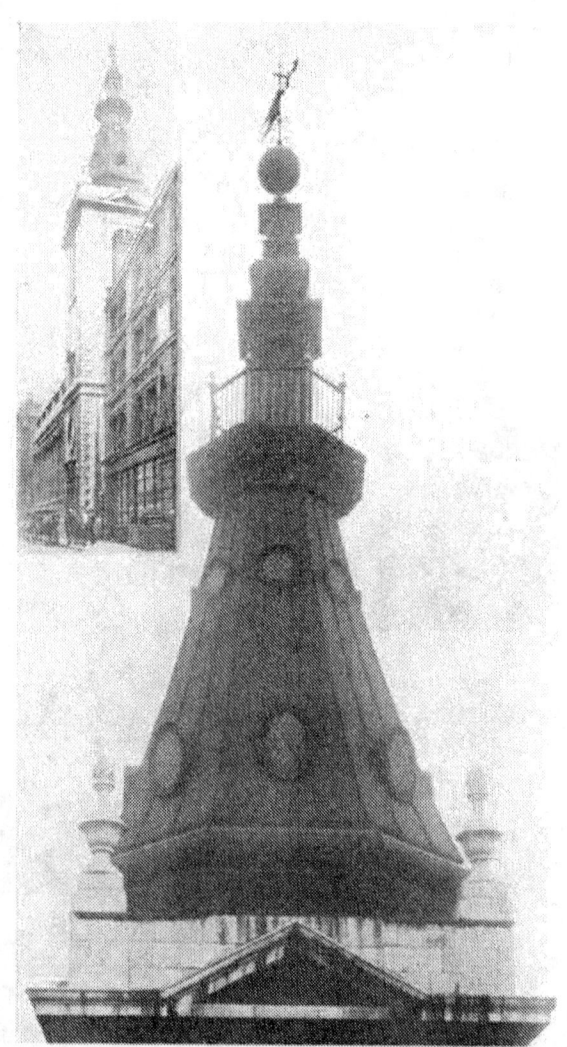

FIG. 233.—St Nicholas, Cole Abbey.

Externally there are five soldered dots to each sheet, but inside there are in addition a large number of secret tacks, two to each face of the octagon, spaced 2 feet apart vertically. The welts at the angles are 1¾ inches projecting 1½ inches. The moulded stiles of the panels are 10 inches wide inclusive of 2¾ inches moulding, while the depth of the panel on the face is 1¾ inches, and there are three clips to each panel. The lead is dressed over the pediments of the spire-lights, but there are no lead coverings to the louvres. About two years ago two new sheets were put up, and an inscription says that the spire was re-leaded in 1834, but this can hardly apply to the whole work, for some of it seems contemporary with the spire. The timbering is on the same general lines as at St Swithin's, but the central post only comes down from the apex as far as the level of the top tier of spire-lights. The angle posts are 9 inches by $5\frac{1}{2}$ inches put flat-wise with bevelled faces, and the sides are framed and cross braced, the latter being 7 inches by 5 inches, and of St Andrew's Cross form. Many of the old iron straps remain, but some further cross ties and braces have been added in modern times. The boarding is 9 inches by $\frac{3}{4}$ inch, spaced 3 inches apart.

The obelisk of St Margaret Lothbury (Fig. 213) is framed on four 9-inch by 9-inch posts, 3 feet 6 inches apart, which come down on to two 12-inch by 12-inch beams which

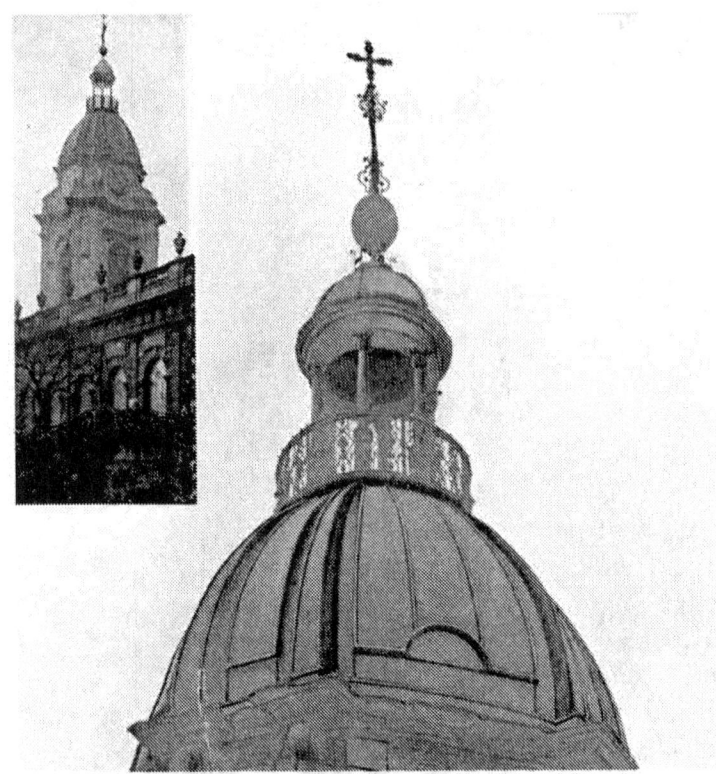

cross the top of the tower and rest on wall plates. Diagonal beams and braces run from the junctions of posts and main beams to the corners. The round and hollow curves of the spire outline are formed by cradling from this central core. In this respect the construction is analogous to St Paul's because the obelisk really runs on through the apparent ogee outline which supports it. The curved ribs are 5 inches by 3 inches, and 14 inches apart, with close boarding instead of open as at St Swithin's. The oval spire-lights have 3 feet by 2 feet openings. The details of the leading cannot be seen, as there is no door to the outside, and could be inspected only from cradling or scaffolding.

FIG. 234.—St Philip's, Birmingham.

The lantern of St Benet, Paul's Wharf (Fig. 230), is peculiarly interesting. There are eight posts to the lantern, 9 inches by 4 inches, spaced to give openings 14 inches wide, and the attached consoles between project 12 inches at the bottom and 4 inches at the top. They are sheathed with lead all round, with welts at the edges of the console and one at the back, inside the lantern. The work has been freely nailed with lead-headed nails, but many of these have gone. The inside diameter of the octagon is 5 feet, the height of the console to the entablature 5 feet 9 inches, and the entablature about 1 foot 9 inches, with a projection of 10 inches. The wood mouldings are covered with lead throughout. The horizontal sheet joints are arranged so as to give a drip at the bottom edge of the top fillet of the moulding. The dome below this lantern has tapering ribs projecting about 2 inches, with two angle welts giving a fluted face. Between the ribs are three sheets showing two welts. The welts above the lantern are worked in the same way, indeed fat welts are the great feature of this steeple and give its rich appearance. The lead sheeting of the dome is carried on battening 3 inches by 1 inch, 2 inches apart, the chief interest of which is that it is placed diagonally. The eight posts of the lantern rest on as many inclined 8-inch by 6-inch braces secured at the feet by a framed floor over the top of the tower. The curvature of the dome is formed by $2\frac{1}{2}$-inch segmental cradle pieces on the back of the braces with a greatest projection of about 18 inches.

The strength and simplicity of these Wren spires is no less admirable than the design. It is hoped that these details of their construction and lead covering may be found instructive, and may lead to more attention being given to the subject, particularly when repairs are undertaken.

Before leaving London's leaded steeples a point of colour is worthy of note. In the country the tendency of lead is to weather to a silvery grey, and sometimes so brightly that spires look as though they have been whitewashed, whereas in many cases the stone tower has weathered to a dark hue. In London the precise opposite is the case. The Portland stone has remained white, while the lead of the spires has been blackened by smoke and impurities. How white the church towers of London can look may most sensitively be realised in Westminster on a November day. The black fog will sometimes hang over the Thames long after the sun has driven it from the north and west, and against this heavy background the sun-lit western towers of the Abbey take on a snowy whiteness. On one observer, at least, the effect has been so to magnify and ennoble these not too beautiful towers, as to convey somewhat the impression that Coleridge took from the architectural dreams of Piranesi.

By way of comparison with Wren's treatment of leaded domes and lanterns, Archer's tower of St Philip, Birmingham (Fig. 234), is illustrated. The tower proper

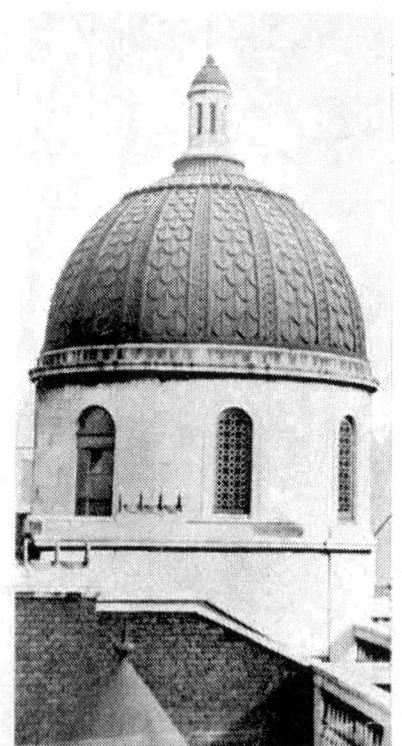

FIG. 235.—National Gallery.

is certainly the finest part of this fine composition, but the dome is a very notable achievement. It may be felt that the columns supporting the small cupola are a little attenuated, and the balcony railing rather trivial in detail, but, taken altogether, the work bears comparison with all but Wren's best work. The detail of Archer's leadwork is careful, but a little undersized for the bold rococo character of the tower. The columns supporting the cupola are cased in lead, which is heavily seamed at the joints. The capitals have elaborate acanthus leaves in gilt cast lead, and the bases are cast in rings and fitted round the columns. St Philip's is altogether a notable church in a city not too notable for architectural beauty. Archer's Garden Pavilion at Cliefden has a leaded cupola that will also repay study.

The leaded dome of the National Gallery (Fig. 235) is very different but distinctly

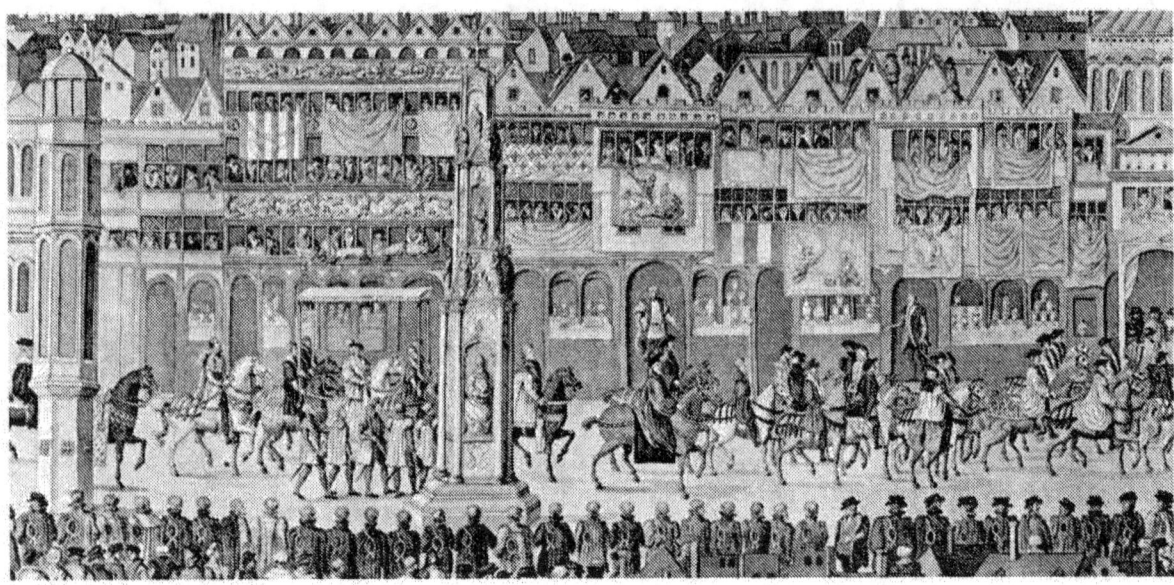

FIG. 236.—The Progress of Edward VI. (part of engraving) showing Goldsmiths' Row and Cheapside Cross.
(Reproduced by permission of the Society of Antiquaries.)

interesting. Built as late as 1839 by Wilkins, the dry classic detail of the leadwork is almost as far removed from Wren's straightforward rather thoughtless manner as from the luxuriant crocketting of the best mediæval work. It shows an appreciation of the value of pattern on bold curved surfaces, even if it fails altogether of an understanding of the right treatment of lead roofs. It is doubtless inspired by the classic idea of a bronze scale roof. It is hardly necessary to do more than mention the steeples of St James, Piccadilly, and of St Ann's, Soho. Both are disfigured by clocks. Wren was not responsible for the first; S. P. Cockerell was for the second, of which we may say, with John Timbs, that it is a "whimsical and ugly excrescence."

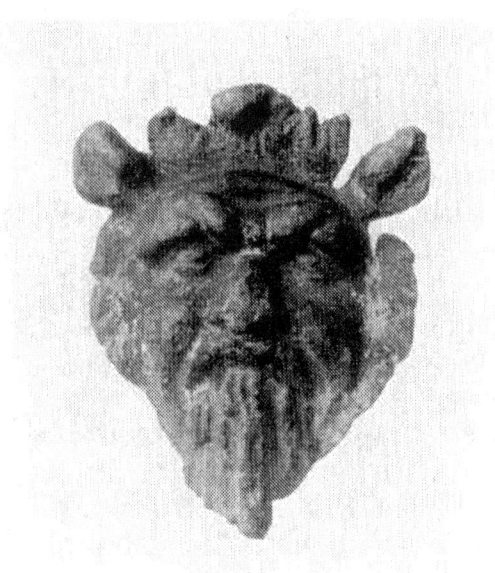

FIG. 237.—Aberdeen Mask.

We have so far dealt with lead coverings for spires, domes, and lanterns. There remain roofs and walls. With simple roofing it is not proposed to deal, as the many interesting points raised are mainly questions of technical detail and not of ornamental treatment. One delightful little decorative detail, however, may here be noted. The little mask (Fig. 237), about 3 inches long, is one of eight fixed at the ends of piend rolls (of lead) of a small octagonal larder at Scotston House, Aberdeen. It is probably of about 1800.

Of lead coverings for walls in Britain there is little history. Mr Lethaby has quoted the case of the Saxon church at Lindisfarne. Eadberht, bishop of that place in A.D. 638, took off the thatch and covered it, both roof and walls, with lead. Mr J. Park Harrison claims that this church is to be identified with a building which is shown on an illuminated MS. in the library of Corpus Christi College, Cambridge. If this be true the lead was clearly in the form of tiles or shingles and not in the form of sheeting as in the case of a mediæval spire. Unhappily the great leaded timber buildings are in the limbo of history, and there are gaps and uncertainties in building records which make it difficult accurately to establish uses. Mr Starkie Gardner, in his admirable paper on " Lead Architecture," sought to prove that the chief glory of Nonsuch Palace was in the decorative leadwork, and rather scoffed at the idea that the modelled panels which appear in Hoefnagel's view were of any sort of plaster. Mr Maurice B. Adams, in a note in the *R.I.B.A. Journal*, says that " Pepys describes the building as *sheeted* with lead." That is hardly the case. Pepys' own words are now set down in parallel column, with the description of Nonsuch by a much more competent observer, John Evelyn.

PEPYS' DIARY.	EVELYN'S DIARY.
1665. *Sept. 21.*	1666. *Jan. 3.*
" . . . Walked up and down the house and park ; and a fine place it hath heretofore been, and a fine prospect about the house. . . . And all the house on the outside filled with figures of stories, and good painting of Rubens' or Holben's doing. And one great thing is, that most of the house is covered, I mean the posts, and quarters in the walls, covered with lead, and gilded.	" I supp'd in None-such House . . . and tooke an exact view of the plaster statues and bass relievos inserted 'twixt the timbers and punchions of the outside walles of the Court ; which must needs have been the work of some celebrated Italian. I much admired how it had lasted so well and intire since the time of Henry VIII., expos'd as they are to the aire : and pitty it is they are not taken out and preserv'd in some drie place ; a gallerie would become them. There are some mezzo-relievos as
" I walked into the ruined garden . . ."	big as the life, the storie is of the Heathen gods, emblems, compartments, etc. The Palace consists of two courts, of which the first is of stone, castle-like, by the Lo. Lumlies, the other of timber, a Gothic fabric, but these walls incomparably beautified. I observ'd that the appearing timber punchions, entrelices, etc., were all so cover'd with scales of slate, that it seem'd carved in the wood and painted, the slate fastened on the timber in pretty figures, that has, like a coate of armour, preserv'd it from rotting."
(NOTE.—Nonsuch Palace, near Epsom, was in sufficiently good repair at this time for the Exchequer to be moved there during the Great Plague. It was Exchequer business which took Pepys to the Palace.— L. W.)	

These two extracts should be read together. Pepys only claims lead-covered posts, and is quite silent about lead panels. There is no evidence that his story of Rubens and Holbein providing the exterior paintings contains a word of truth ; but, in any case, it is evidence for something very different from cast-lead panels. Evelyn is definite about the plaster statues and reliefs, and his " scales of slate " abolish lead covering even for the main timbers.

Where there is a conflict of testimony, we must consider credibility of witnesses. Pepys was an acute observer, but of men and manners rather than of buildings. Evelyn's architectural taste was highly trained by long residence in Italy, and his general accuracy of observation and his detailed description of Nonsuch may make us hesitate to reject his evidence.

It would be pleasant to give leadwork the benefit of any doubt, but even if we accept the leaded posts and quarters of Pepys, and assume a slate-like, scale-like treatment for their leading, we must reject any idea of lead statues and reliefs.

The evidence from Stow as to the lead panels on Goldsmith's Row, Cheapside, is explicit. In the view reproduced in Fig. 236, the "Goldsmithes armes and the likenes of woodmen in memory of his name (Thomas Wood's) riding on monstrous beasts, all which is cast in lead, richly painted over and gilt," are unhappily covered by the draperies hung out for the royal festivities, but the two long panels with scroll ornament (to the left of Cheapside Cross) may be taken to have been of modelled cast lead. Thomas Wood was sheriff in 1491.

English Lead Fountains—The Great Example at Windsor.

Professor Lethaby in " Leadwork" devoted a chapter of one and a half pages to fountains, a measure of the poverty of English leadwork in this direction. In the chapter on lead statues generally there are described various figures which do service as fountains, but they had no characteristics which seemed to call for their segregation in a separate chapter, and it is best to include here (for want of a better place) some account of a great lost example. Had any reasonable drawing remained of the fountain that once stood in the Upper Court of Windsor Castle it would have justified special treatment, but the little sketch in Norden's view of Windsor Castle in the reign of James I. is obviously inadequate when compared with the descriptions in the building accounts. Either the fountain was re-modelled between 1555 and 1607 (the date of Norden), or we must accept his sketch as only a vague indication.

The particulars given in Tighe and Davies' "Annals of Windsor" are full enough to indicate how serious was the loss to the history of leadwork when that splendid structure was destroyed.

A plan by Hawthorne makes it clear that the base was octagonal and of stone. That the stonework was an important element is obvious, for Roger Amice, surveyor, was paid £3 "for viewing and appointing stone at Reading for building of the fountaine." It was railed about with wood, for which work carpenters were duly paid.

On the stone base was a great tank, which may probably be identified with "the great lead panne," for the carrying of which from London to Windsor 1s. 4d. was paid. Norden's view suggests that the lead tank was covered in by stonework on the outside, that the pillars were also of stone, and the lead confined to the ogee roof and its ornaments. The dragon is shown gilt and standing in the tank. There is no sign of the other royal beasts mentioned in the accounts.

It was on 9th October 1555 that the pipe conveying the water from Blackmore Park was brought into the Upper Court, and "there the water plenteously did rise 13 feet high." The fountain was part of a reservoir scheme, and "of curious workmanship."

By collating the fragmentary indications in the building accounts with Norden's sketch, it would appear that the fountain in general form resembled that of Trinity College, Cambridge, which was built only forty-six years later, but its detail doubtless retained more of the Gothic spirit. The making of wood patterns for the plumbers was a considerable item. The carpenters made the "greate mould in the plombery," also "cisterne cases and other necessaries for ye fountaine." Carvers wrought "scouchions in wainscott to make patterns for the moulds of the scotcheons and badges to garnish the cisterne and topp of the fontaine."

The chief decorations of this fine structure were the six "beasts royall, viz., the eagle, the lyon, the antilop, the greyhound, the gryffith (varying between 5 feet and 6 feet high), and the dragon with his base (13 feet 4 inches high)." The carvers were paid 6s. 8d. a foot for carving them; and it would seem that another item, "founders casting paternes in metall to garnish the cesterne and top of the fountain," shows the next step, the casting of the beasts in lead.

The harte is mentioned later in the painting account, and is necessary to complete the scheme, but must have been carved at some other time.

There seem to have been escutcheons and coats of arms in stone on the lower part of the fountain and in lead above. Carvers were paid for "carthowges and scouchions" (carthowges and cartushes are both delightful spellings for cartouches), and plumbers for "sodering the armes about the fountaine."

It was the work of the latter to "lead the lavatory," and that the leadworker was the main craftsman on the work is clear from the following:—"To John Puncherdon, serjant plumer, and Henry Deacon, for finishing and garnishing of the fountaine in great, as it was agreed between the Lord Treasurer and them, £60."

The painters' accounts give us the final touches, and indicate the gay and splendid work that Puncherdon completed.

They painted and gilded one great vane with the King's and Queen's arms with a great Imperial crown, and did the same for the lion and eagle that held it up. They painted the gryffon (the gryffith of the earlier reference), harte (not mentioned in the carving accounts), the greyhound, and antilope, holding up four compartments with four badges crowned within them, and finally we read of the painters working on the "top of the fountaine with all cartushes, pedestals, armes, beasts, pendants, compartments, pillars, cornice, architraves, and friezes."

The fountain must have had eight pillars, from which sprang arches, probably round. Above the cornice there was a roof of ogee outline, and standing on the cornice were the royal beasts with their gilded vanes flashing in the sun.

CHAPTER VIII.

LEAD PORTRAIT STATUES.

Fairfax—Charles II.—William III.—Marlborough—Prince Eugène—Queen Charlotte—Sir John Cass—George I.

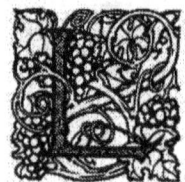 EAD portrait statues do not need an apology, but it may fairly be said of lead in this connection that it takes the place of bronze for reasons economical. It is hoped that the next chapter will not only justify the use of lead for garden figures of a more or less trivial and purely decorative character, but establish for it a fitness peculiar to the garden atmosphere.

In the case of the *Marlborough* and *Eugène* figures (Figs. 242 and 243), though they are portrait statues of a portraiture quite serious, they are also, in their present home at Glemham Hall, garden ornaments.

In the case of the Queen Square statue (Fig. 245), it also stands in a garden, as do the Hoghton Tower *William III.* and the Wrest Park *William III.*

When we come to the equestrian figures a defect must be admitted. The weight of the horse's body and of the rider is a heavy stress on the horse's three lead legs, and in the case of the Petersfield *William III.* a stay rod has been summoned in aid, an addition frankly disturbing. Yet even here no worse has happened than in the case of some stone equestrian figures, which have also needed support.

The portrait of the great Lord Fairfax (the earliest in order of date) is not only a fine achievement in sculptured likeness of a strong type, but is probably the oldest lead portrait bust in England.

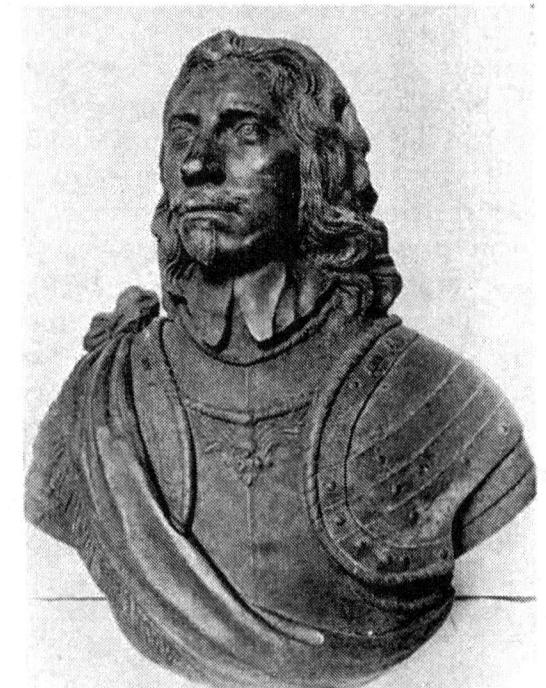

FIG. 238.—Fairfax.

It is in the Council Chamber of the York Philosophical Society, by which Society it was bought in 1879 at Sheriff Hutton near York. It had belonged to Mr Leonard Thompson, whose family bought the Park Estate from the Ingrams of Temple Newsam in the reign of Charles II. So far we are on solid ground and have a grasp of facts, but the information is not very fruitful. To what artist may we attribute this very notable bust, for whom and when was it modelled? We are obliged to fall back on conjecture and comparison. No local will mentioning the bust has yet come to light; it is impossible to say, therefore, whether the original possessor of the bust was "Black Tom" himself or some member of his family.

At Leeds Castle, Kent, which once belonged to the Fairfaxes, there is a bronze bust of which the York lead bust is an exact replica. For the lead bust there may safely be claimed the greater claim to interest. Though the epithet "unique" is a dangerous one, it is fair to apply it to a lead portrait bust of the middle of the seventeenth century, and the same cannot be said of bronze.

The questions of authorship and date are bound up together. There is no signature or other mark on the York bust, and we turn, therefore, to the evidence of its portraiture.

In 1644 was fought the battle of Marston Moor, out of which Black Tom came with a wound in his left cheek. This scar appears in the bust as in most of the portraits, and the bust cannot, therefore, be earlier than 1644. After Naseby, in November 1645, an enamelled jewel incorporating a portrait of Fairfax and made by Bordier was presented to him by his Parliamentary admirers, and he wore it round the neck on a chain.

This jewel, known as the Naseby enamel, which was in the possession of Thoresby, the famous Yorkshire antiquary, and at his death was bought by Horace Walpole, appears in portraits by Bower and others. It is likely that the Naseby jewel would have appeared in the York bust if Fairfax had possessed it when the bust was modelled. The year 1645 may

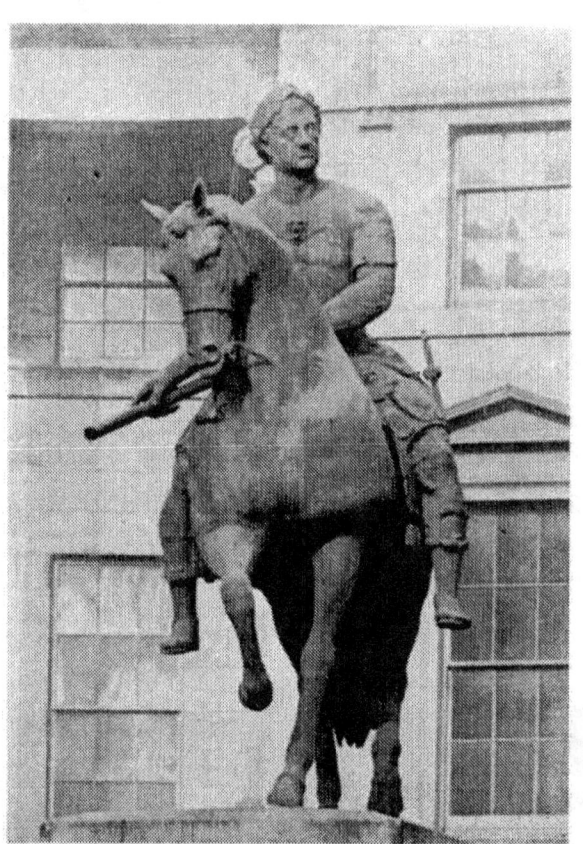

FIG. 239.—Charles II., Edinburgh.

be taken as the most notable of Black Tom's career. Aged thirty-five, he had won the supreme military position on the Parliamentary side by sheer capacity, and, as has happened to other successful generals, there was a rush to immortalise his features. In this year Thomas Simon executed four medals of Fairfax, and these are very similar to the York bust in armour and cast of features.

Abraham Simon, the brother of Thomas Simon, and the "virtuoso fantastical" of John Evelyn's phrase, was a modeller of large portraits, and it seems very likely that towards the end of 1645 Fairfax entered on a debauch of sitting for his portrait—to Thomas for the medals, to Abraham for the bust, and to Bordier for the Naseby jewel. The attribution to Abraham Simon of the bust is nothing more than a guess, but it seems a reasonable one.   Andrew Karne was in York somewhere between 1633 and 1638, but we do not know of his being there as late as 1645.   He is a possible but unlikely author of the bust.

In Parliament Square, Edinburgh, is an equestrian lead statue of Charles II. as a Roman general (Fig. 239).   The face has that saturnine look (not inappropriate to Saturn's metal) which is shared by the "shaven" portrait of the Merry Monarch by Sir Godfrey Kneller.   The horse and rider are about 10 feet in height, and on the back of the tunic there is a winged cherub as an ornament, a little inappropriate to the Roman guise.   The legs of the horse are unfortunately splitting somewhat and need repair.

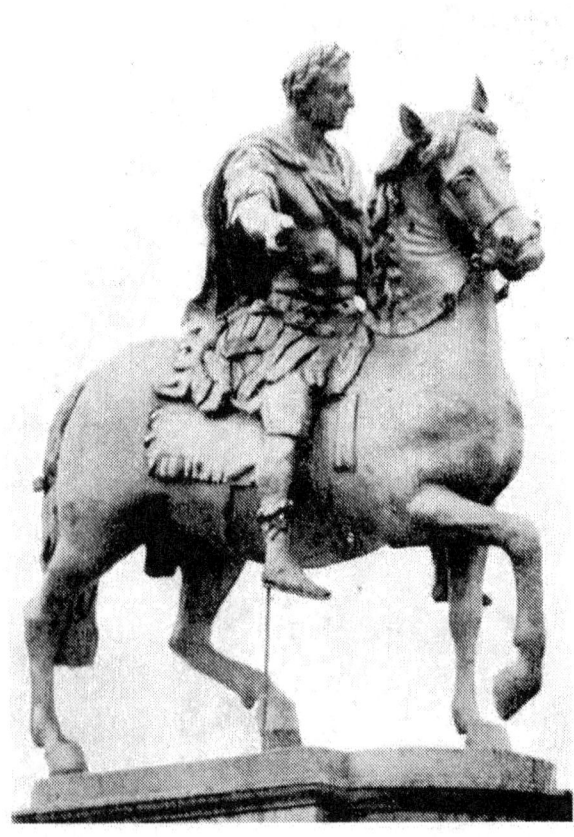

Fig. 240.—William III., Petersfield.

King William III., however, is the king of leadwork.   At Dublin, in College Green, his statue has been the sport of contending factions.   Warburton, White-law, and Walsh in their "History of Dublin" incorrectly describe this figure as being of bronze, and they go on to say, "By an effusion of more loyalty than taste, both statue and pedestal get a new coat of paint every year."   The Corpora-tion of Dublin no longer paint the pedestal, which is of stone, and is orna-mented with trophies of arms in the marine store style of decoration, but the figure is still painted brown to imitate bronze.   One good feature, appropriate to leadwork, remains.   The trappings of the horse, the cross gartering of the King's Roman legs, his laurel wreath, and parts of his tunic are gilt.   Being Roman, he abstains (as do Charles at Edinburgh and William again at Petersfield) from using stirrups.

Redgrave was mistaken in attributing the Dublin *William III.* to van Nost. The Corporation muniments record that the commission was given to Grinling Gibbons, and he received payment for the statue, which was set up in 1701.   A pasquinade on artists who worked in Ireland, by the vitriolic John Williams, says that the younger van Nost was the son of the van Nost of Piccadilly who made lead garden figures, and that he went to Dublin in 1750.   It is perhaps not too speculative to suggest that van Nost *père* did the actual casting of the statue for Grinling Gibbons, and that the connection with Ireland so established led the younger van Nost to decide on an Irish career.

Van Nost *fils* was also a maker of lead statues, for the Corporation of Cork invited him to Ireland some time after 1780 to make a statue of one Mr Lawton.   Most, however, of his work is in stone, but in the upper yard of Dublin Castle there are lead figures by him of *Justice*, *Peac*, and *Mars*, which were put up in 1753. The figure of George II. in St Stephen's Green, Dublin, he did in 1758 for the Corporation. They advertised for designs and selected van Nost as "the most knowing and skilful statuary in this kingdom," but he elected to do George II. in stone, not in lead. In an old Dublin newspaper of 1765, among the London intelligence there is the following note : "Mr van Nost, an eminent statuary from Dublin, is lately come over to take a model of His Majesty for a lead statue which is to be erected in the Exchange about preparing in that metropolis."

There is extant an advertisement by the younger van Nost of casts of a bust of King William, which he originally did in marble. These busts were probably in lead, and it would be most interesting to know if one of them survives.

He died in Mecklenburgh Street, Dublin, in 1787.

Of the authorship of the equestrian *William III.* at Petersfield, nothing is known (Fig. 240). It stood originally in front of the house of the Jolliffe family. When the house was demolished it was moved to the square at Petersfield. The drapery of the figure is of a freer type than the Dublin example. The outstretched arm gives it more action, but at the loss of some dignity. Both are inferior to the splendid brass statue of William III. at Bristol by Rysbrack. The Bristol horse is a particularly fine creature.

It would be satisfactory to find some justification for labelling a *William* lead statue with the name either of Rysbrack (1693-1770) or Roubiliac (1695-1762), but there is not a tittle of evidence. That Roubiliac worked in lead we know ; that he learnt it from Sir Henry Cheere (1763-1781), to whom the *Queen Charlotte* is attributed

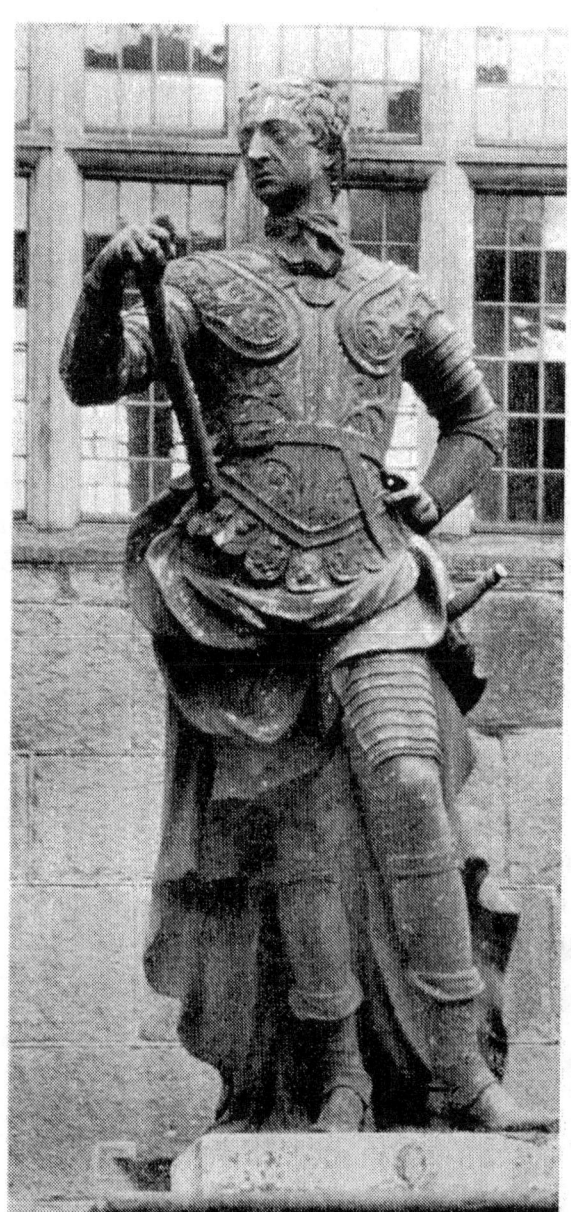

FIG. 241.—William III., Hoghton Tower.

later, we may guess. He left Cheere on securing a commission from Jonathan Tyers for a figure of *Handel* to stand in Vauxhall Gardens. For this same Tyers he did a *Milton* in lead "seated on a rock, in an attitude listening to soft music," and his *Cass* is described later. It is, however, to some competent artist of the calibre of Rysbrack or Roubiliac that we must look for the authorship of the lead figure of William III. now at Hoghton

Tower, Lancashire (Fig. 241).   The portraiture is strikingly good, and the easy pose
of the figure bespeaks an artist of no little
ability.   One detail is amazing, the absence
of a wig.   There is no portrait among the
scores of engravings at the British Museum
where this is lacking.   In one emblematic
engraving, where Britannia offers William
the sceptre and an angel is crowning him, he

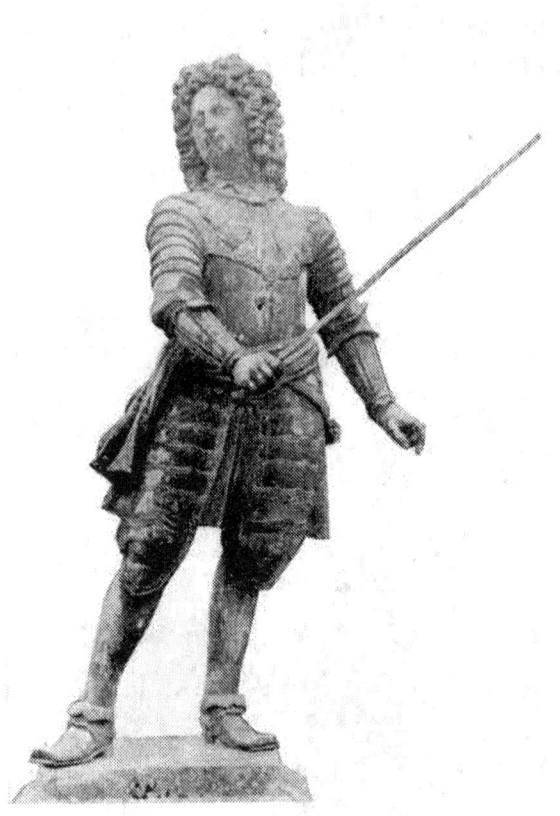

FIG. 242.—Prince Eugène, Glemham Hall.

wears costume in all respects Roman save
for the ridiculous addition of a wig.   In
other engravings where he is made to look
somewhat ethereal, and is crowned with
laurel, he pertinaciously retains his wig.
Even as a little boy he is bewigged.
Everywhere a wig but in this statue.   No
complaint is made of this notable absence
as of something indecent, but it is clear
that here we have evidences of a statuary
who disregarded the conventions.   Had
William been represented as at Dublin,
Petersfield, and Bristol in Roman costume,
the absence of the wig would wring no
withers, but at Hoghton Tower the cuirass
indicates the military dress of his time,
and his arms are not bare in the Roman
manner.

There is a directness and simplicity

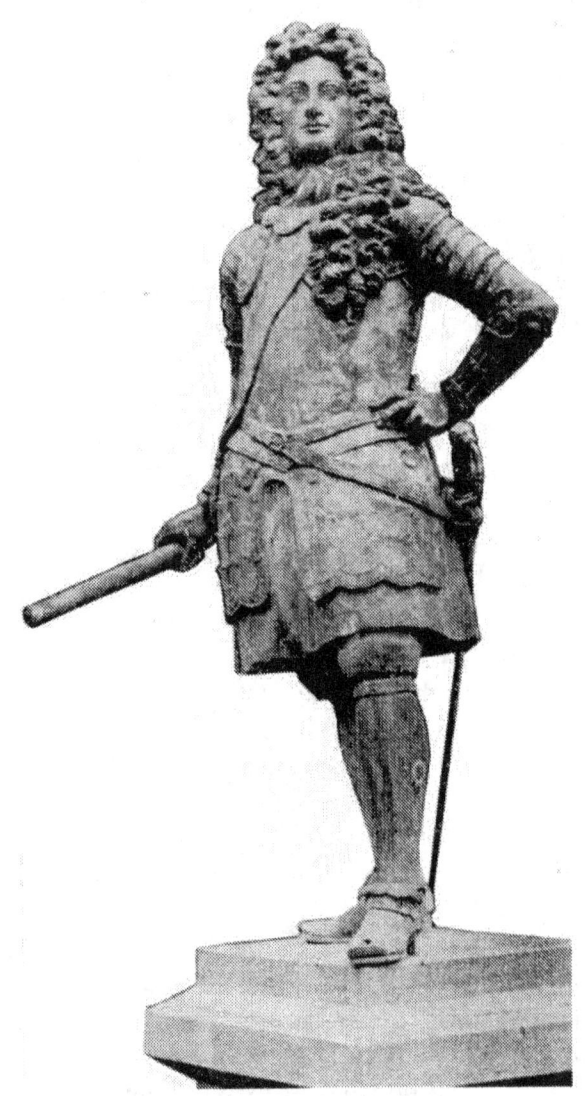

FIG. 243.—Marlborough, Glemham Hall.

about this work which perhaps suggests it was done by an Englishman rather than by a foreigner.

When Henry, Duke of Kent, laid out the grounds of Wrest Park, an avenue was planted in honour of the Revolution of 1688, and a lead statue of William III. set up in front of the Pavilion. It faces up the lake towards the house, and the pedestal is inscribed to the King's "Glorious and immortell memory." The sword which is seen in Fig. 244 resting against the pedestal is ordinarily carried under His Majesty's right arm. As, however, it has obviously nothing to do with the statue, the author removed it before photographing. The treatment of the mantle, &c., is closely akin to that of Grinling Gibbons' bronze statue of James II., which, after much travel, is now in front of the west elevation of the new Admiralty block. The detail is, as becomes lead, somewhat coarser. The name of the sculptor is lacking, but the statue is clearly from a very competent hand.

At Glemham Hall, Suffolk, are two delightful lead figures of Prince Eugène and of John Churchill, first Duke of Marlborough. The *Eugène* shows him with drawn sword, in a slightly theatrical attitude, wearing a bulky wig and the collar of a Knight of the Golden Fleece. He lived from 1663 to 1736. The best way to date Eugène is by the fatness of his face. There is an engraved portrait of 1701 (when he would be thirty-eight years old) which resembles our statue. A portrait of 1712 shows him with his face longer and thinner, and in another of 1735 this development of gauntness is very marked indeed. Most of his portraits, notably that by Sir Godfrey Kneller, shows him with his marshal's baton in his hand. There seems to be none with a drawn sword.

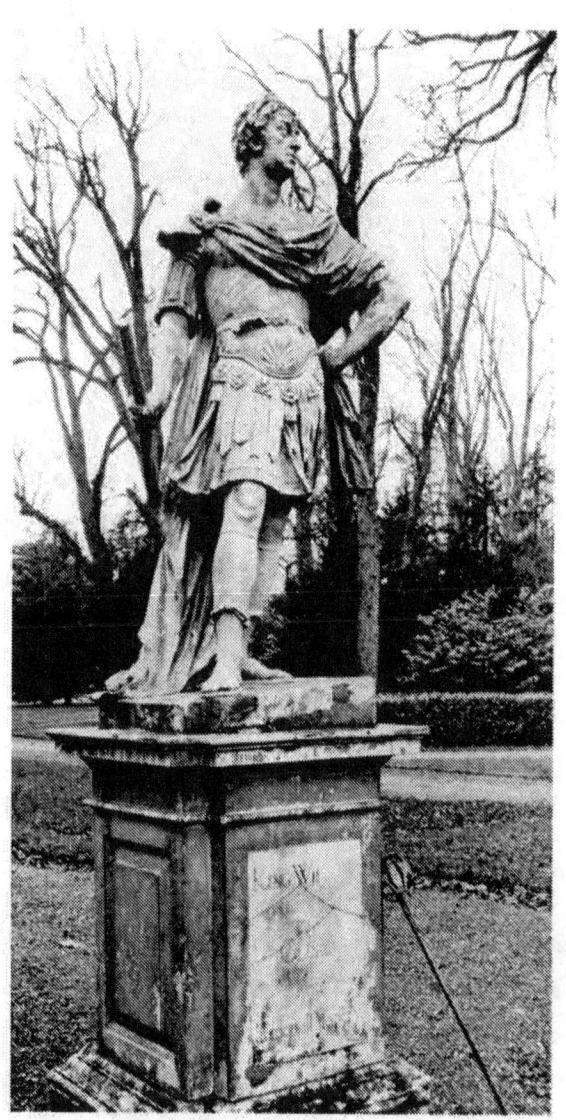

FIG. 244.—Statue of William III. at Wrest Park.

The *Marlborough* is a splendid figure of great ease and nobility of pose. The wig is luxuriant, and while the duke carries his baton he wears no order. He looks rather younger in the statue than in the Kneller portrait of 1705, but otherwise the statue as a portrait is excellent. It is perhaps not impertinent to remark the continuing faithfulness to type of the Churchill family.

As to the authorship of the Glemham Hall figures there are no facts to give. Rysbrack did the monument of Marlborough at Blenheim, but these statues are probably earlier, and it seems safe to date them as c. 1700 if they were modelled from the life.

The lead statue of a queen in Queen Square, Bloomsbury, has been variously described as of Queen Anne, and of the consort of George I., Queen Charlotte. It presents some difficulties, but the evidence seems to be in favour of Queen Charlotte. Mr Henry B. Wheatley in his "London Past and Present" is on the side of Queen Charlotte, and says that the statue was presented by General Strode. Strode does not appear in any biographical dictionary, but he seems to have been a kind of Carnegie of public monuments. The equestrian statue of the Duke of Cumberland, modelled by Cheere (of whom more hereafter) and set up in Cavendish Square in 1770, was given by Lieutenant-General William Strode. It is not recorded whether this was of bronze or of lead. It was taken down to be repaired in 1868, and incontinently disappeared. The need of repair and the subsequent vanishing point to lead rather than to bronze. Strode also set up in Stratford Place a pillar, which made haste to fall down a few years later. Assuming, therefore, that Strode gave the statue in Queen Square, it is more likely to have been of Charlotte, who was pursuing her dull and decorous course as consort in 1770, the date of the *Cumberland*. Strode was apparently a courtier, and would have been less interested in Anne, who was even then unquestionably dead. The giving of the *Cumberland* statue is strong evidence in favour of Charlotte. Sir Henry Cheere was the most notable modeller of lead statues then flourishing. As Strode was his customer for the *Cumberland*, what more natural than that he should go to him for the *Charlotte*?

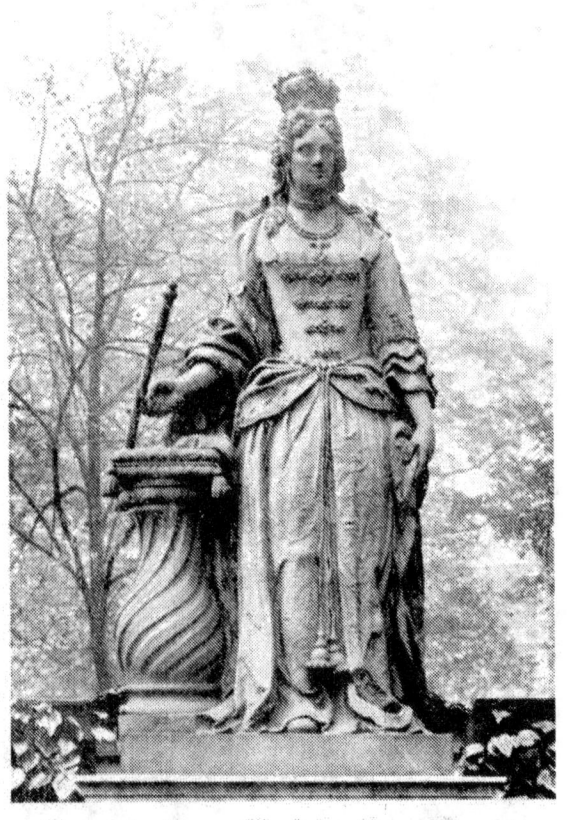

FIG. 245.—Queen Charlotte, Queen Square.

The evidence of the figure itself is puzzling, but the balance is in favour of Charlotte. She carries a sceptre in her right hand, wears a crown, and carries no orb. Her robes are of the ordinary coronation type, and she wears no orders. All this suggests Charlotte.

Every engraved portrait of Queen Anne wearing a crown, of the scores examined (except one), shows her also with the collar and star of the Garter and the George. The one exception is a fanciful sketch, from which a formal ornament like the collar might not unnaturally be omitted. The portrait statues of Anne in Queen Anne's Gate, at Blenheim, and in St Paul's Churchyard, not only have the Garter ornaments but also the orb. Were the Queen Square figure of Anne, it would certainly have the orb and the collar and star of the Garter. The portrait of Charlotte by Reynolds shows her seated in coronation robes similar in general character to those of the Queen Square statue in respect of the corsage and sleeves, and there is a sceptre on a cushion. Here again we find no orb and no Garter ornaments. The chief difficulty of the statue is in the hair. It is arranged in heavy curls hanging down over the neck, and is very similar to that of the *Anne* in Queen Anne's Gate. In the Reynolds picture of Charlotte the hair is done up in the usual late eighteenth-century manner, and only one curl strays on to the neck. In this the engraved portraits of Charlotte agree, save for one

at the age of twenty-three, which shows as many curls as the statue does. It is possible, however, that Francis Bird's statue of Anne, set up in 1712 in St Paul's Churchyard, may have crystallised the long curls into a queenly convention, which the later statuary, who did the Charlotte figure, thought well to follow. The features tell little. Charlotte was very plain, and in life her nose was markedly snub. The Queen Square statue has a non-committal sort of nose, neither Roman like Bird's figure of Anne, nor honestly snub like Charlotte's less flattering portraits. Accurate portraiture, however, was not universal in the statues of those days, *e.g.*, the *Anne* of Queen Anne's Gate has a nose not at all Roman.

This last statue and also the George II. in Golden Square have been included in lists of lead statues, but incorrectly. Both are of stone—the *Anne* of Portland stone, the *George II.* of some more friable and coarsely grained stone, which now shows ominous cracks and is like to perish before long.

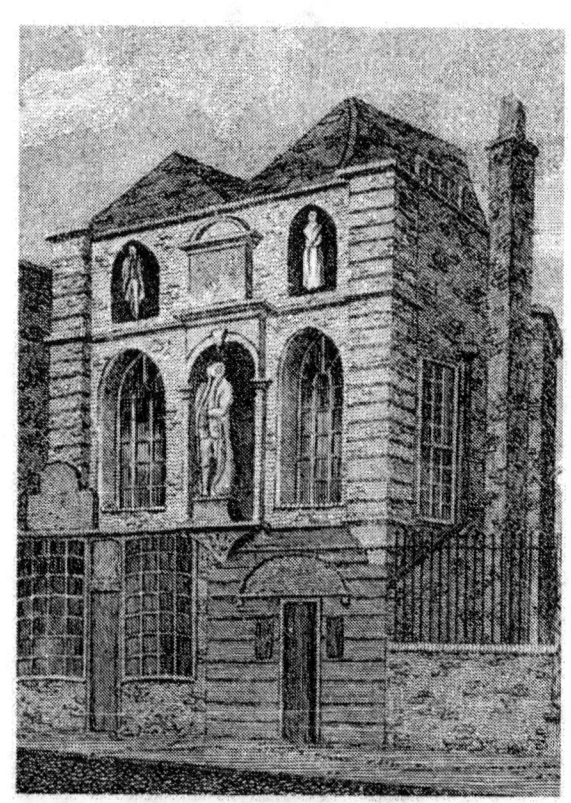

FIG. 246.—The Old Cass School
(destroyed).

The most satisfactory lead portrait statue extant, as far as detailed knowledge of it goes, is that of Sir John Cass. It stands high up in a niche on the new building of the Cass Foundation Institute in Jewry Street, E.C. (Fig. 247).

In 1710 Cass established a school, in 1718 he died, and in 1750 the trustees of the charity "resolved that it be referred to the Treasurer to prepare a statue of Sir John Cass to be made by a skilfull Artist in such manner as he shall be advised, and that the same be erected in the Niche for that purpose in the Front of the sd. schoole."

Sixteen months later Mr Treasurer wrote, "acquainting the Board he had agreed with Mr Roubilliac, statuary, for making Sir John Cass's effigies."

The sculptor borrowed Sir John's picture "to fform the effigies by," and a month later "attended with a modelle," and such of the Trustees present as remembered Sir John Cass in his lifetime gave Mr Roubilliac the best description they could of " Sir John's persone."

In November 1751 the statue was ready to be set up, and the treasurer "was of the opinion it would be proper for some of the Trustees to go and see the Statue at Mr Roubilliac's, in St Martin's Lane."

On the 9th January 1752 it was "resolved that the Treasurer do pay Mr Roubilliac the sum of one hundred pounds."

The minutes of the trustees from which the above extracts are taken are full of detail with one odd omission, the material of which the figure is made.

With the single exception of the lost *Milton* made for Vauxhall Gardens no other lead figure can be attributed to Roubilliac. The engraving (of which part is reproduced in Fig. 246), dated 1810, shows the figure in its original place. The figure is too high in its new position, and should be moved into the board room of the Governors. This statue does not suffer from the fantastic artificiality which is characteristic of so much of Roubilliac's work, notably of the *Nightingale* monument in Westminster Abbey. Sir John Cass is given a calm and dignified pose, very different from the buoyant triviality of the *Shakespeare* at the British Museum. The detail of the robes is exquisitely clean but does not suggest undue effort. There is none of that restless straining after characterisation which appears in the heads that Roubilliac modelled from the life. Among lead portrait statues the *Cass* has no equal except the *William III.* at Hoghton Tower, and that it was modelled *ad hoc* for architectural use gives it an added interest.

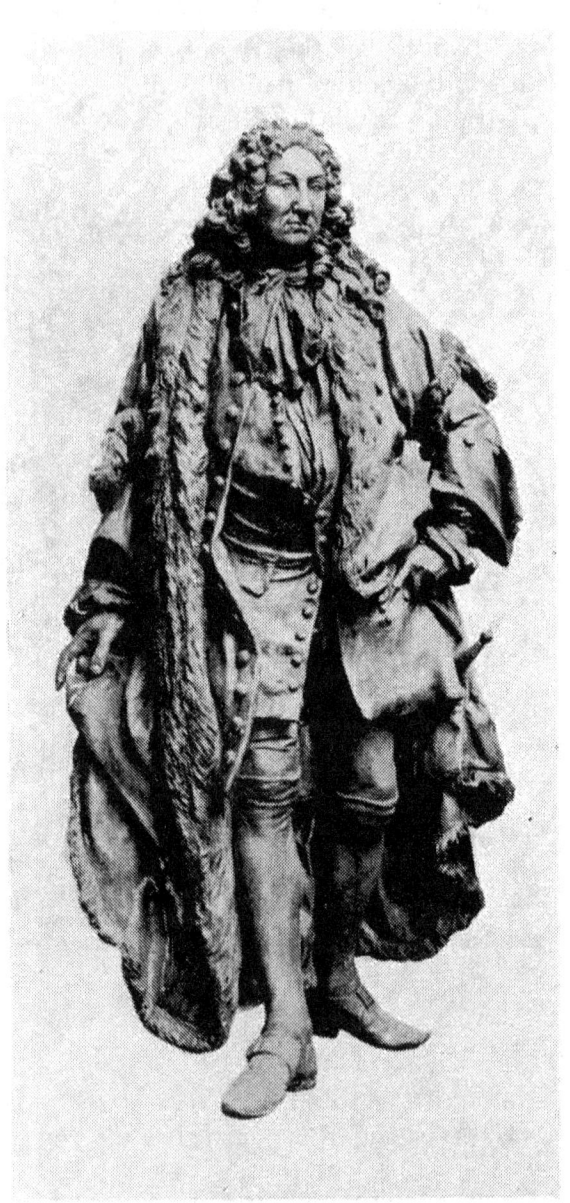

Fig. 247.—Sir John Cass.

J. T. Smith records that the *Cass* was at one time painted various colours to give it a life-like appearance, in the manner of the wax figures at Westminster. Garden figures were often tricked out in the same fashion.

In Leicester Square there stood a gilt lead statue of George I. It was originally made by van Nost for Canons House, Edgware. It was set up in Leicester Square by Frederick, Prince of Wales, to annoy his father, George II. Being in 1872 much damaged, it was sold for £16.

In Grosvenor Square there·was erected in 1726 an equestrian statue of George I., said to have been by van Nost, and, if so, doubtless a replica of the Canons statue. In 1727 this figure, which was "doubly gilt," was defaced by a partisan of the Pretender, and it has since disappeared. Malcolm speaks of Vancost of Hyde Park Corner (doubtless John van Nost) as modelling a statue of George I. from that of Charles I. in 1721, so presumably van Nost thought it safer to follow Hubert le Sœur than trust to his own unaided ideas.

As this chapter was going to press, news came of the sale of some of the Glemham Hall figures, among them the Marlborough and Prince Eugène.

CHAPTER IX.

LEAD FIGURES GENERALLY.

The Cross of Cheapside—Neptune at Bristol—Karne—Melbourne, Derbyshire—Giovanni de Bologna—Harrowden Hall—Wrest Park—Wilton—Nun Monkton—Methods of Casting—Hampton Court—Syon—Castle Hill—Deceitful Figures—Forgers of "Antique" Leadwork—Studley Royal—The Water Note in Leadwork—Eighteenth-Century References to Statues—Hardwick Hall—Glemham Hall—Enfield Old Park—Norfolk Market Crosses—The London Apprentice.

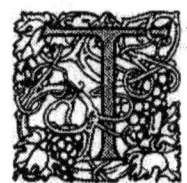

HE antiquary may be pardoned a not unnatural desire to prove early dates, and lead statues would lose some of their importance if no record of them in England could be found earlier than the seventeenth century. Mr Edmund Gosse has complained of the scantiness of the records of sculpture even in the eighteenth century, and one might despair of finding anything in the way of mediæval lead statues were it not for the records of the Cheapside Cross.

In J. T. Smith's "Antiquities of London," there is a rough picture of the destruction of the Cross by the Puritans, and under it the legend :—

"The 2d of May 1643 the Cross of Cheapside was pull'd down. A Troop of Horse and 2 Companies of Foot waited to guard it, and at the fall of the top Cross, Drums beat, Trumpets blew, and multitudes of Caps were thrown in the Air, and a great shout of people with joy. The 2d of May the Almanack sayeth was the Invention of the Cross and the 6th day at Night was the leaden Popes burnt in the place where it stood, with ringing of Bells, and a great acclamation and no hurt done in all these actions."

"Leaden Popes," a very stimulating reference. Now the history of the crosses in Cheapside is shortly as follows :—

The first was a stately cross of stone, built by Edward I. in 1290 in memory of Queen Eleanor. This fell into disrepair, and was rebuilt in 1441 at the expense of the City of London. Henry VI., in connection with this second cross, granted to John Hatherley, Mayor, licence "to re-edify the same in more beautiful manner." Hatherley "had licence also to take up two hundred fodder of lead for the building thereof and of certain conduits and a common granary." Two hundred fodder represent roughly 200 tons, and possibly some of this lead went to the making of the "leaden popes" that were burnt in 1643 in the place where the Cross had stood. It was building from 1441 to 1486, and Stow mentions that it was "at the charge of divers citizens (notably John Fisher, mercer) curiously wrought." By 1581 people had come to be irritated by emblematical figures, and the Cross was almost demolished, and the images defaced, but it was repaired. Incidentally the Philistines of that day wanted to move it to make a street improvement.

In 1599 the timber of the Cross at the top "being rotted within the lead," the top was taken down, but the Privy Council ordered repairs.

After a year's delay, and more commands from Queen Elizabeth, a cross of timber was framed and set up (in 1600), covered with lead and gilded, but the image of Our Lady was again defaced. On the accession of James I. it was railed in, repaired, and beautified. Its final downfall has already been described, a destruction which Evelyn witnessed, "I saw the furious and zelous people demolish that stately Crosse in Cheapside."

Several illustrations of the Cross remain. It was of a purely monumental type, not practically a building, as was Paul's Cross. Among the Thomason Tracts at the British Museum is one entitled "The Downe-fall of Dagon," which was doubtless published in or soon after 1643. It is a delightful publication, and purports to be not only a description of the Cross, but also its last will and testament dictated by itself, and its epitaph, "Dagon," being a puritanical pet name for it. In the will we find, "Item, I give to the Red-Coate souldiers all the lead which is about me to make bullets if occasion be ; if not, I give it to the Company of Plummers to make cisterns and pumps with."

The illustration shows three of the figures bearing pastoral staves, and though it may be claimed that these would be bishops not popes, there is other evidence. In the Crace Collection of prints is one of Cheapside Cross as it appeared in 1547, with part of the procession of Edward VI. on his way to his coronation at Westminster. This print shows, in the lowest tier of figures, one with a triple crown. In another print, a *Representation of the Demolishing of the Cross*, one figure wears a mitre, but there is none with a triple crown. In the Pepysian Library, Cambridge, there is a picture of the third Cross built of leaded timber in 1600, and in the Crace Collection a copy of the drawing as well as an engraving after it. Here again in the lowest tier of figures is one with a head-dress which is certainly not a mitre, and though it is not an accurately drawn tiara, it is differentiated from the next figure, which wears an obvious mitre, and may fairly be claimed as the triple crown. Stow says, "The lowest Images . . . being of Christ, his resurrection, of the Virgin Mary, King Ed. the Confessor, and such like." "Such like" is not very definite. So much for the "popes." Now as to the "leaden." We have established the very large use of lead by John Hatherley. To quote again from "The Downe-fall of Dagon," "Some report divers of the Crownes and scepters are silver." Now silver ornaments are much more likely to have been applied to lead than to stone statues. There is also the evidence of the frequent regilding of the Cross on the occasion of royal progresses, &c. Lead statues are much more likely objects so to be gilt than stone figures. From the somewhat rude sketches of the Cross which remain, the figures which decorated it seem to have been about twenty in number. The evidence suggests that John Hatherley adorned the second Cross with these figures, in lead, and that the statues were of popes and saints. The date 1600 is a very unlikely one for the production of ecclesiastical figures of this character. Probably the rebuilding of 1600 consisted merely of placing on the leaded timber framework the "leaden popes" that came to so untimely an end in 1643. We may turn now to the later work, where we are on more solid ground.

It is an unhappy thing that, with the exception of the *Neptune* of Elizabeth's reign at Bristol, there is no English lead statue of the sixteenth century or earlier, at least none has been recorded. Of mediæval lead statues there must have been plenty, but in England they have not survived.

The *Neptune* of Fig. 248 stands in the street at Bristol, in the shadow of the leaning tower of the Temple Church. The figure (if a local tradition recited on the pedestal has any value) has an historical interest which gives it an important place among English lead statues. The story has it that the pumps from captured ships of the Spanish Armada provided the material, and that it was given by a Bristol plumber to celebrate the great defeat. Even if this story is not true, the figure is certainly old, as lead statues go, and it may be accepted as sixteenth-century work. Mr Lethaby thinks "the limbs are contorted with too much life," and it is certainly a coarse piece of modelling, but it is the most interesting figure in Bristol.

We come next to the leadwork done by Andrew Karne (or Kearne), variously described as a Dutchman and a German.

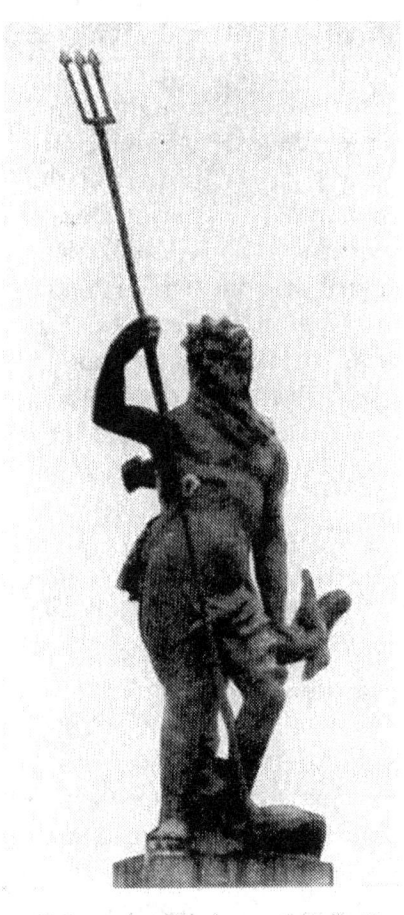

Horace Walpole relates of him that he was brother-in-law of the sculptor Nicholas Stone the Elder, for whom he worked. At Somerset Stairs he carved the river-god which answered to the Nile, carved by Stone, and a lioness on the water gate of York Stairs. He died in England, and left a son who was living after 1700. The date of his birth seems unknown. The most definite and interesting fact about him is contained in Sir Henry Slingsby's Diary. About 1625 Slingsby began to build the Red House, Marston Moor, and writing in 1638 of the oak staircase (which in 1861 was removed to the chapel), he says: "Ye staircase yt leads to the painted chamber was furnished ye last year by John Gowland. Ye stair is about five feet within the sides in wideness; ye posts eight inches square; upon every post is a crest set of my especial friends and my brother-in-law, and upon that post yt bears up the half pace . . . yt leade to the painted chamber, there sits a blackamore cast in led by Andrew Karne, a Dutchman, who also cut in stone ye statue of ye horse in ye garden. The blackamore sits holding in either hand a candlestick to set a candle in to give light to ye staircase."

The "blackamore in led" sits there still (Fig. 249), and is the earliest lead statue in England to which an exact date can be given, for there is no documentary evidence as to the Neptune at Bristol. The black boy's candlesticks have unhappily gone, and one arm with them, but he is still a pleasant boy.

FIG. 248.—Neptune at Bristol.

The majority of lead garden statues are the product of Georgian times, but the seventeenth century saw their use well established in the pseudo-classic atmosphere in which they chiefly flourished.

C. G. Cibber was born in Flensburg, Holstein, in 1630, and, in Colley Cibber the dramatist, had a son more famous than himself. He was originally employed by John, the son of Nicholas Stone.

Peter Cunningham says of him that "his residence in Rome and the general favour extended to classic subjects . . . induced Cibber to carve allegories and gods. He performed for the vista and the grove what Thornhill and La Guerre did for the ceilings and the walls. Neptune with his Tritons appeared in the midst of the pond, Diana and her nymphs in the recesses of the grove, Venus adorned some shady arbour, and Minerva or Apollo watched by the portico." From this one would suppose that Cibber was the first to use gods in the garden, but Nicholas Stone the Elder (1586-1647), the father of Cibber's employer, was engaged in 1632 on statues of Cupid, Venus, Ceres, Hercules, and Mercury for the Paston family, and one may assume some of these were for the gardens of Oxnead. Mr A. E. Bullock, who has written so fully of Nicholas Stone, has found no reference to his having worked in lead.

Careful search has also failed to identify Cibber with any lead figures. He delighted in freestone, for it is easily worked, and god after god could be turned out rapidly to satisfy the urgent demands of the *cognoscenti* of his day. A few years of rain and frost, and the insidious creeping of lichen, produce in a freestone statue an air of desolation and decay. Hence the recourse to lead for

> " Homer, Cæsar, and Nebucadnezar,
>   All standing naked in the open air,"

for frost, which will split a stone figure, leaves lead unhurt.

It is interesting to note that Pepys had a word to say about garden statues, as indeed about most things that minister to the pleasures and graciousness of life.

He spent a Sunday afternoon at Whitehall with Hugh May, who was near to getting the post of surveyor to Charles II., but happily lost it. It was given to Sir Christopher (then Dr) Wren.

Hugh May was doubtless, as Pepys says, "a very ingenious man," but one trembles to think what we should have lost if he had been the architect of St Paul's and the City churches.

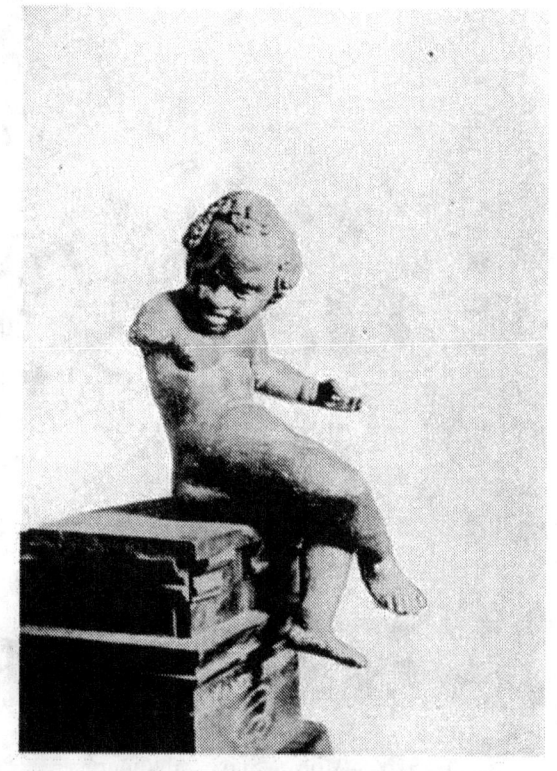

Fig. 249.—At the Red House, Marston Moor.

About gardens May seems to have been sound, and told the diarist that "we have the best walks of gravel in the world, France having none, nor Italy, and our green of our bowling alleys is better than any they have. So our business here being Ayre, this is the best way, only with a little mixture of statues or pots, which may be handsome, and so filled with another pot of such or such a flower or greene as the season of the year will bear."

While "a little mixture of statues" is here admitted as being part of the "best way," Hugh May unfortunately did not enlarge on the question of material, or refer to the subjects he thought fit for such figures. However, "our business here being Ayre" is

a delightful English touch, for which we may well be grateful, and forgive him for
omitting to descant on the charms of statues and pots when of lead, or the statues which
came up to his standard of "handsome."

Most of our knowledge of the makers of lead statues comes from the antiquarian
writings of J. T. Smith. He has been quoted at large by Mr Lethaby, so the bare facts
only need be here set down.

John van Nost, a sculptor who came to England with William III., started the
first lead yard for the regular supply of garden figures in Piccadilly. We are told that

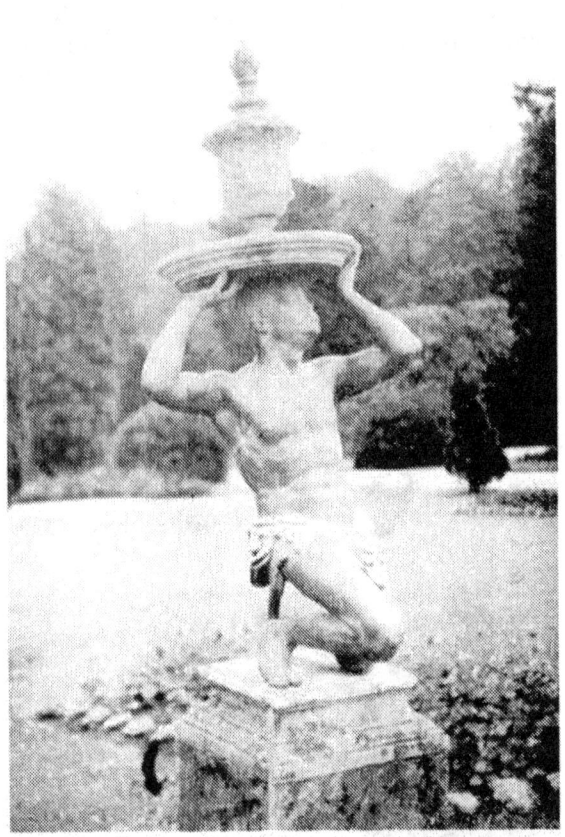 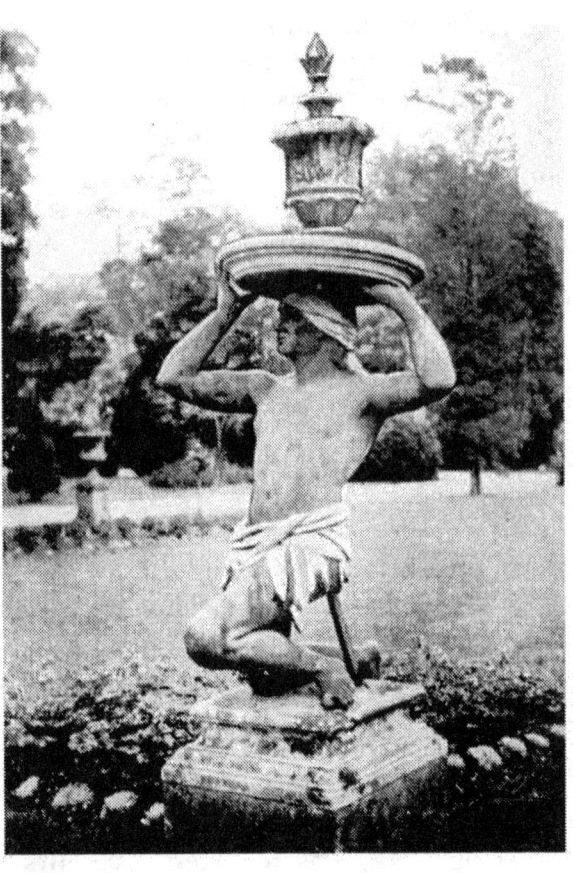

FIG. 250.—African Slave, Melbourne.   FIG. 251.—Indian Slave, Melbourne.

there was a sale of his effects in 1711, but this was doubtless a temporary reverse only,
for John Cheere did not take over the van Nost yard until 1739. We will trace his work
as far as we may, in face of the difficulty that there are few subjects so deplorably
lacking in documents as the history of sculptors and sculpture of the seventeenth and
eighteenth centuries, or one which would better repay careful research.

The gardens of Melbourne, Derbyshire, which were remodelled by Henry Wise, are
a mine of leadwork. The figures, or many of them, came from John van Nost early in
the eighteenth century, and the accounts are preserved. There is an item of "Young
Triton with brass pipe in middle, £6. 9s. 0d." Perhaps this is the delightful boy of

Fig. 253, though Triton seems hardly a proper description. However, there is no Triton of the fishy sort, and the brass pipe which makes him a fountain is possibly enough to identify him.

There are two *Kneeling Slaves* in the upper garden, Figs. 250 and 251. They were until lately painted black with white waist cloths, but when recently mended the paint was fortunately removed.

These slaves are the most common of lead garden statues. One is markedly negro in hair and lips, and has always been called "the Black-a-Moor," the other is a turbaned

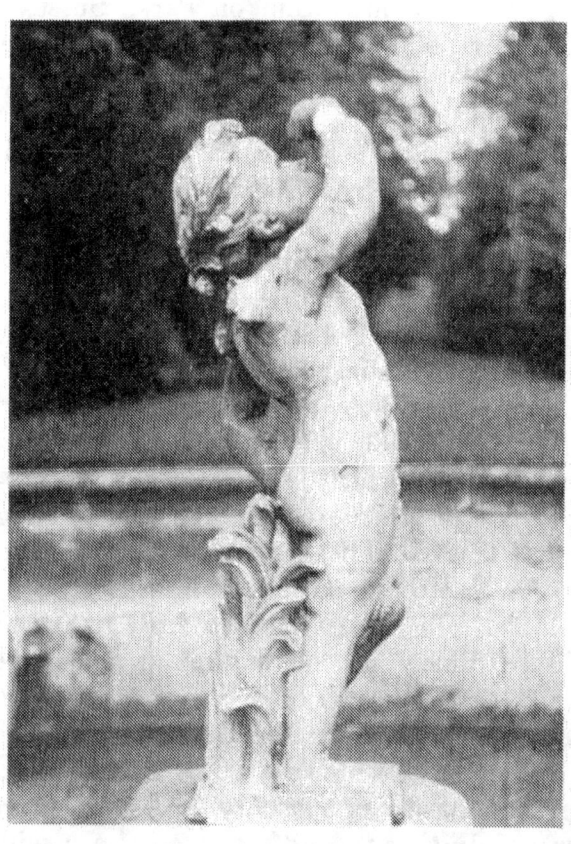

Fig. 252.—Melbourne.

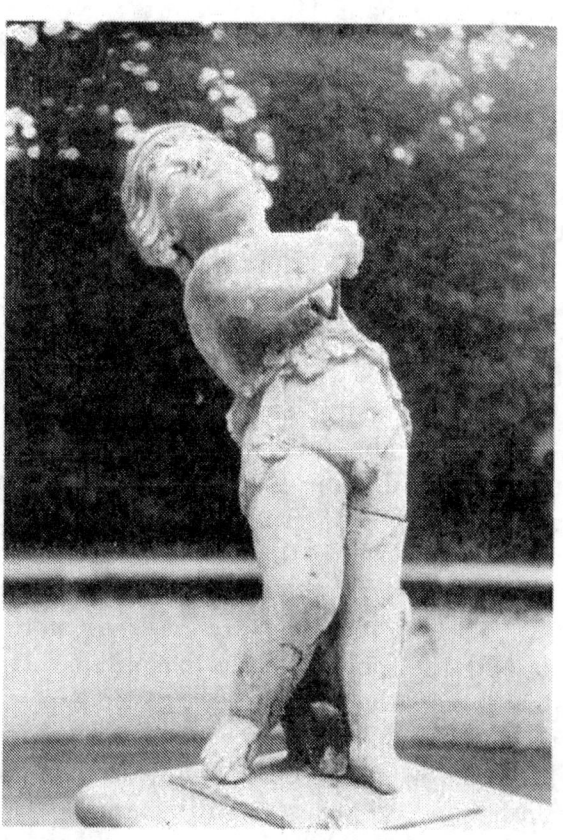

Fig. 253.—Melbourne.

figure of Indian type. Both are about 3 feet 6 inches in height to the top of the tray. They cost £30 the pair.

At Melbourne they carry stone trays, and on them lead vases. Sometimes they carry sundials. The pose is admirable. The tracing of the supply of these figures is not without interest. There is a pair at Glemham Hall, Suffolk, which came from Campsey Ash, when it belonged to the Shepherds. The best known example is the Black-a-Moor in the gardens of the Inner Temple. It is dated 1731, and its former home was Clement's Inn, where once the following verses were found attached to it :—

"In vain, poor sable son of woe,
      Thou seek'st the tender tear ;
From thee in vain with pangs they flow,
      For mercy dwells not here.
From cannibals thou fledst in vain,
      Lawyers less quarter give ;
The *first* won't eat you till you're dead,
      The *last* will do't alive."

Lord Algernon Percy has another slave at Guy's Cliffe. There was one in the gardens of Sandywell, Gloucestershire, now laid waste. There are others at Knowsley, Arley, Aldenham House, Herts ; Norton Conyers, Yorkshire ; Slindon Park, Sussex ; Purley Hall, near Pangbourne ; Ockham Hall, Surrey ; and Mr Philipson-Stow has one which came from Cowdray. Reference will be made later to a variant in which the Black-a-Moor's face is that of a boy, but the figure and pose the same.

It has been suggested that this figure is after one by Pietro Tacca, who modelled the wonderful group of galley slaves at Leghorn. No evidence of this is, however, to be found.

Van Nost must have found the lead slave trade brisk and remunerative, for the list is doubtless far from complete. Replicas must have perished in scores when formal gardens were abandoned for what Mr Lethaby delightfully calls "mean productions in the cemetery style, an affair of wriggling paths, little humps and nursery specimens." In such futile parodies of gardens the lead statue was an offence and a hissing.

The Melbourne *amorini* are chubbily pretty, and the story of quarrel and reconciliation, told in the four groups of two, gives a dramatic touch which is pleasant. Figs. 254-256 show the progress of the quarrel, which arose out of a struggle for a garland. The fourth group shows them healing their quarrel with kisses. These groups were modelled by van Nost in 1699, and were supplied in 1706 for £42 the four.

The single figures are perhaps more admirable. The artist had no story to attend to, and the modelling has benefited. It would be difficult to find figures of a happier grace than those of Figs. 252 and 253. The pose of the boy of Fig. 252 is very like that of a bronze Cupid of the school of Andrea del Verrocchio in the South Kensington Museum, while the other is a little reminiscent of the Boëthos figure of a *Boy with a Goose.* Both stand on pedestals in the middle of large sunk basins of masonry, and gaily spout up water through brass tubes. Their brothers of Figs. 258 and 259 were busy with archery. Though the bows have perished, and the arrows have long since found their mark, the look of mischievous intent remains, and they doubtless smote some lingerers in these gardens in anacreontic fashion, μέσον ἧπαρ ὥσπερ οἶστρος. The crushed look of the right leg of the boy of Fig. 259 is due to the partial collapse of the lead.

The tendency of sculptured *amorini* is to a (not unnatural) liveliness of limb which is of less happy effect in lead than in bronze, but the quiet action of some of these boys makes them rank high in their race. The youngster of Fig. 257 is the most lively of the whole series, and not without sufficient reason. He has disturbed a nest of hornets in the hollow of a tree stump, and they are working their vengeance on him. One is on his right hand, another on his face, and his fat little person is paying the toll of interference. His features are screwed into an ecstasy of pain, but the sense of artificiality remains to spare us the discomfort of too genuine a sympathy.

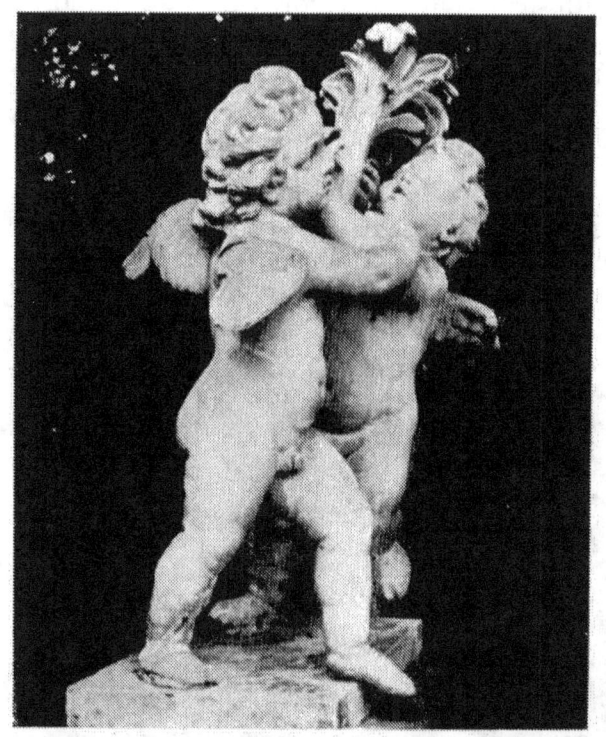

FIG. 254.

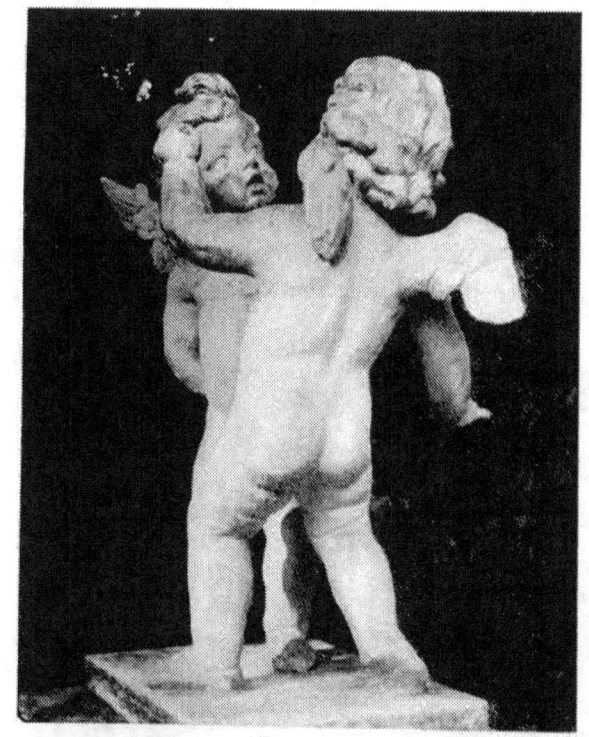

FIG. 255.

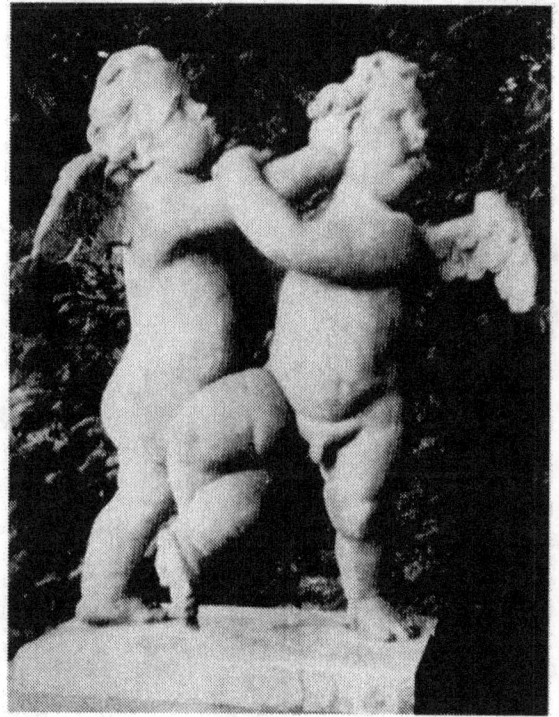

FIG. 256.

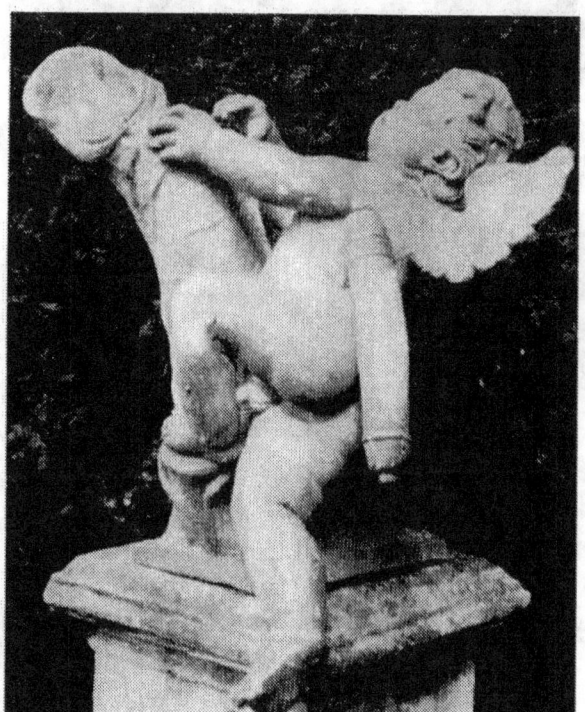

FIG. 257.

AMORINI, MELBOURNE, DERBYSHIRE.

If these Melbourne *amorini* are compared with such figures as Andrea del Verrocchio's bronze *Cupid with Dolphin*, it will be seen that the sense of merry elfish agility which Verrocchio's figure suggests is not only absent from the Melbourne figures, but would be misplaced in lead.

The question of muffled detail is particularly noticeable in the wings. In Verrocchio's figure the feathers are distinct, at Melbourne they are little more than suggested. There is, of course, the inferiority of the artists in lead as modellers. It would seem, however, that in many cases the figures have been modelled with an intentional roughness, appropriate to lead, which would be coarse in bronze. Compare, for example, the bronze *Cupid*

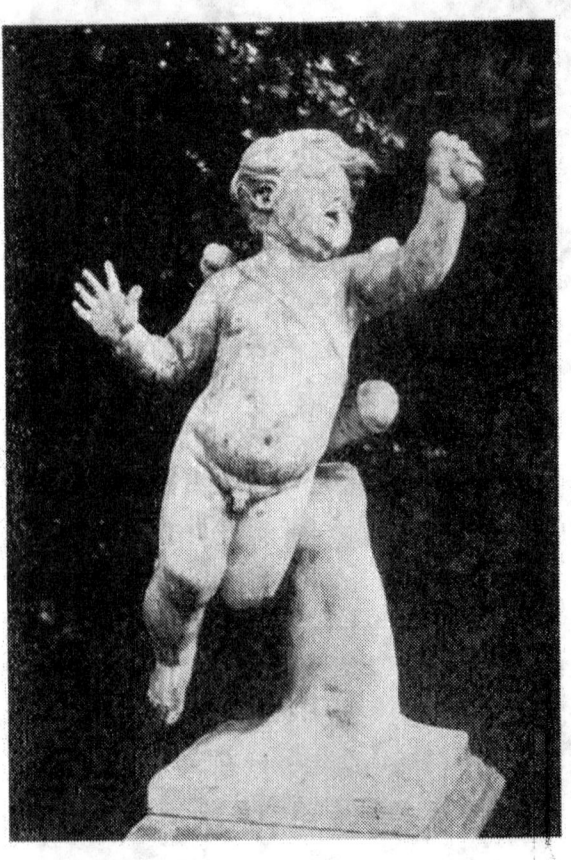

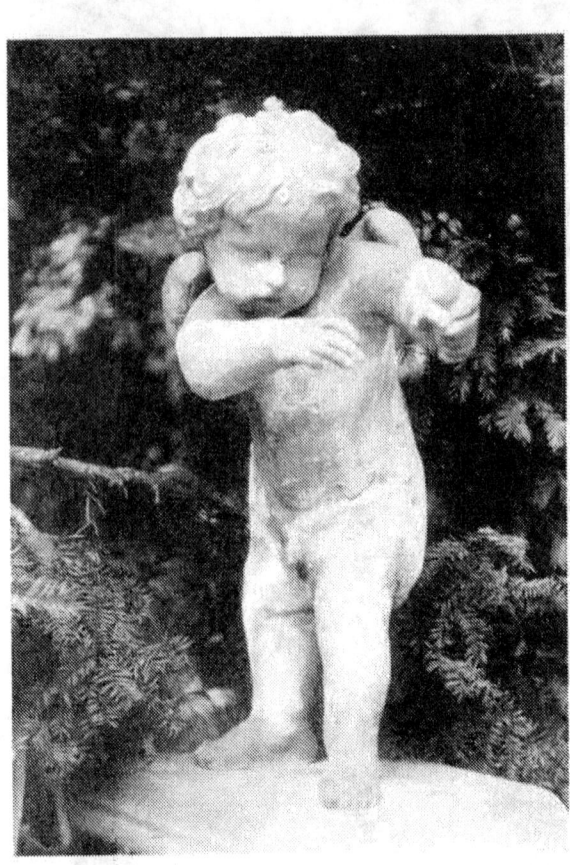

Fig. 258.—Melbourne.   Fig. 259.—Melbourne.

by Donatello which is in the National Museum at Florence, with the lead *amorini* at Melbourne. The fine lines and detail of the Donatello would lose if reproduced in lead. Even if attempted, they would soon be blurred by the battery of time and gently effaced by lichens. Impossible, too, in lead, that exquisite delicacy of expression which Donatello gave to his bronze, the impish gaiety which a surface defect would destroy. The Melbourne *amorini* are from 2 feet 3 inches to 2 feet 6 inches high.

At the bottom of the Melbourne gardens, one on each side of the "Birdcage," an exquisite garden-house of open ironwork, stand *Perseus and Andromeda*, facing the fish pond. They have been painted white, and have a ghostly look against the back-

ground of yew. Perseus Fig.)
260) is holding out an affrighting
Medusa head, and turns away
with a rather unconcerned
manner, not devoid of swagger.
His clothing is somewhat nonde-
script, and looks Roman rather
than Argive, but the artist has
been careful to give him the
winged sandals and the helmet
of Hades. He is a heavy figure
compared with such a *Perseus* as
the Canova in the Vatican, or the
Benvenuto Cellini. Andromeda
is rather more interesting (Fig.
261). She is chained to the rock

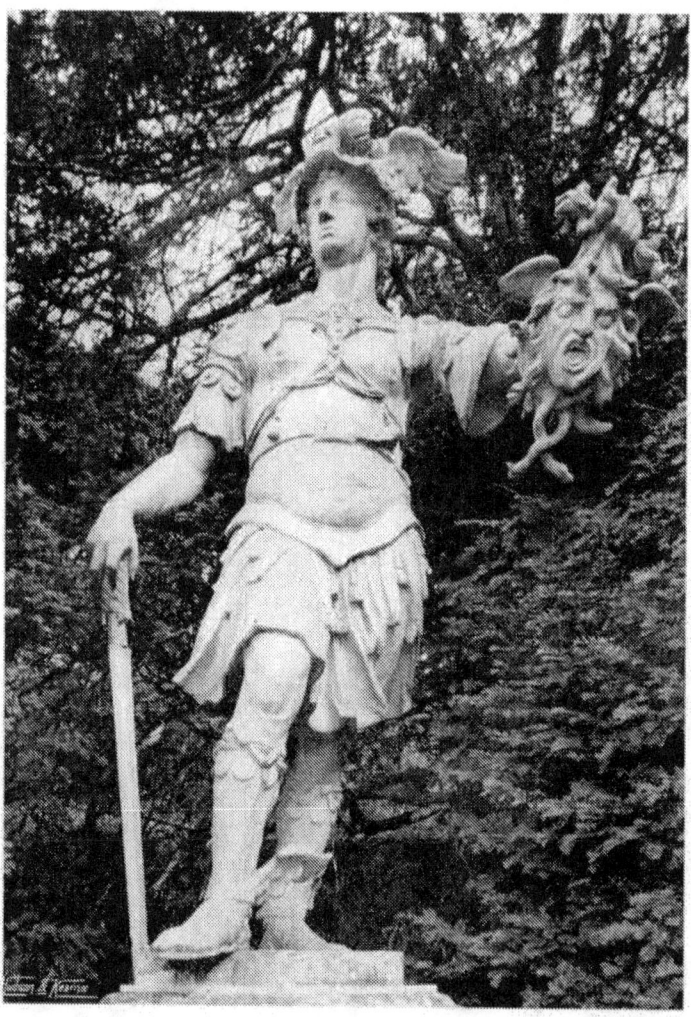

FIG. 260.—Perseus, Melbourne.

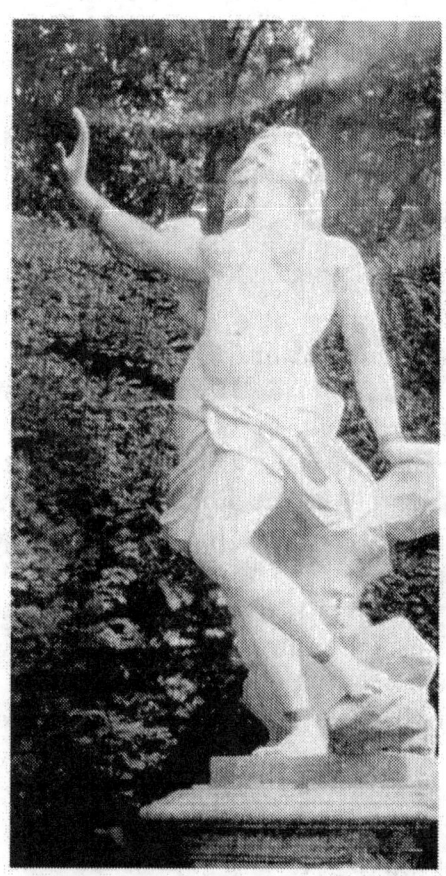

FIG. 261.—Andromeda, Melbourne.

in orthodox fashion, and the pose of persecuted
maidenhood waiting and crying for deliverance is
tolerably convincing. The accounts show them as
costing £25 for *Perseus*, and £20 for *Andromeda*.

These two, after all is said, are merely classic
personages as the eighteenth century understood
them. They are ornamental, and give a pleasant
academic flavour to a garden which is reminiscent
of courtly manners and a sedate, if not very
intelligent, affection for the arts of life.

When we turn to Fig. 262 we have a figure
which we recognise as properly a bronze figure.
There is another at Holme Lacy, and its photograph forms the frontispiece. Giovanni de
Bologna was a prime favourite with the lead founders of Piccadilly. As he was a

Fleming, from Douai, despite his Italian name, the Dutchman van Nost, who copied his figures, would doubtless be drawn to his work as that of a fellow Low Countryman.

Not only is there this *Flying Mercury* at Melbourne, but the *Rape of the Sabines* in lead at Painshill, Surrey (the original is in marble in the Loggia dei Lanzi). The *Cain and Abel* which used to stand in Brasenose Quadrangle was after Bologna's *Samson slaying a Philistine.* It was set up in 1827, and removed and destroyed in 1881. The original was presented to Charles I. at Madrid, and is now in the possession of Sir William Worsley at Hovingham Hall. Other replicas of this remain at Wimpole, at Harrowden Hall, at Chiswick House, and at Drayton House, Northamptonshire.

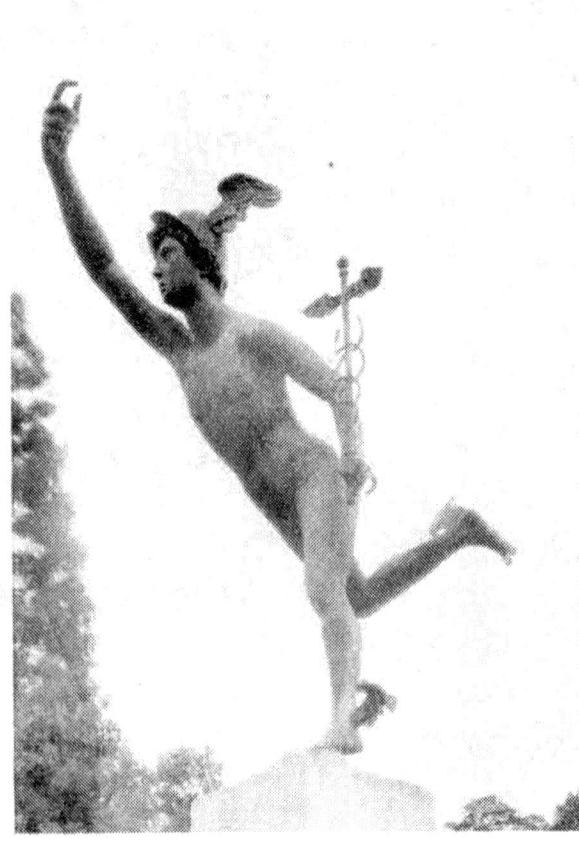

Fig. 263 shows the *Samson* at Harrowden Hall, and Fig. 264 another pair in the same gardens. Originally there were four groups, but one pedestal now stands empty. The *Wrestlers* of Fig. 264 are after the same original as those at Studley, illustrated in Fig. 303, but with enough small differences to make it possible that they came from different lead yards. The *Samson* at Drayton Park was cast by Peter Scheemakers (1691-1769), an important sculptor, from whom Sir Henry Cheere learnt his business. The other *Samsons* doubtless came from him, and he must have been the modeller of many other of the statues now illustrated, but the building accounts of the great English houses need to be examined before attributions can be made with any certainty.

There was a lead *Mercury* at Christ Church, Oxford; but, by a curious conjunction of metals, the head was of bronze, and is now preserved in the library. The late Mr Vere Bayne rescued the head from a stonemason's yard. The figure was presented about 1695 by Canon Radcliffe, and removed from the fountain (it is said during a "rag") some seventy years ago.

FIG. 262.—Bologna's Mercury at Melbourne.

The only excuse for the Melbourne *Mercury* being in lead, apart from its cheapness (for it and a figure of "Syca," now disappeared, cost only £50 the pair), is the exquisite patina which lead takes on when it weathers. This is a charm peculiar to leadwork, and it is of a simple graciousness which makes the figures harmonise with the domestic dignity of English formal gardens in a way that stone never does.

There are comparatively few large groups in lead, but four at Wrest Park make an imposing series. The subjects are not altogether clear, but that of Fig. 269 may safely be described as *Æneas Rescuing Anchises,* of Fig. 267 as another tableau from the story of Troy, and of Fig. 266 as the *Rape of the Sabines*. The last is markedly less heroic in treatment than Giovanni de Bologna's work. The four groups stand well in front of the early nineteenth-century house, which replaced, but on higher ground, the original building, and help to realise the description which has been given to Wrest Park of a "miniature Versailles." They certainly accord better with the spirit of English gardens than the chilly white marble figures which have been added of late years. One group is illustrated as it stands on its pedestal to show the general setting, though at the expense of the figures appearing to a smaller scale (Fig. 269).

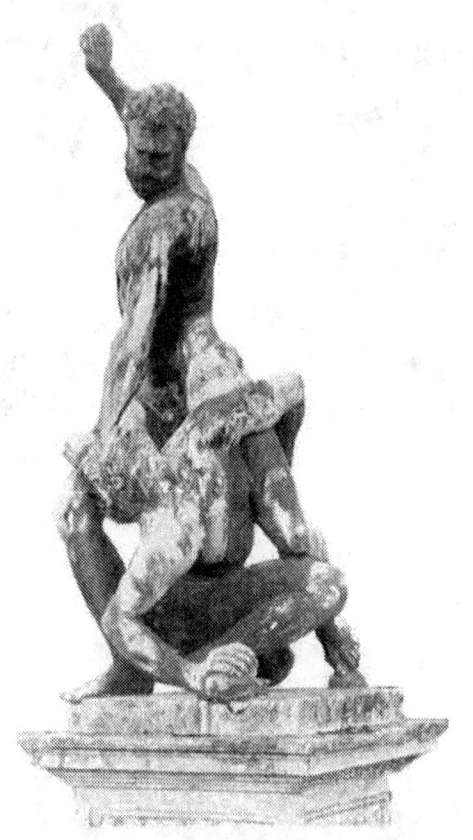

FIG. 263.—Samson Slaying the Philistine, Harrowden Hall.

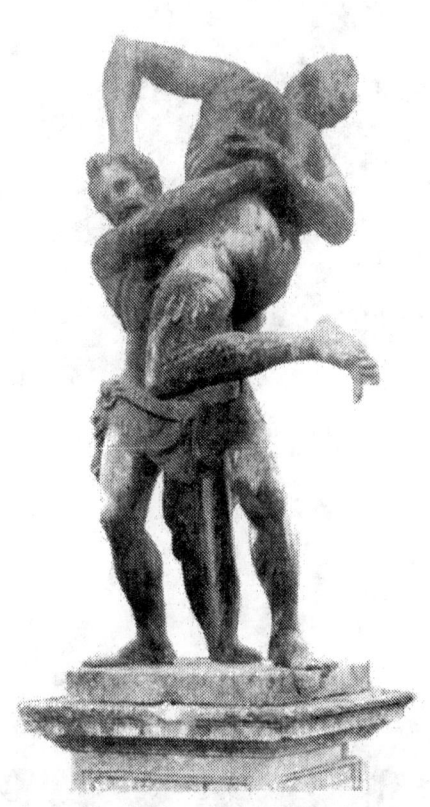

FIG. 264.—The Wrestlers, Harrowden Hall.

The gardens at Temple Dinsley have some agreeable little boys in lead, but the best figure is *Old Time* (Fig. 268). The scythe is not of lead.

The sky-line of Wren's Hampton Court has been altered not a little by the loss of four colossal lead figures which once adorned the south front. Many years ago they were taken down and deported to Windsor. Two were brought back and now stand behind the railings on the south front, but are deceptively painted brown, and look more like terra-cotta than lead. One is a *Roman Soldier*, the other a *Hercules*.

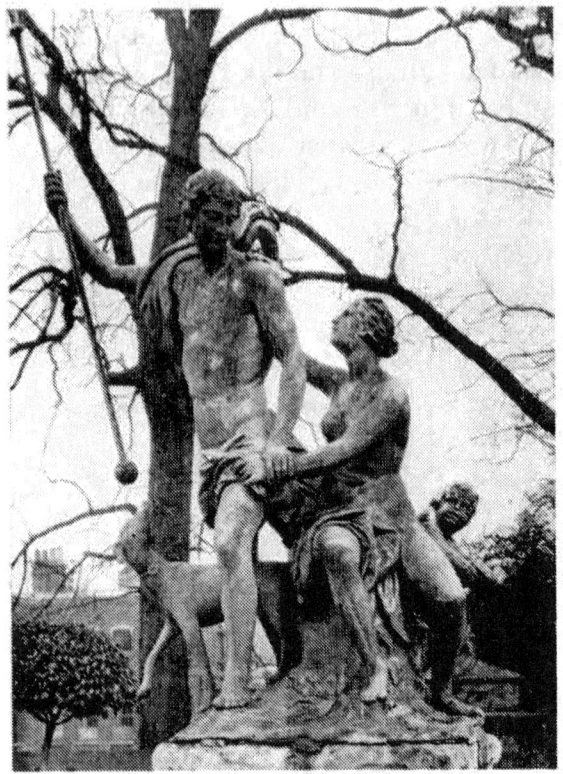

FIG. 265.—At Wrest Park.

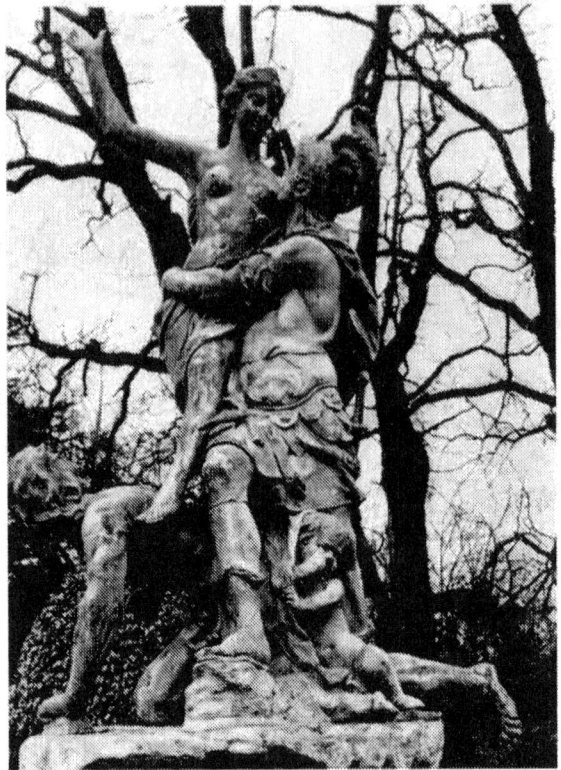

FIG. 266.—Rape of the Sabines (?), Wrest Park.

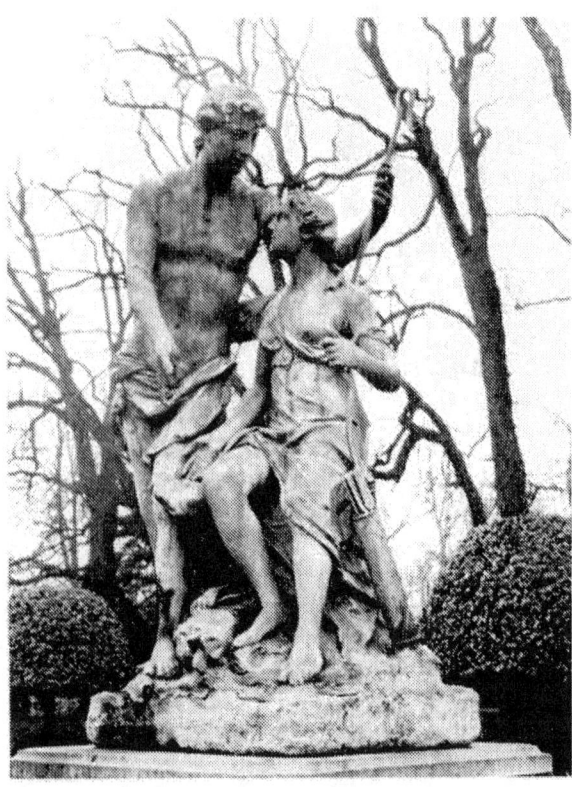

FIG. 267.—At Wrest Park.

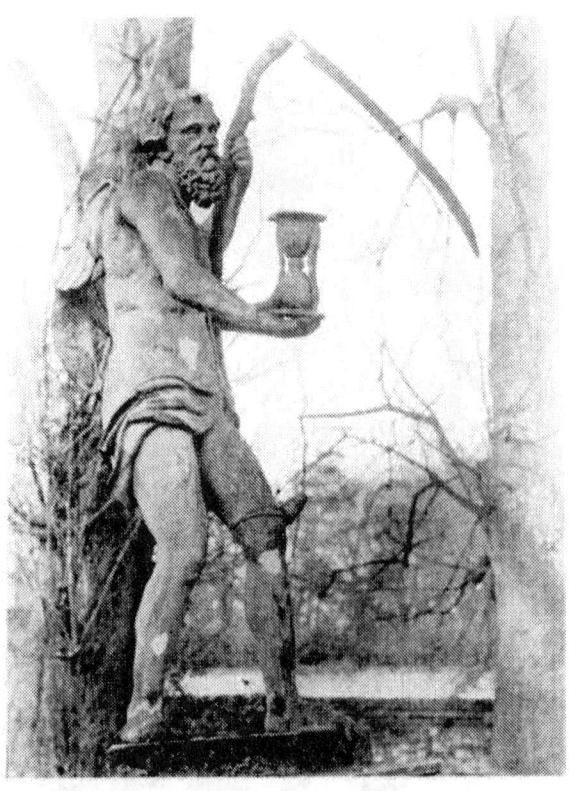

FIG. 268.—Father Time at Temple Dinsley.

John Thomas Smith, when referring to the "despicable manufactory" of lead figures, says "they consisted of Punch, Harlequin, Columbine, and other pantomimical characters, mowers whetting their scythes (Fig. 278), gamekeepers shooting (Fig. 291), and Roman soldiers with firelocks ; but, above all, an African kneeling with a sundial upon his head found the most extensive sale." The African we know well, and two others, to the illustrations of which references are given above. The author has not met Harlequins, but there is a memorial of their presence in the name of some semicircular arbours at Wrest Park once called "My Lady's Alcoves" and also the "Harlequin's Half-houses." The latter odd title they got from once having sheltered leaden Harlequins, but unhappily the figures have disappeared.

J. T. Smith calls the products of the Piccadilly yards, "these imaginations in lead," and mentions Dickenson as a maker as well as van Nost, Cheere, Carpenter, and Manning. Of the productions of the four last we have traced examples, but so far Dickenson has eluded search.

From the fact that the *Cupid making his Bow* at Wilton (Fig. 271) is cast from the same pattern as one at Melbourne, it is reasonable to assume that the Wilton lead-work came from the yard of van Nost or his successors. The right hand boy of the pair in Fig. 270 has so benevolent a forehead that he looks unduly elderly, and his brother with the bowl-shaped hat is a little half-hearted in his gesture. The Wilton *amorini* alternate with delightful lead vases (illustrated in a later chapter) round the formal garden. The most important leadwork at Wilton is, however, the equestrian statue of *Marcus Aurelius* on the arch designed by Chambers. It is very similar in general character to the *William III.* at Petersfield.

On the front of the house which looks towards the river and the Palladian Bridge,

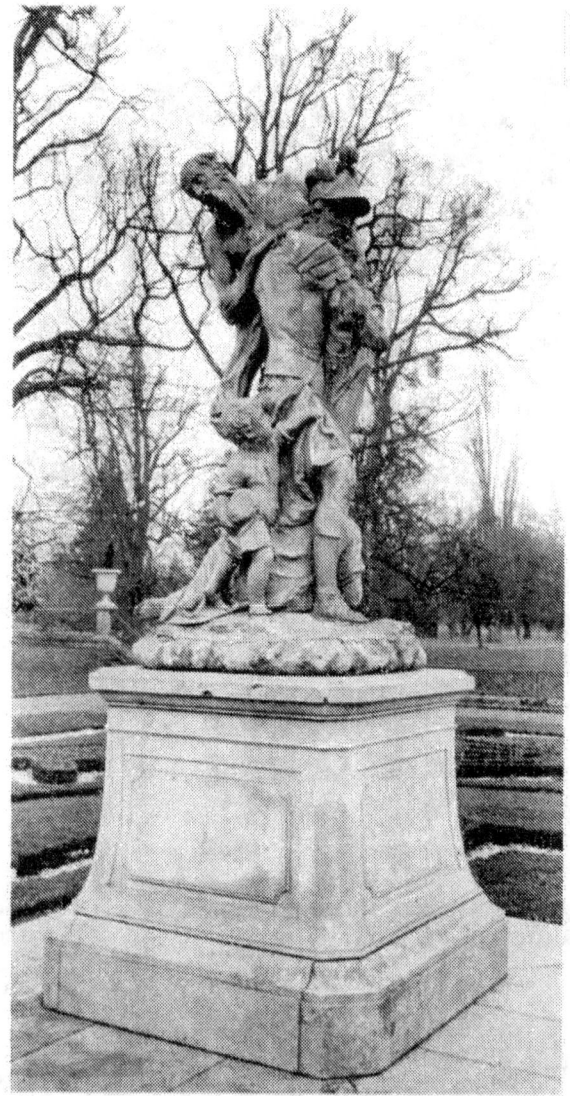

Fig. 269.—Æneas and Anchises, Wrest Park.

and sitting high on the parapet, is a lead figure of a woman (Fig. 272), which was certainly added well after the time of Inigo Jones, and is frankly a somewhat disturbing element.

Of Charpentière (or Carpenter), who died in 1737, being then over sixty, we have rather more information than of John van Nost.

He had been his assistant before setting up in business for himself. He supplied in

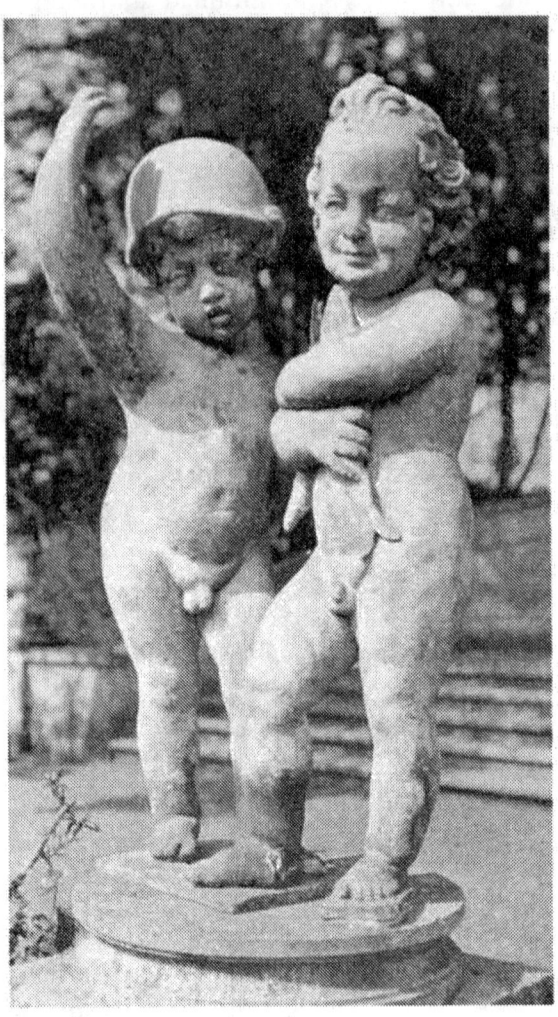

FIG. 270.—Wilton.

1722 and 1723 to Ditchley, Oxfordshire, the seat of Viscount Dillon, the lead figures of *Fame* (Fig. 273) and a *Roman Soldier*, which stand on the parapet. The bills for them amounted to £35 and £20, and the figures are 7 feet 3 inches high. *Fame* is trumpeting lustily, and has a spare instrument in her left hand for emergencies. The *Roman Soldier* might easily have been deadly. His uplifted arm became loose, and was recently for safety's sake removed and replaced by a wooden arm. As the lead arm weighed 40 lbs. the precaution was wise. *Fame* seems to have been a favourite subject with Georgian statuaries, for the *Fame* in the gardens at Nun Monkton is a

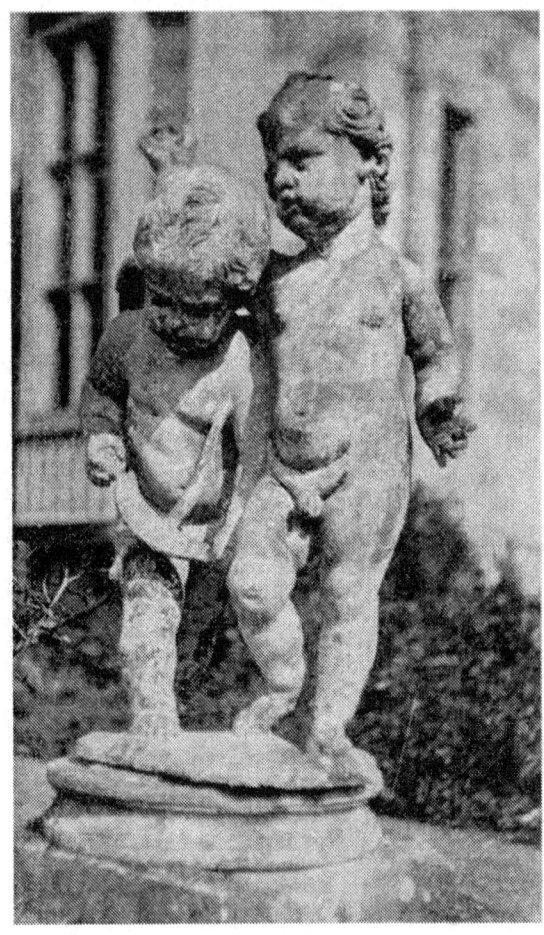

FIG. 271.—Wilton.

different figure, and the late Mr F. Warre had a small *Fame* 3 feet 2 inches high.

In 1702 Carpenter must have been well known, for we find Thoresby writing in his diary: "Sat up too late with a parcel of artists . . . Mr Carpenter, the statuary, and Mr Etty, the painter, with whose father, Mr Etty, sen., the architect, the most celebrated Grinling Gibbons wrought at York."

In 1714 (11th May) Thoresby again "walked to Piccadilly to Mr Carpenter's, the carver's," and saw "curious workmanship of his in marble and lead."

Walpole tells us that Carpenter was much employed by the Duke of Chandos at Canons, and apparently shared the Duke's work with his old chief, for van Nost certainly did the statue of George I.

The presence of *Fames* and *Roman Soldiers*, though not from the same models, both at Ditchley and at Nun Monkton, makes it appropriate to illustrate the latter figures at this point, though nothing is known of their origin.

The Nun Monkton collection of figures is particularly fine, and is of especial interest as nearly all the types of eighteenth-century garden sculpture are represented. In addition to *Fame*, plump and trumpeting, already mentioned (Fig. 274), there is a graceful young woman masquerading as a soldier (Fig. 276), and affecting a most unmilitary pose. There is also a real male Roman soldier.

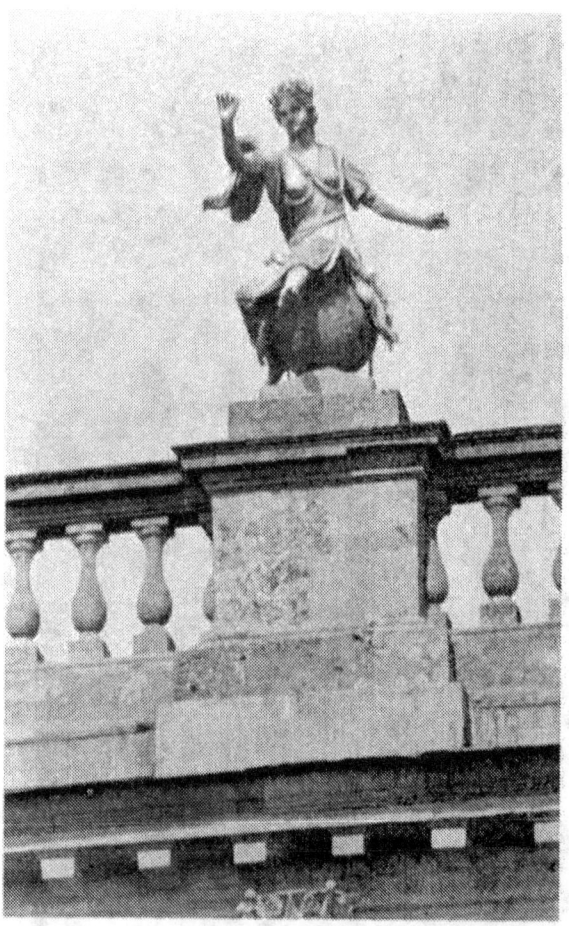

FIG. 272.—On the Parapet, Wilton.

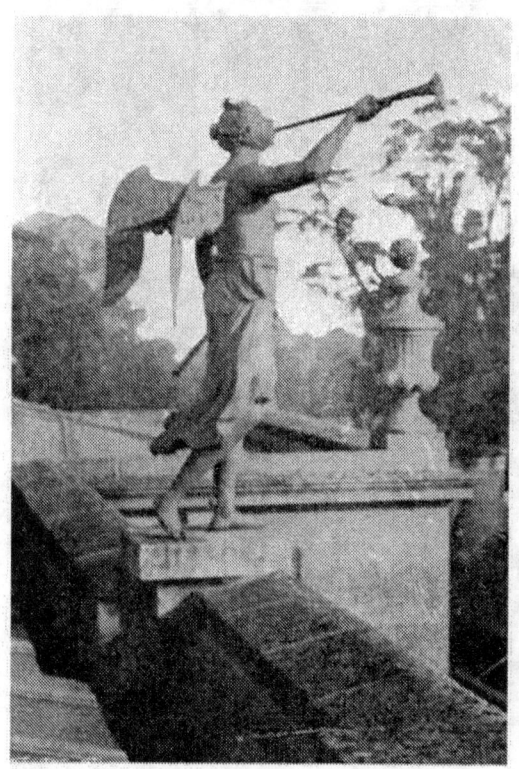

FIG. 273.—*Fame*, Ditchley.

Another figure is a rustic maiden (Fig. 275) regarding some fruit with a languid air, and, best of all, a really vigorous gentleman of buccaneering aspect (Fig. 277) pledging the garden world with the contents of his little barrel. He is rather Dutch than English, which is hardly astonishing when it is remembered how many sculptors from the Low Countries settled in England.

At Bicton, Budleigh, are four figures of the same character as those at Nun Monkton. There is a girl very like the rustic lady of Fig. 275, but cast from a different model, a vigorous figure of a *Mower* (Fig. 278), the pretty shepherdess of Fig. 279, and an elegant young man in knee breeches, most elegantly laying his hand on his heart, doubtless for the

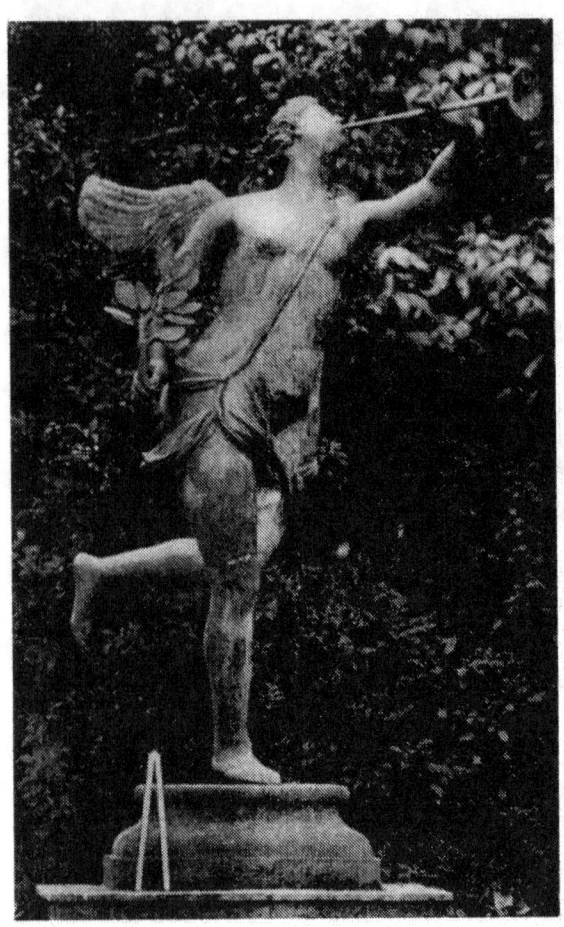

FIG. 274.—*Fame*, Nun Monkton.

benefit of the shepherdess. At the Bridge House, Weybridge, are a *Cymbal Player* and an *Apollo*.

The statues of Nun Monkton stand on both sides of a shady walk, and look altogether charming. The right placing of figures in a garden is their justification.

In the *Annual Register* of 1764 William Shenstone, the poet, unburdened himself of some "Unconnected Thoughts on Gardening," which are marked by excellent sense. These thoughts are reprinted in Volume II. of his works published in 1777. They were doubtless the outcome of musings in his garden at Leasowes.

For lead statues the poet pleads with judgment, and, amongst much that is delight-

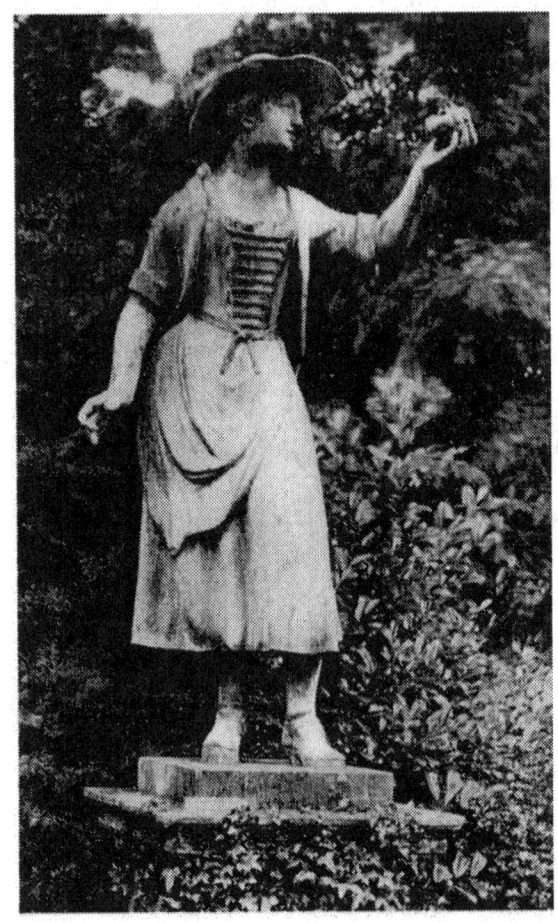

FIG. 275.—At Nun Monkton.

ful, writes: "By the way, I wonder that lead statues are not more in vogue in our modern gardens. Though they may not express the finer lines of an human body, yet they seem perfectly well calculated, on account of their duration, to embellish landskips [*sic*], were they some degrees inferior to what we generally behold. A statue in a room challenges examination, and is to be examined critically as a statue. A statue in a garden is to be considered as one part of a scene or landskip; the minuter touches are no more essential to it than a good landskip painter would esteem them were he to represent a statue in his picture." This excellent good sense is the more notable when it is borne in mind that by 1764 lead garden statues had fallen into some disrepute, and the palmy days of the Piccadilly lead founders had gone for ever.

Of the making of lead statues a word may here be added. All the English examples seem to have been cast. For cast figures one of two methods would be employed : for figures of which one only was wanted, the lost-wax process ; for stock patterns like the *Kneeling Slaves*, a set of casting patterns and core stocks. There are no modern methods of making a lead statue to supplant the old. When one turns to bronze and copper, there is the elasticity of electrotyping in copper as an alternative to casting in bronze.

It is not perhaps generally known that some large statues which appear to be bronze are, in fact, built up from thin copper electrotypes on an iron skeleton framing. This is analogous to the building up of lead figures from hammered sheet lead. This method was employed in mediæval France. The lead was beaten out on a model of carved wood, and the edges of the adjacent pieces either soldered or lapped. An internal framing of a main rod with struts ensured rigidity. For such figures as angels with wings outstretched, the repoussé method is obviously the best, as it makes for a convincing lightness of appearance, while strength need not be sacrificed. In England it never found favour. Nor is the omission confined to statues. On pipe-heads repoussé work was but slightly employed. The beating-up of patterns in relief seems to have been avoided, except on some of the

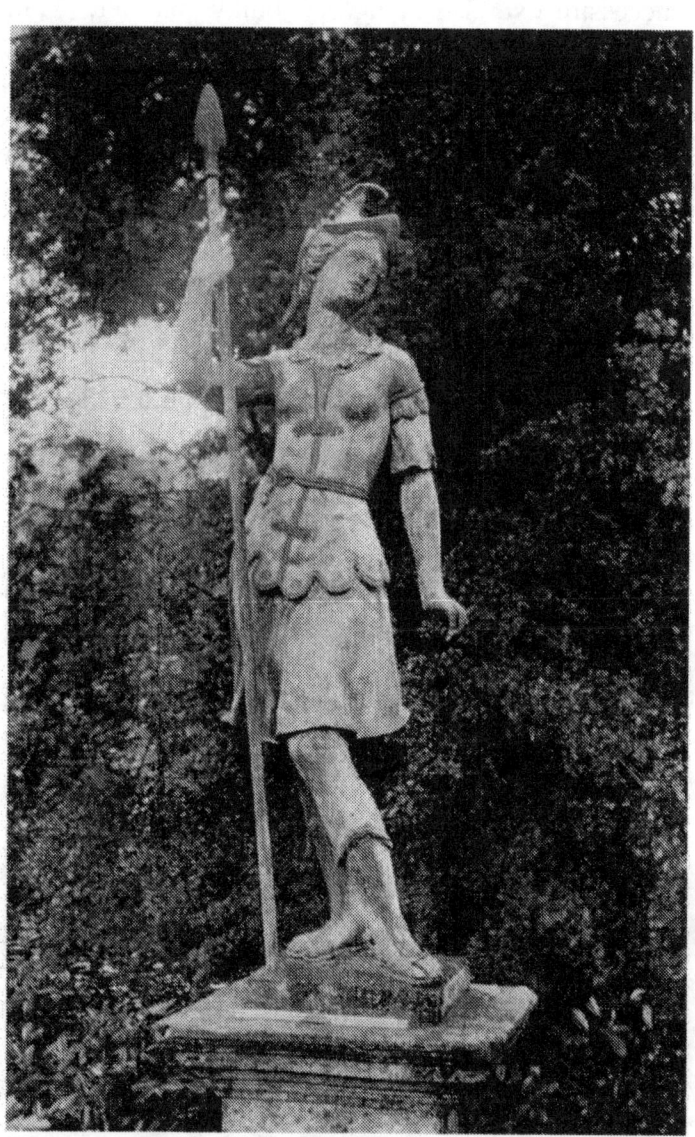

FIG. 276.—The Military Girl, Nun Monkton.

eighteenth-century vases where the type of decoration often called particularly for repoussé work.

Giacomo Leoni, an architect imported by Lord Burlington (and employed as the "ghost" of that ingenious nobleman), showed some forty statues on the elevations of the palace which he designed for Thomas Scawen at Carshalton Park, but, perhaps fortunately, never built.   It is evident that one of the figures was to have been the same *Gladiator* that we find at Burton Agnes (Fig. 280).   The entrance gates and a little bridge are the only features of this pretentious scheme that ever took shape.   As the two statues on the stone piers that flank the gates are of lead, it is not unreasonable to suppose that the other forty would have been of the same material.   One may regret the lead statues, but the house was best unbuilt, as it was a ponderous and not very successful exercise in a very bulky manner.   The two statues on the gate piers are of

Diana (Fig. 281) and *Actæon*, and give an added interest to a range of admirable wrought ironwork. The carving of the very fine stone piers has been attributed to Catalini, and the statues to van Nost.

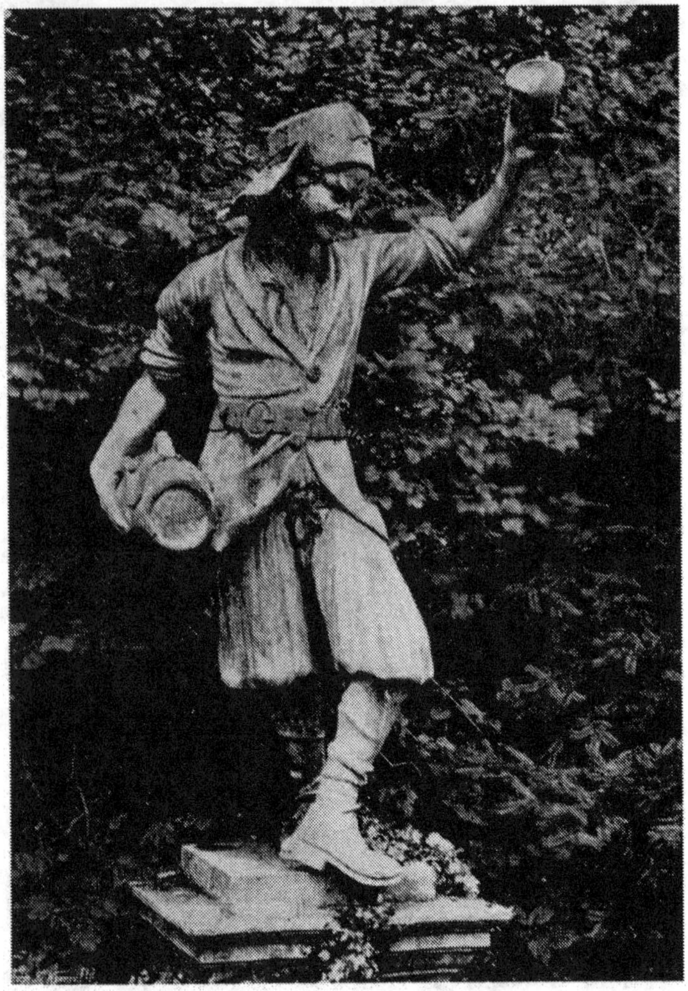

There is perhaps no more delightful use of lead figures than in the middle world where garden craft and architecture meet, the entrance of a great park.

The groups of three charming boys upholding trophies of fruits give its name to the *Flower-Pot Gate* at Hampton Court (Fig. 282), and are perhaps the most completely successful terminals ever devised for gate piers. This gate was part of the improvements carried out by London and Wise about 1700, under the supervision of William III. himself. Probably of the same period are the *Lion and Unicorn* and *Trophies of Arms* in lead that crown the piers at the main entrance which leads to the Wolsey part of the Palace (Figs. 283 and 284). These gates, and their ornaments, appear, though very minutely, in Kip's view, which was published between

FIG. 277.—Buccaneer, Nun Monkton.

1706 and 1710.   The piers and trophies were there in 1700, but the shield of arms supported by the royal beast is that of George II., and was perhaps substituted for an earlier shield of William III.

The Hampton Court lion is a more convincing beast than the pair of lead lions at the Bar Gate, Southampton, who are sitting up in a rather comic pose (Fig. 286).   They are a pleasant example of the strange efforts of the eighteenth century to devise new Gothic trimmings for old buildings.

The great lead lion, weighing three tons, which once stood on the summit of the street front of Northumberland House, at Charing Cross, now occupies a similar position at Syon House, whither it was removed by the sixth Duke of Northumberland in 1874. The lion is after a model by Michael Angelo, and stands on a *Chapeau d'honneur*. Redgrave says that it was modelled by Laurent Delvaux, an assistant of Bird and a partner of Scheemakers, but erroneously states that it is of bronze. It has also been attributed to Thomas Carter of Knightsbridge.

At Syon House there was also a statue of Flora, about double life size. It unfortunately fell with fatal results. The lead was only about three-sixteenths of an inch thick (a significant commentary on the economic tendencies of the eighteenth-century lead yards), and the statue was filled with brick rubbish, &c., held together by cement. The bust, however, survives, also an arm and hand holding a wreath. The arm was strengthened by an iron bar, and the wreath is covered with repoussé leaves. The figure had not been painted, and what remains bears patches of silvery patina.

The smaller lead lion at Syon is the one that Robert Adam set up on the Lace Gateway (Fig. 285). The best feature of the gateway is, however, the pair of *Sphinxes* (Fig. 288). They are admirably modelled. The *Stags* (Fig. 287) at Albert Gate, are also in lead, and have this in common with the Syon sphinxes, that they came from an Adam building, the Ranger's Lodge in the Green Park, which was built in 1768. Syon was in Adam's hands in 1761-62. It is possible that John Cheere was the maker of these, for the "despicable manufactory" (as J. T. Smith calls it) of lead figures was rather on the wane by 1768, and some of the lead yards were closing. In 1778 he made the lead sphinxes which are high up on the back of the Strand front of Somerset House, and got £31 each for them. They are markedly inferior to the Syon sphinxes, as are those on the gate piers of

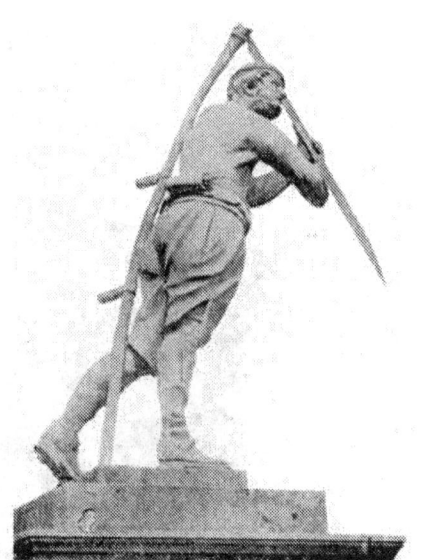

FIG. 278.—Mower, Bicton.

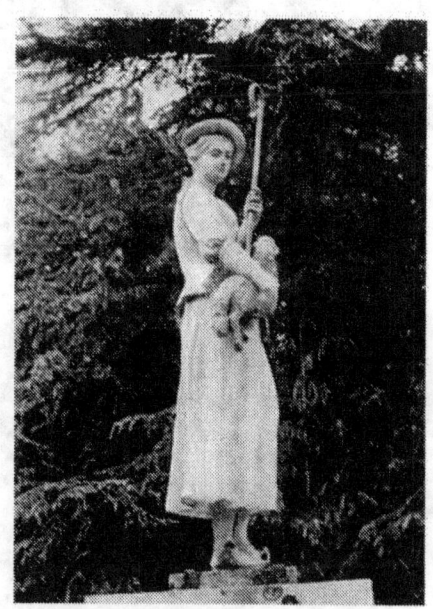

FIG. 279.—Shepherdess, Bicton.

Devonshire House, Piccadilly, which came from the Burlington Villa at Chiswick, where there remains another pair of replicas, of which one is in stone. At Chiswick there is a stone goat signed *Rysbrack*, but it is hardly possible that this able sculptor can have done the very poor Devonshire House sphinxes. The Chiswick villa was built in 1729, and as we again meet the same sphinx (Fig. 298) in lead at Castle Hill, it seems reasonable to

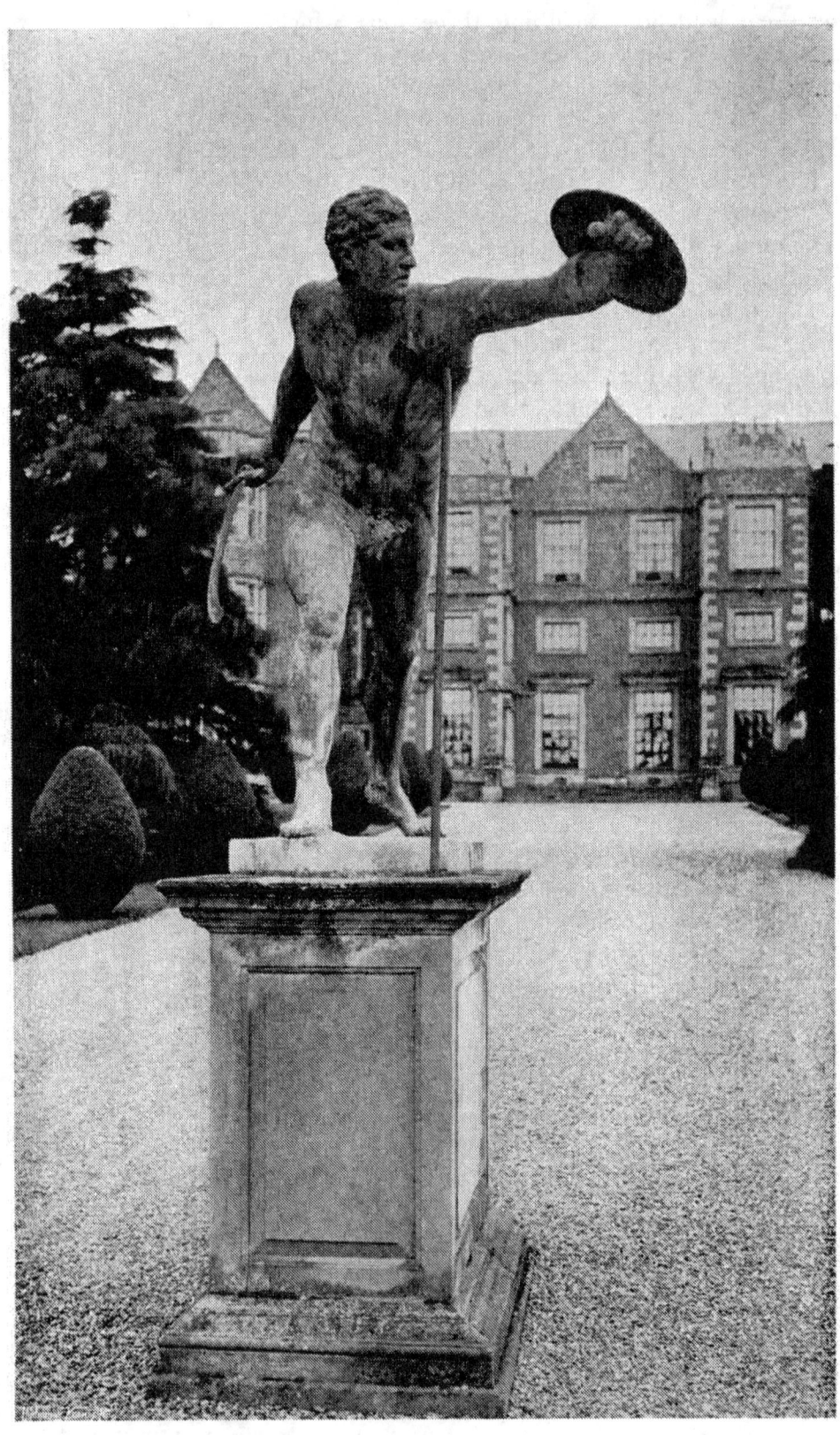

FIG. 280.—GLADIATOR, BURTON AGNES, YORKSHIRE.

suppose either that all the Castle Hill figures are
of early in the eighteenth century, or, if they were
set up when Chambers was working there in
1770, the *Sphinx* which Kent used at Chiswick,
and the *Cymbal Player* which he used at Rousham, were popular over a period of forty years.

Amongst the many figures at Castle Hill,
Devonshire, the residence of the Earl of For-
tescue, there is a bust of Pan (Fig. 293) of quite
extraordinary interest. It stands on a stone pillar
which slopes down to its base, and against a
background of trees is a very incarnation of the
woods. Grapes are in his hair, and above his

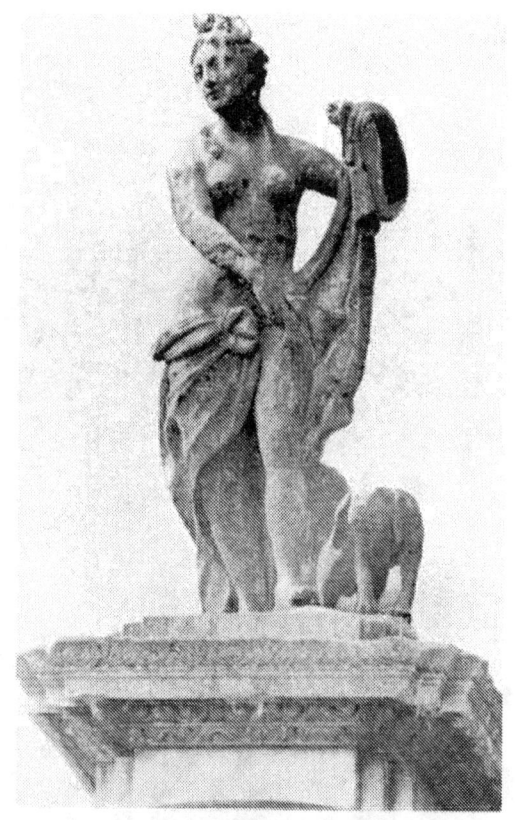

FIG. 281.—Diana, Carshalton.

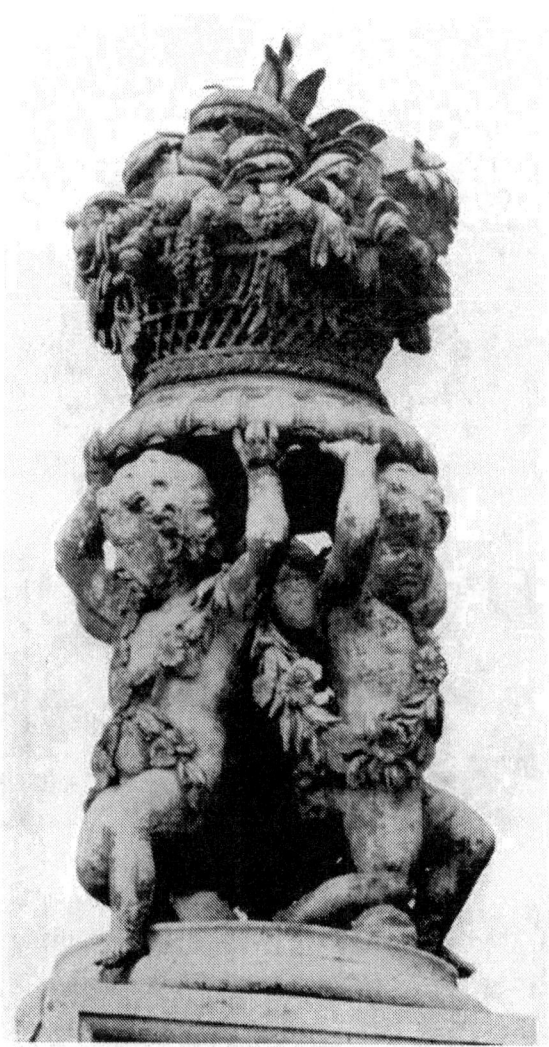

FIG. 282.—On the Flower-Pot Gate, Hampton Court.

wicked ears the horns are seen. His appearance in the wood would scarcely bring panic
fear to the wayfarer. He may not be benevolent, but he is not alarming, and there is
much subtlety in the look of smiling, quiet
lust on his lips. It is a hypnotising face,
libidinous and cynical, and one may well
hope that the authorship of this fine work
may later be established. It was a fantastic
wit that put him in the same garden with the
sphinx (Fig. 298). She is cold, unamusing,
and one is convinced, little friendly to the
bust of Pan; chastely glad, perhaps, that
the artist gave him no goat's feet to set him
dancing, as statues will of nights, in any
wisely peopled gardens. The sphinx has a
wonderful headdress; even Pan would take
no liberties with such severity.

The *Cymbal Player* is also at Rousham,
but the Castle Hill *Venus* (Fig. 290) is very

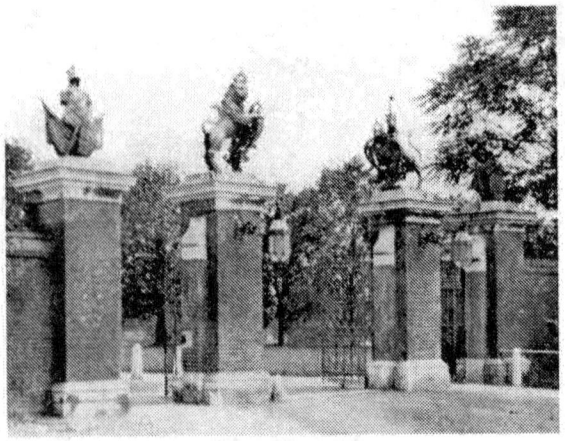

FIG. 283.—Entrance Gates, Hampton Court.

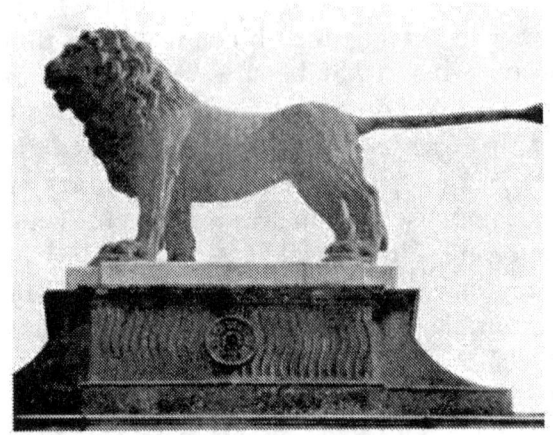

FIG. 285.—Lion Lace Gate. Syon House.

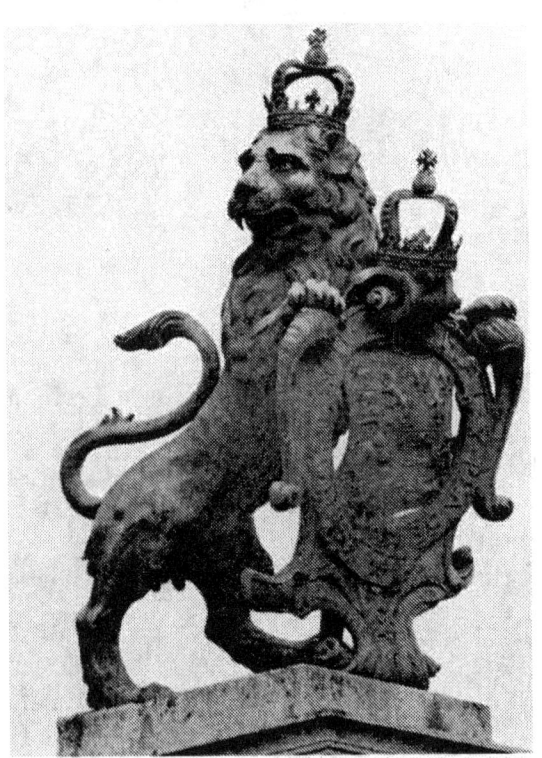

FIG. 284.—On Entrance Gate, Hampton Court.

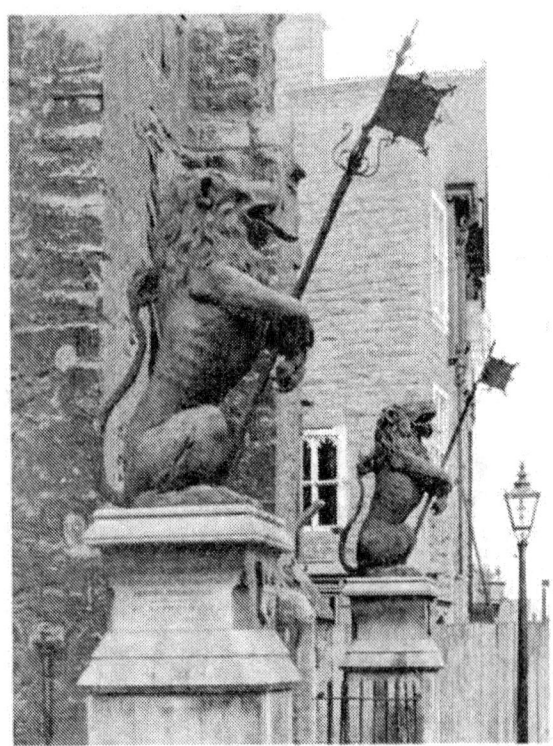

FIG. 286.—Lions at Southampton.

different from the Rousham *Venus*, which is the Medici figure.   The Castle Hill statue has the pose of a clumsy *ballerina*, and must be an altogether eighteenth-century product.   It is a good example of how bad art makes the nude naked.

Also at Castle Hill are to be found *A Lion, a lioness, and a greyhound* (Figs. 296, 297, 299).   The first is not remarkable, but the lioness has a powerful head, and is vigorously modelled. The greyhound is a quite convincing hound, and the artist has managed to give him the look of wistfulness which is so attractive in life.

There is a formality about these beasts lying on their stone pedestals which one does not always find in the lead fauna of gardens. Sometimes the base of the casting is let into the lawn. In one case of a *Fox* stealing away with a fowl, at Weald Hall, Brentwood, the figure ceases to have anything to do with art, and becomes an illusion in lead, a theatrical trick far removed from the spirit of the formal garden. Perhaps the most amusing example of this type is the lead *Cow* at Biel House, Haddingtonshire (Fig. 292). Could any landscape produce a more convincing cow? At Biel, too, is a lead *Gamekeeper* standing on the grass, and leaning forward to aim with a long fowling-piece (Fig. 291). Mr Hamilton Ogilvy also possesses at Winton Castle, in the same county, a *Kneeling Hercules* supporting a sundial on his head. It seems a plagiarism of the *Kneeling Slave*, and is far inferior in modelling. It was taken to Winton from Bloxham Hall, Lincolnshire.

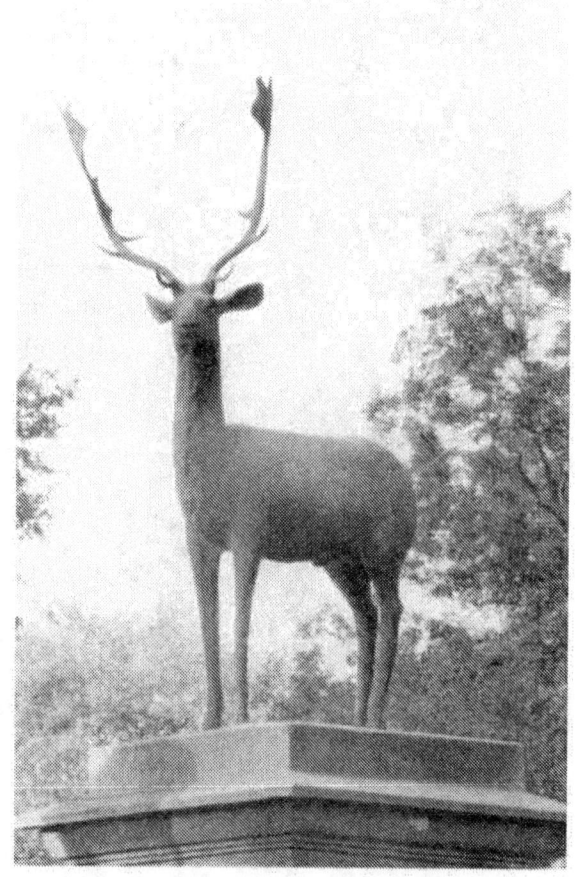

FIG. 287.—Stag on Gate Pier, Albert Gate.

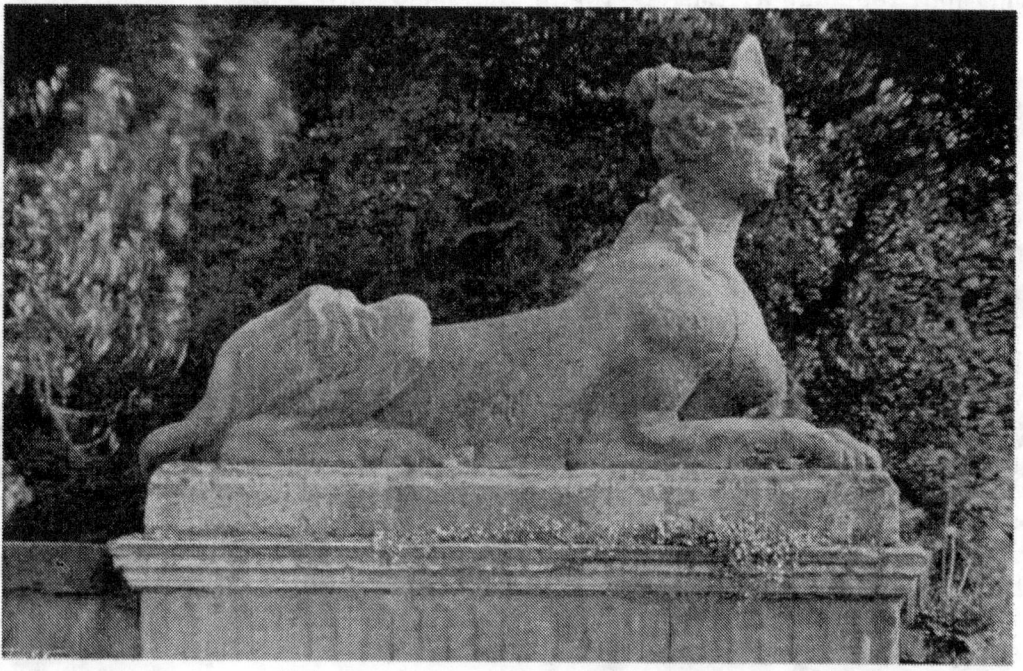

FIG. 288.—Sphynx, Lace Gate, Syon House.

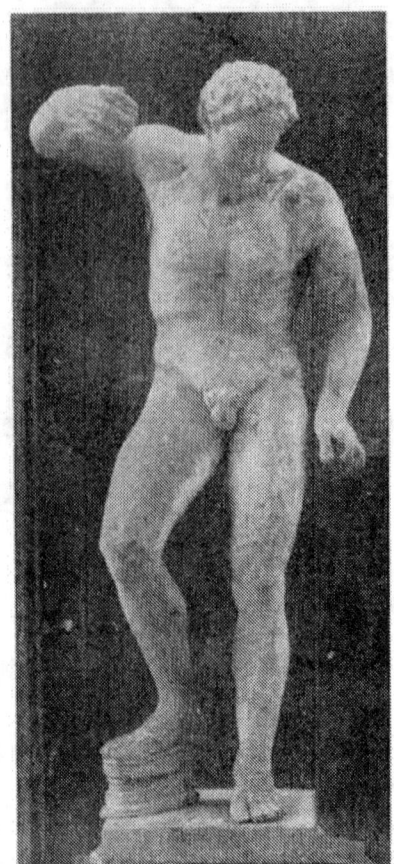

FIG. 289.—Cymbal Player at
Castle Hill.

The *Cow* at Biel is said to have come from Holland,
but it is more likely to be the work of a Dutchman
working in London, perhaps of van Nost.   The *Fox* at
Weald Hall has an appropriate neighbour in the same
Gamekeeper that we find at Biel.

Among the greater houses of England, Rousham,
near Oxford, is very little known, far less than it deserves, both for its gardens and pictures.   Kent took
the former in hand, and there is a good deal of garden
architecture in his solemn classical manner.   To him,
too, may probably be attributed the niches in that Gothic
manner of his, which Mr Reginald Blomfield justly calls
barbarous.   The niches are provided with lead statues,
and one is the Venus de Medici, a chilly monument.

Much more satisfactory, indeed altogether delightful, are the two *Cupids on Swans* (Fig. 294).   One is
unhappily much battered.   The other figures include a
Cymbal Player, as at Castle Hill, a *Boy with Dog*, as
at Studley, a *Flying
Mercury*, and five others
of a classic sort holding
rather dreary revels
beneath overshadowing
trees.

Purely architec-
tural in its use is the
bust of Fig. 295, which

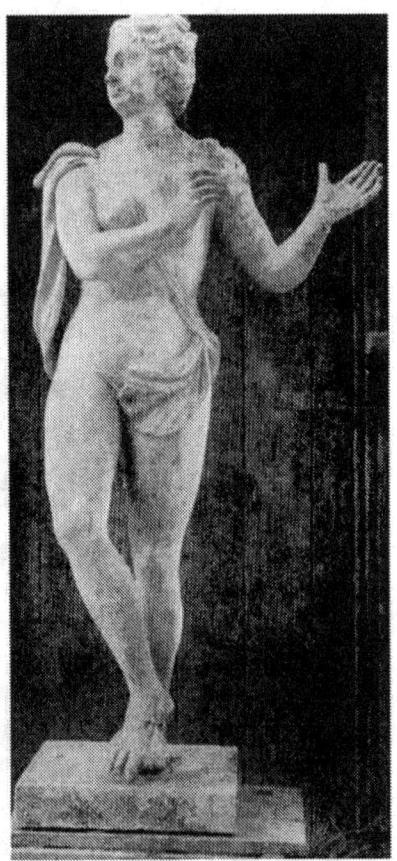

FIG. 290.—Venus at Castle Hill.

is built into the wall at Castle Hill in the same way
that the long series of classical busts is employed on the
front of Ham House, Petersham, which was built in
1610.   This fashion was set by Wolsey at Hampton
Court, but his Italian artists worked in terra-cotta.

At Castle Hill the designer of the gardens had
more than a fondness for leadwork.   It amounted almost
to obsession.   The seat illustrated as tailpiece to Chapter
X. is of lead, and of a riotous ugliness.   The swag has a
fat amorphous lonely look which is positively grotesque.
White marble seats in an English garden are inappropriate enough, for they grow green and have a cold and
dank look; but this lead object is an equally good example
of how not to make a garden seat.

In Fig. 300 the figure of *Paris* adjudging the apple
(South Kensington Museum) is shown as a good example of a type of statue which is not suitable for
reproduction in lead.   The original is in marble at the

Louvre, and was by Nicolas François Gillet (1709-1791). There is no record as to the date of this lead reproduction, but, judging from the terra-cotta pedestal on which it stands, it is probably of late in the eighteenth century. It is a little figure 2 feet 10 inches high, and the subject seems altogether too delicate for lead. If the original material (marble) were abandoned for metal, the smooth feeling of the figure seems to call for bronze ; lead has too much texture ; but whatever the material, the figure is graceful and charming.

Another Cupid is illustrated in Fig. 301. Life is more serious to him than to them of Melbourne. He carries a sundial, and has no time for archery. He differs markedly from the

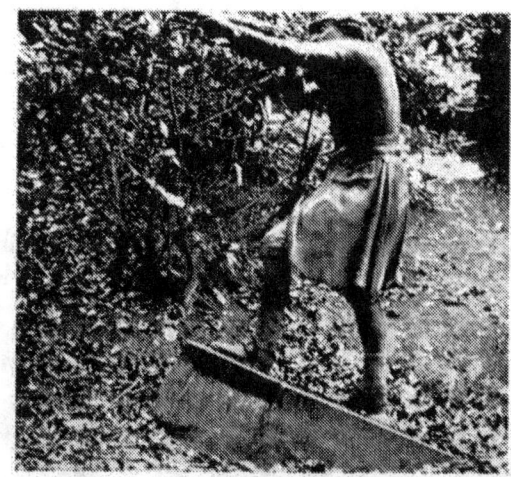

FIG. 291.—The Gamekeeper of Biel House.

Melbourne family in his wings, which are folded, but are large and practical for flying.

The modelling is poor, and one does not see why this figure has been more extensively chosen than any other for copying and sale as "antique." It crops up incessantly in sales of garden ornaments with such labels as "from an old garden near Bath." The last indignity was reached when it appeared among the weeping angels of white marble in a tombstone yard in the Euston Road. Poor Cupid, to have fallen among such dismal company !

The methods of the makers of "antique" lead figures and vases are not without interest. The great purpose is to achieve the silvery patina, which is so delightful a feature of the old work that has honestly weathered. The commonest method is as follows :—The lead figure is first heated and washed over with hydrochloric acid. It is then, while still hot, brushed with water and dried. The patina so obtained can, however, be rubbed off with the finger, and appears in the crevices, whereas true patina comes on the raised surfaces. This method is so quick that a statue has been cast, treated, and sold as an "antique" in one day. *Caveat emptor.*

Another method is more efficient and difficult to detect. The work is buried in wet lime long enough for the surface of the lead to be eaten away somewhat. After washing it is buried in old tea leaves or other wet herb stuffs that will give the brown tinge that is often found on the old work. A third trick is to paint the figure with a thin oil colour, and after with

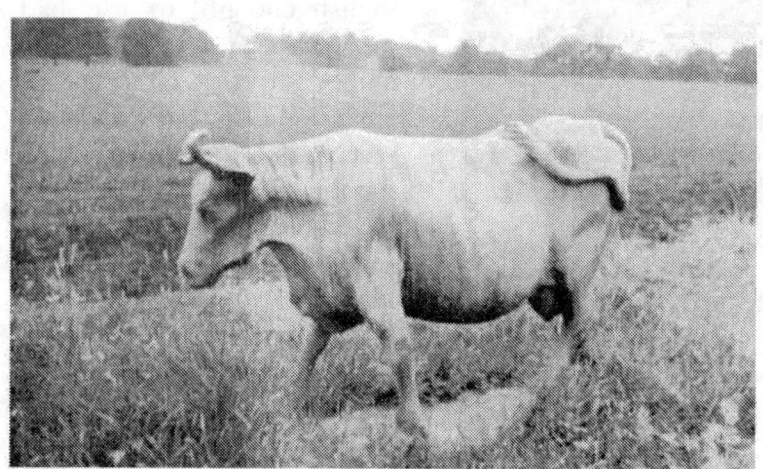

FIG. 292.—Lead Cow.

a solution of copperas. The lead is then scorched, painted again with one or more coats of dirty colour, and scraped and scratched. As most of the genuine work has at some time been painted, the deception is often more complete than attempted patina.

As to the casting itself, the cheapest method is to cast in sand, without the use of cores, the patterns being handled much as in the practice of brass casting. After pouring, the lead is allowed a few seconds to cool, and the casting frames tipped, which releases the molten lead through the pouring hole.

Sometimes a "chill" is made, for which castings can be turned out in dozens. For single copies the "lost-wax" process is used, clay sometimes being used instead of wax, and the mould is generally made in a mixture of plaster and sand. It is a melancholy fact that the *Kneeling Slave* has been reproduced, and in one case known to the author, the first casting method described above was employed.

Generally, however, the forgers of "antiques" are foolish enough to use bad models. A common example is a *Girl with a Rabbit*, but other worthless stucco futilities have been employed, and they ought to deceive neither the elect nor the comparatively ignorant.

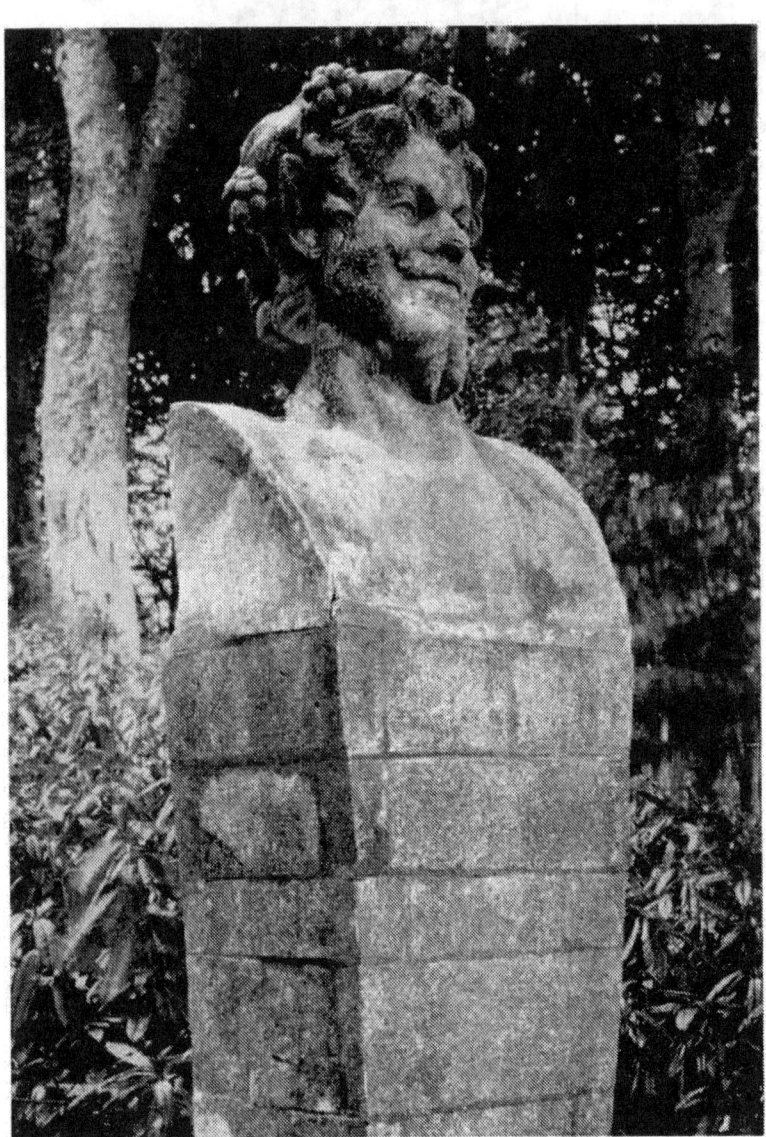

FIG. 293.—Pan, Castle Hill.

Reference will be made later to the *Neptune* at Studley Royal, the Yorkshire seat of the Marquess of Ripon. Close by the moon and half-moon ponds are several statues, all in the classic manner, and among them two pairs of *Wrestlers*, of which one is shown in Fig. 303. The other is the famous group at Florence, which has so important a place in the history of sculpture. Another is a *Faun with a Dog*. He carries a trophy of fruit, and is strongly stayed with iron bars, another example of a subject unsuitable for execution in lead. Away from the water and near the church is a *Pan*. He was making

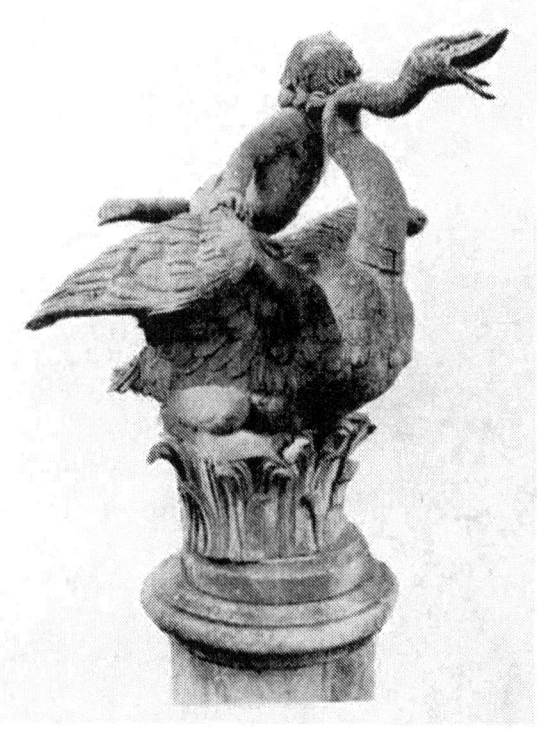

Fig. 294.—Cupid and Swan, Rousham.

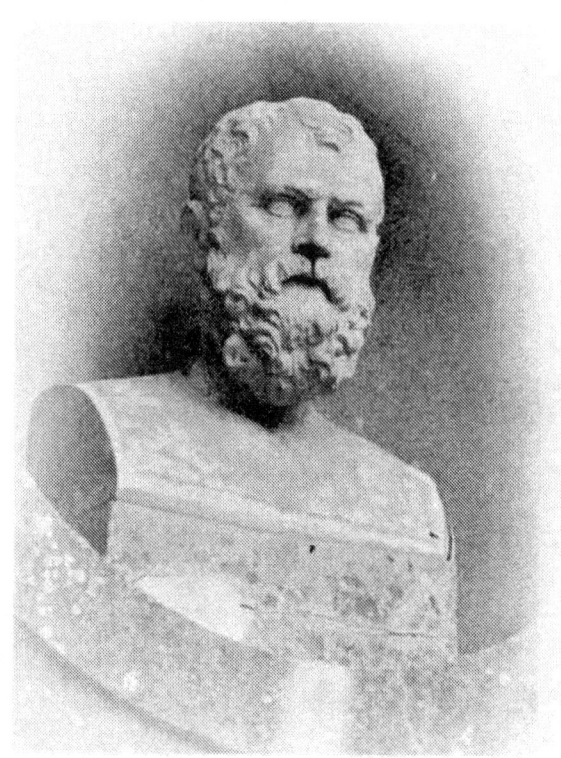

Fig. 295.—Bust at Castle Hill.

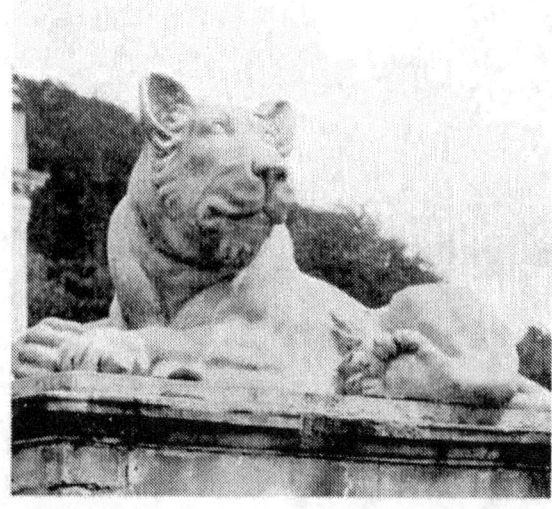

Figs. 296 and 297.—Lions at Castle Hill.

music, but his pipes have gone. One figure the author found lying battered in the brake; the lead was only three-sixteenths of an inch thick.

The modelling of these figures, which group so charmingly with the lake and woods, and with the stately Temple of Piety, is partly veiled by successive coats of paint which have cracked and give a false air of decay. Where the paint has gone the natural silvery patina is revealed, and one may hope that some day this unpleasant shroud may be altogether removed.

Studley Royal shows that lead figures are of particular value when used in connection with ornamental waters. The watery garden had a great impetus when Dutch artists and gardeners came to England in great numbers at the Revolution, and stimulated the Dutch note in English gardencraft. A typical Dutch lead *Triton* in the State Museum at Amsterdam is shown in Fig. 304. It was evidently at one time a point of freshness in a formal garden. With this before us it is easy to see the source of inspiration of many of the figures turned out by the Piccadilly lead founders.

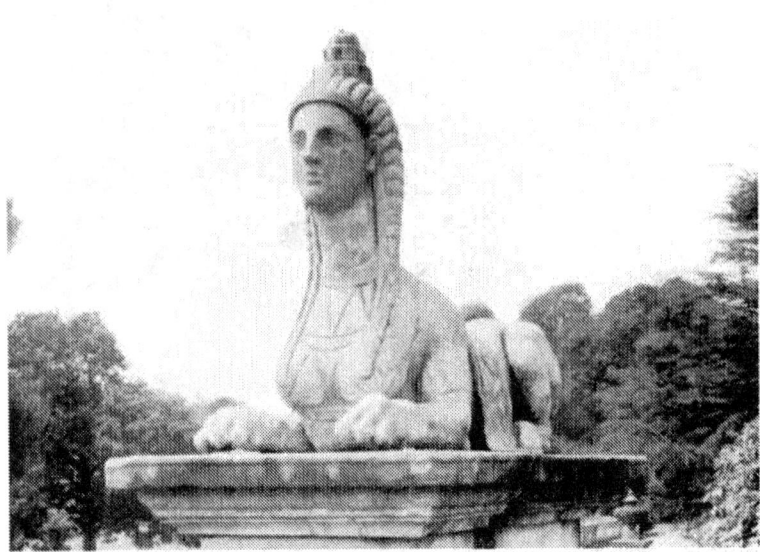

FIG. 298.—Sphinx, Castle Hill.

Pierre Husson's "*La Theorie et la Pratique du Jardinage,*" published at The Hague in 1711, leaves no doubt as to the Dutch attitude towards water in the garden. He tells us that "fountains and waters are the soul of gardens; they make their chief ornament and enliven and revive them. How often it is that a garden, beautiful though it be, will seem sad and dreary and lacking in one of its most gracious features, if it has no water."

Husson is all for "eaux jaillissantes, celles qui s'élevent en l'air au milieu des bassins,

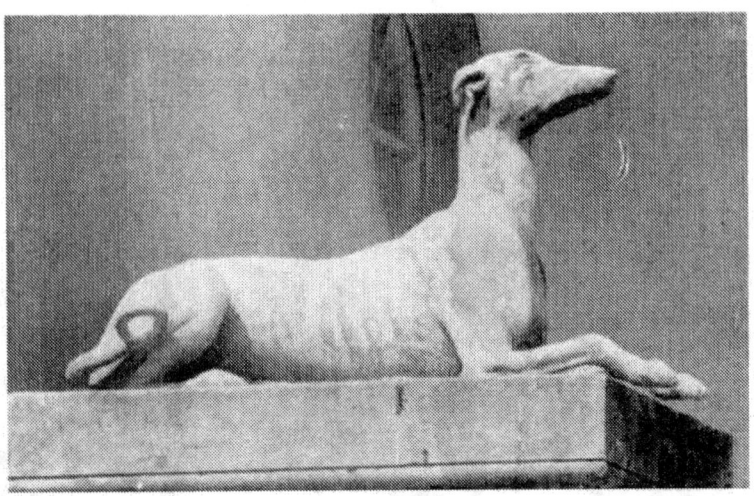

FIG. 299.—Greyhound, Castle Hill.

forment des jets, des gerbes, des bouillons d'eaux." He gives practical instructions for lining basins with lead, but warns his readers that folks are apt to steal the metal. For the figures which adorn the fountains he recommends marble, bronze, and lead gilt or bronzed. Bronzed lead is a puzzling suggestion.

When all is said of fountain statues, however, we must go back to Versailles, which, doubtless, exercised a greater influence on English and indeed all gardencraft

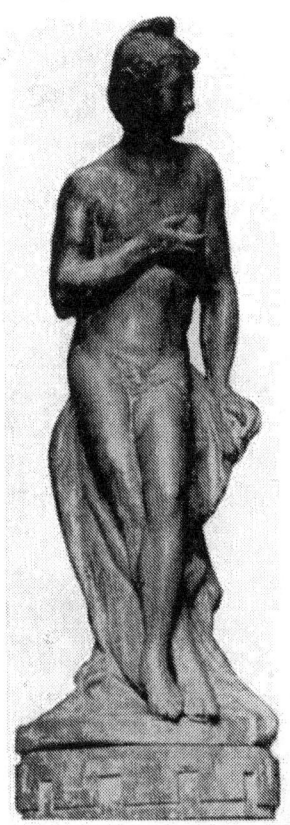

FIG. 300.—Paris, South Kensington Museum.

than all the Dutch gardeners to-gether. In those supreme gardens lead more than won its share of the honours, and chiefly in the water schemes. The *Neptune* at Studley rather shrinks when compared with Sigisbert Adams' group at Versailles in the *Neptune Fountain* (1740). In England there is nothing one can compare with the lazy grace of the *Tritons and Sirens* after Tubi and Le Hongre. Still less can one find anything like Girardon's "Fountain of the Pyramid" (1672).

In 1889 M. Toni Noel recon-stituted from old views the restless group of the *Fountain of Dragons*. This subject has not always been so violently treated. At La Granja, the *Dragons* fountain is a single composition, but among the twenty-five other fountains with which Philip V. of Spain beautified the Palace of San Ildefonso are many of the *Dragons* type at Versailles, notably

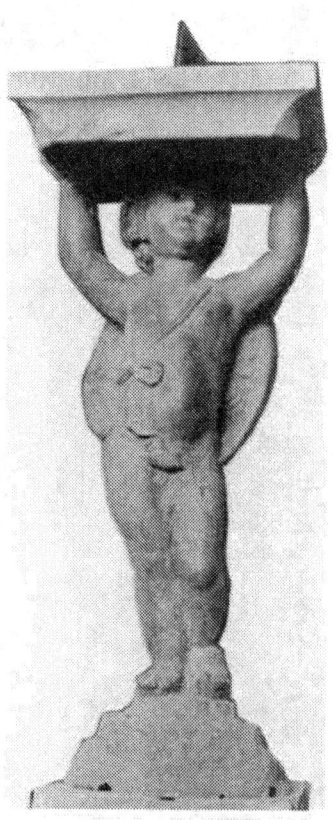

FIG. 301.—Cupid Sundial.

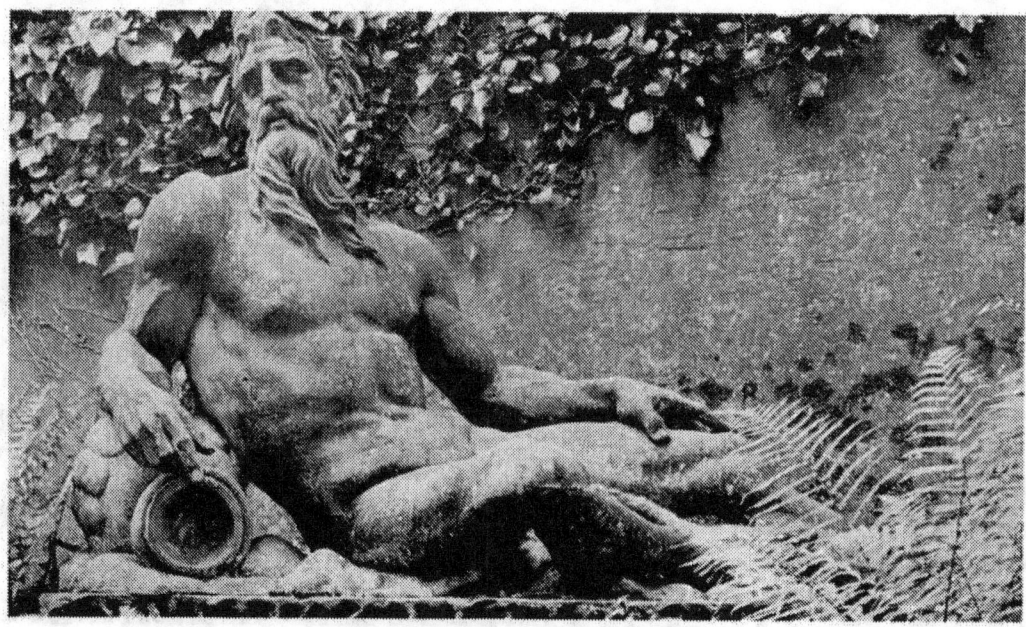

FIG. 302.—River God, Parham, Sussex.

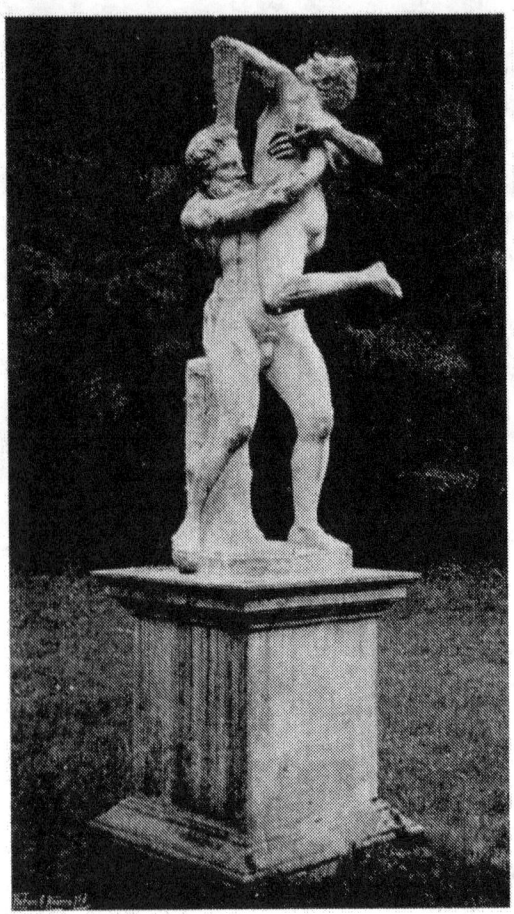

FIG. 303.—Wrestlers, Studley Park.

the Fountain of the *Horse Race*. Whether Philip followed Versailles in his use of lead, as he perhaps excelled it in the wild magnificence of his schemes, this writer knows not.

At Versailles, lead was not used only for those figures which played in the waters, but also for such grave sculpture as Tubi's *Fountain of France Triumphant* (1683, restored in 1883).

Le Gros did a half grotesque *Æsop* in lead, and Tubi a *Cupid* (both in 1673). Bouchardon's fishy creatures, Lemoyne's old god, Hardy's gay children sporting on their islet, Gaspard Marsy's vast and horrible Titan, and Tubi's team of the Sun-god all go to form a splendid tribute to the

perfect adaptability of lead for the varying types of garden and fountain sculpture.

The point which it is important to emphasise is the use of lead by the greatest sculptors of the Grand Monarch for the supreme decorations of his gardens. We may be sure that André le Notre would not have permitted the use of lead if he had regarded it simply as a cheap metal, as a makeshift for bronze. M. Pierre de Nolhac writes of one of the fountains: "The work was once gilt, as was all the lead at Versailles; but time, which has effaced the gold, has made the lead more beautiful, and has left it with tones whose gracious

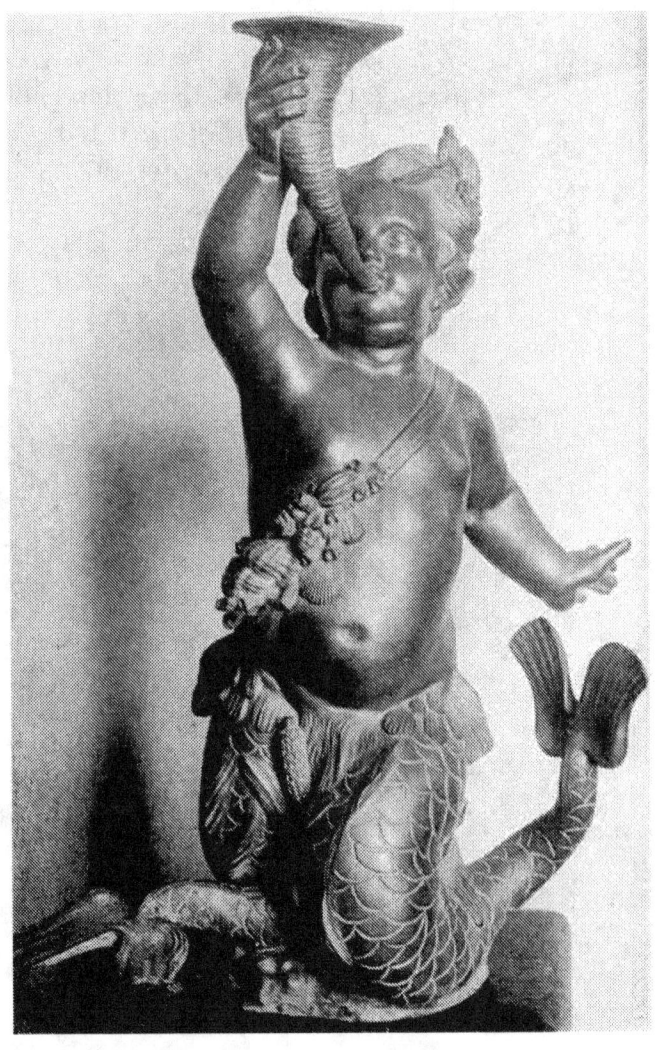

FIG. 304.—Triton, Amsterdam

harmony we must at all costs preserve." It is probable that Jean Jacques Keller, the King's Founder, who looks so imposing in Rigaud's portrait, was responsible for the casting of the lead statues as he was for the bronze.

But return must be made to the less ambitious efforts in our English gardens.

An admirable example of the water note in lead figures is the *River God* at Parham, Sussex (Fig. 302), in the Roman manner.

J. T. Smith in his "Life of Nollekens" tells of a visit he paid with Nollekens and

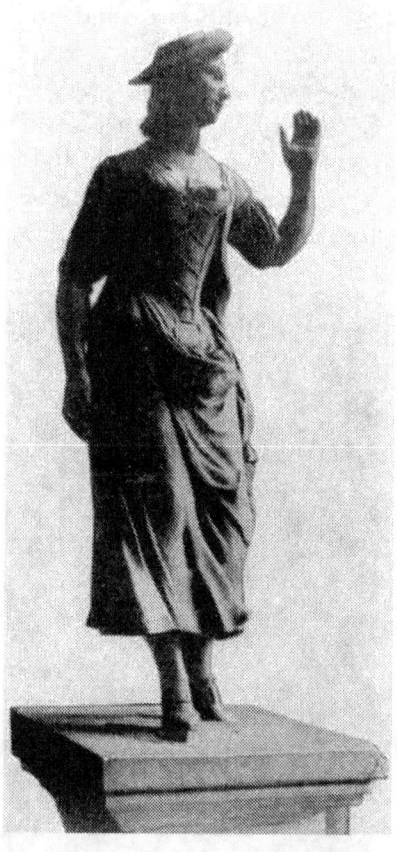

Fig. 305.—Shepherdess.

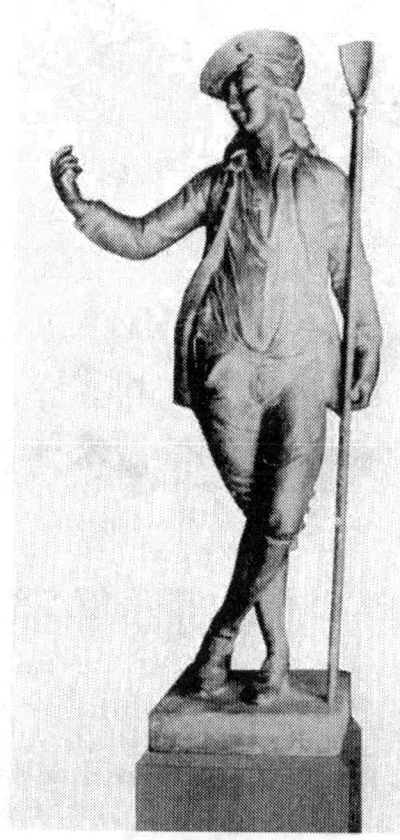

Fig. 306.—Shepherd.

his wife to an old lady, "quite of the old school," who lived near Hampstead Heath. "Her evergreens were cut into the shapes of various birds, and Cheere's leaden painted figures of a *Shepherd* and *Shepherdess* were objects of as much admiration with her neighbours as they were with my Lord Ogleby, who thus accosts his friend in the second scene of the 'Clandestine Marriage': 'Great improvements, indeed, Mr Stirling, wonderful improvements! The four Seasons in lead, the flying Mercury, and the basin with Neptune in the middle are in the very epitome of fine taste; you have as many figures as the man at Hyde Park Corner.'"

John Cheere was the man at Hyde Park Corner. About his work my Lord Ogleby in the play is very informing. The *Flying Mercury* we have met at Melbourne. The great vase at Melbourne bears emblems of the four Seasons, but four charming boy figures in a garden at Bishopthorpe, York, seem better to fit the reference. They are emblematically clothed (as far as their scanty clothing goes) to represent *The Four Seasons*, and are said to have come from the gardens of Nun Appleton, York. Doubtless they are from the same models as those to which my Lord Ogleby referred, and one is illustrated in Fig. 308. It is perhaps worth noting that Evelyn in his diary for 22nd October 1644 mentions *The*

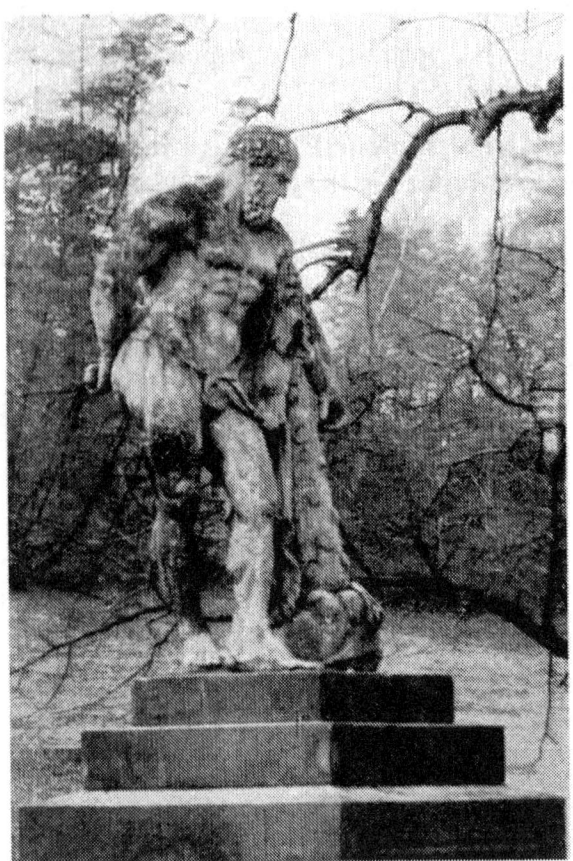

Fig. 307.—Hercules, Shrewsbury.

Four Seasons in white marble on a bridge at Florence.

As to the item of "The basin with Neptune in the middle." Studley Park perhaps provides the answer.

In the middle of the big ornamental water a lead *Neptune* remains and carries on a tradition much older than the eighteenth century and Mr Cheere, for Evelyn notes in 1643, "the Pont St Anne (Paris) is built of wood, having likewise a water-house in the midst of it, and a statue of Neptune casting water out of a whale's mouth, of lead."

"The Clandestine Marriage" (Colman and Garrick, 1766) is a mine of information on some of the more foolish gardens of the middle of the eighteenth century, when lead figures had very undesirable neighbours in Chinese bridges, Gothic dairies, and paths "all

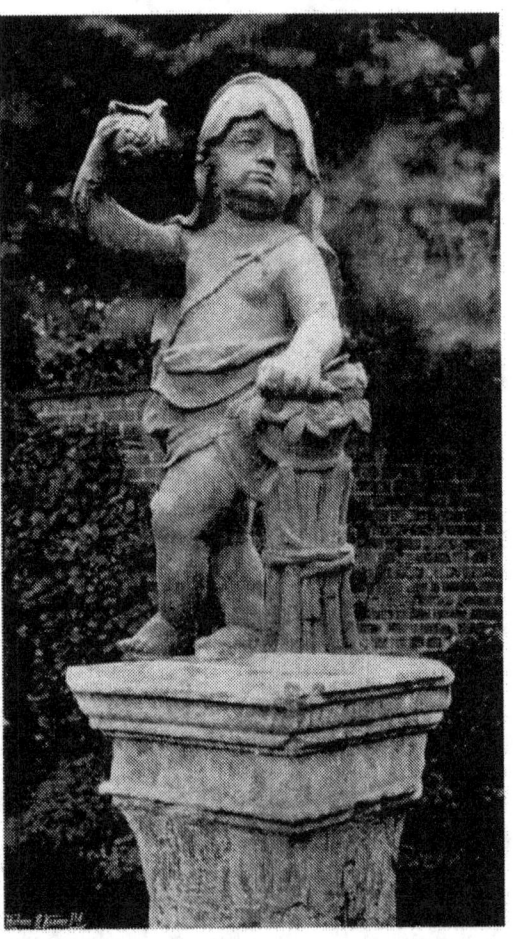

Fig. 308.—One of the Four Seasons, York.

taste, zigzag, crinkum crankum, in and out, right and left, to and again, twisting and turning like a worm, my lord."

The *Shepherd* and *Shepherdess* of the old lady at Hampstead we have no difficulty in identifying with the figures illustrated in Figs. 305 and 306.

Replicas exist of both at Enfield Old Park, and in the South Kensington Museum. Others turn up in the hands of dealers from time to time.

On the question of subjects for garden statues these Arcadian people make one reflect. It is unreasonable to demand too much of a garden statue. In the garden one can be

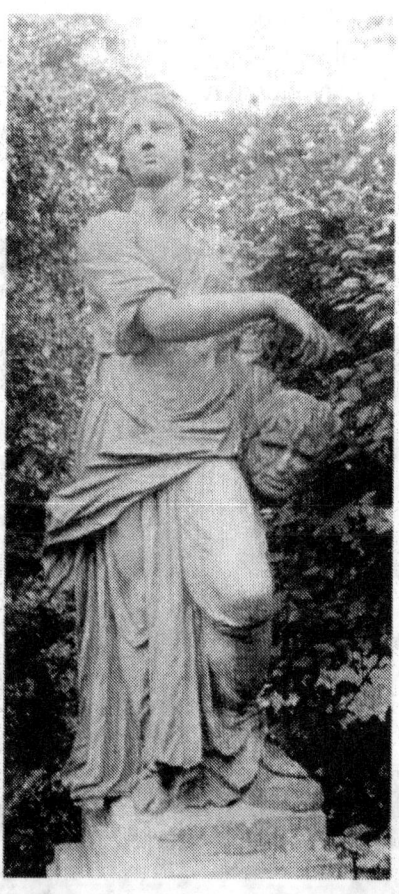 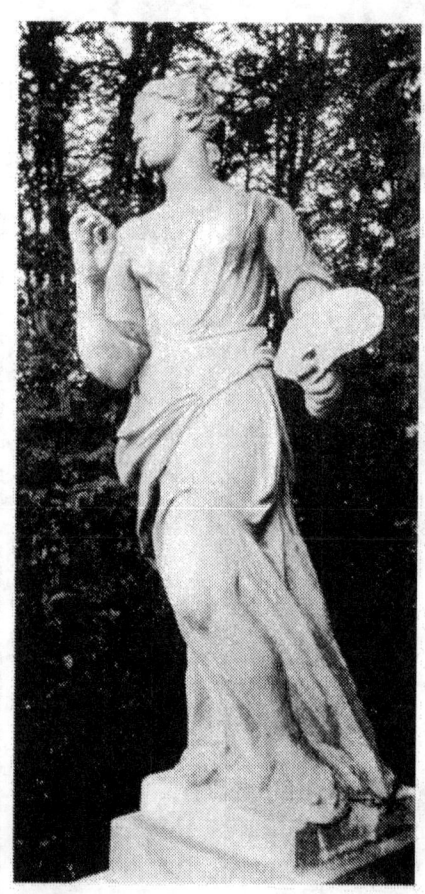

FIG. 309.—Sculpture, Hardwick Hall.   FIG. 310.—Painting, Hardwick Hall.

tolerant, and does not look for masterpieces. To quote Mr Lethaby again (and indeed who in writing of leadwork can resist doing so?), "lead is homely and ordinary, and not too good to receive the graffiti of lovers' knots, red-letter dates, and initials." One cannot, for example, regard seriously these Watteau-like productions. They are merely witticisms in lead, and erect the inappropriateness of material to subject almost into an exact science. Shepherdesses and their swains are so essentially the subjects for the delicacy of Dresden china, that to transpose them into the coarseness of lead and make them 4 ft. high compels amusement. Considering the unfitness of the material, it is noteworthy that the feeling of the figure and the light hang of the shepherd's clothes are so well conveyed.

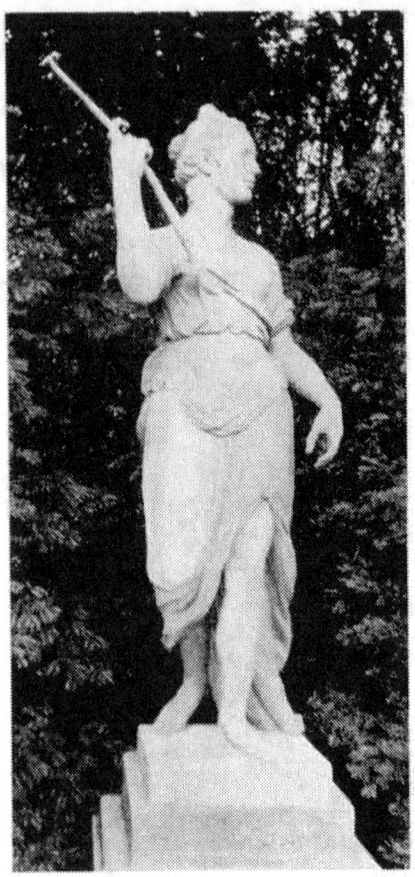

FIG. 311.— Music, Hardwick Hall.

It is the sort of statue that would gain by some touches of gilt. In days past they often went further, and painted the figures all the colours of the rainbow. That seems to be a superfluity of naughtiness. There is a fitness in the gilding of a lead statue. It is a metallic decoration on a metallic ground. It throws up the natural colour of the lead, while painting in other colours (unless they are transparent which illuminate without veiling the metallic feeling) is almost necessarily a mistake.

To return to John Cheere. He died in 1787, and it has always been said that with him the last of the lead yards was closed. This seems inaccurate

in the light of the three female figures in lead on the pediment over the portico of Avington House, near Winchester. It was built in 1789 by James, third Duke of Chandos. One of the figures is a *Flora*. That the use of lead figures never altogether ceased is clear when the pair on the steps leading to the portico of University College, London, is remembered. John Cheere had a long career, for he took over in 1739 the business of the first van Nost. Probably he was more carver and founder than artist, and relied on the stock models of van Nost and the designs of his better known brother,

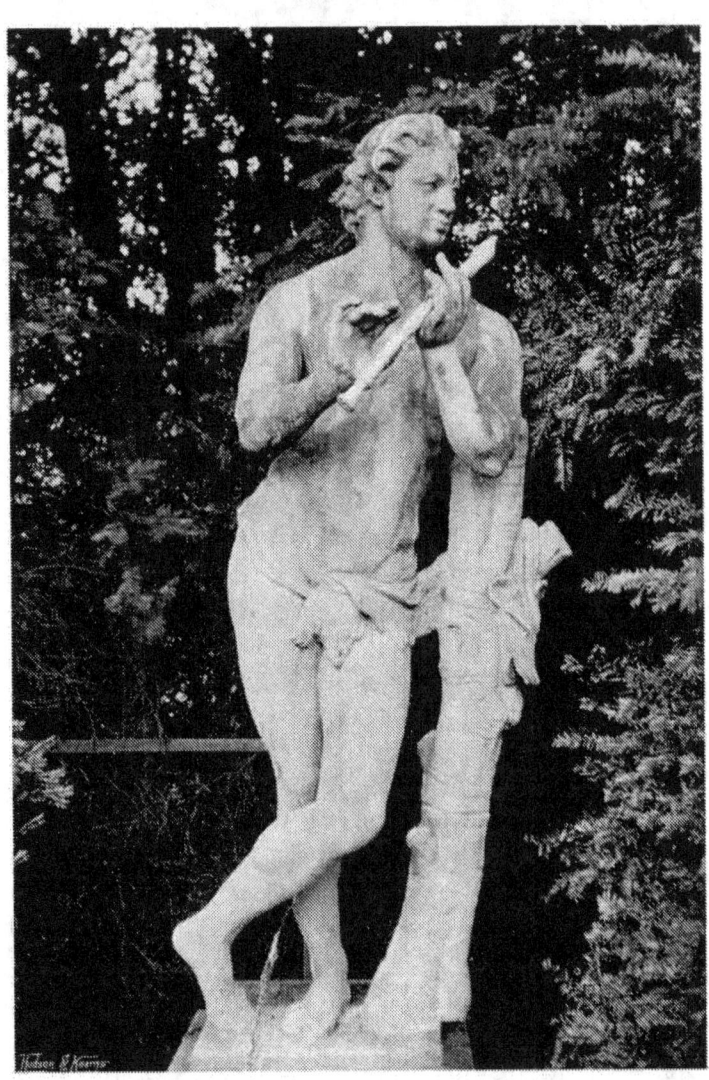

FIG. 312.— Piping God, Hardwick Hall.

Sir Henry Cheere. In the library of South Kensington Museum is a volume of sketches of marble monuments and sculpture generally. It bears no name, but one of the monuments can be identified as by John Cheere. There are also coloured sketches of a pair of charity children, and a pair of old people, evidently designed for an almshouse. These were obviously to be cast in lead, and are likely to have been made by J. Cheere.

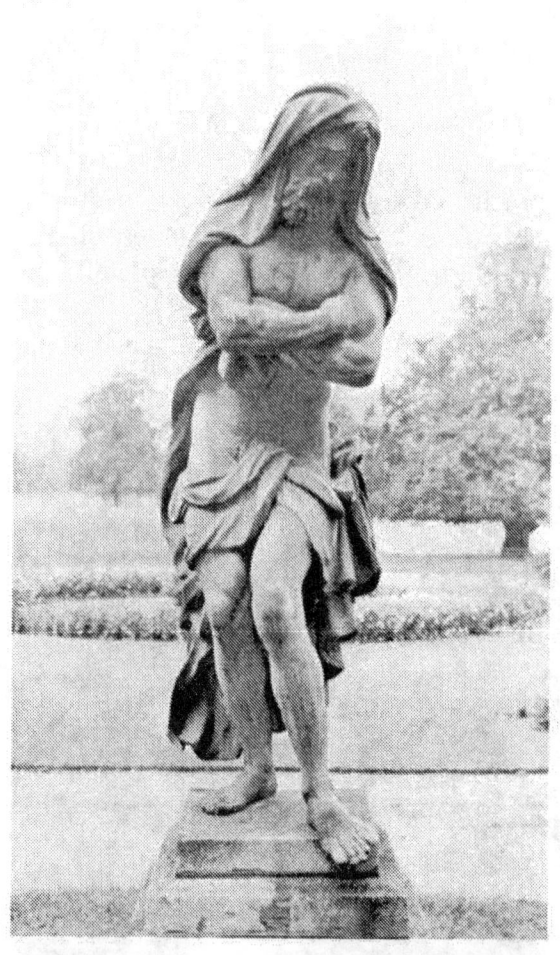

Fig. 313.—Winter, Glemham Hall.

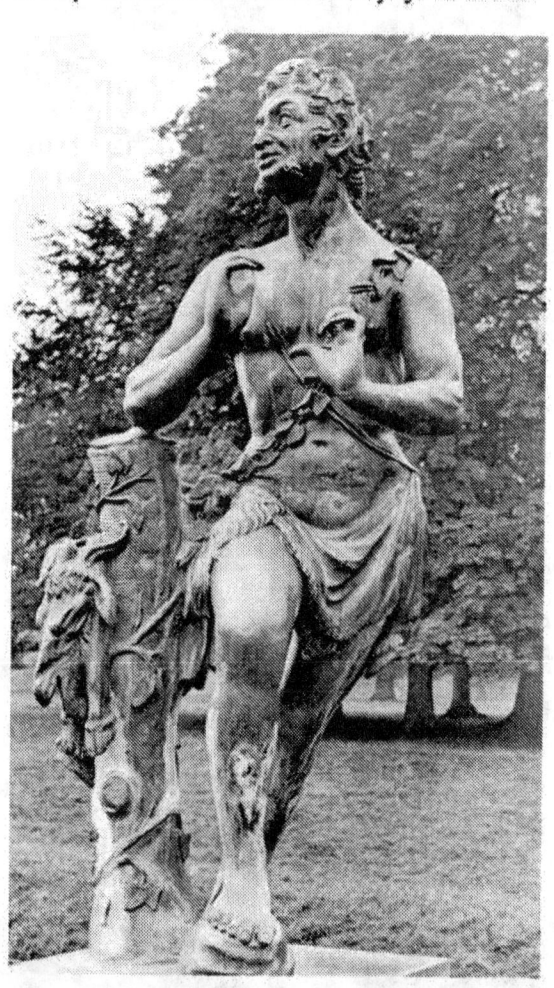

Fig. 314.—Pan, Glemham Hall.

Robert Lloyd in the "Cit's Country Box" also refers to him :

"And now from Hyde Park Corner come
The gods of Athens and of Rome.
Here squabby Cupids take their places
With Venus and the clumsy Graces.
Apollo there with aim so clever
Stretches his leaden bow for ever ;
And there, without the pow'r to fly,
Stands fixed a tip-toe Mercury."

At Aislaby Hall, near Pickering, are four lead figures, *Apollo* (though without a bow),

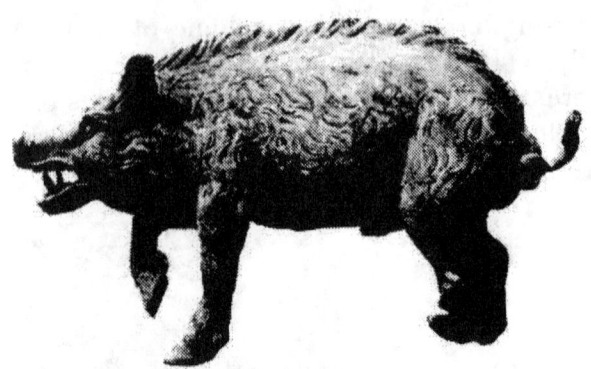

FIG. 315.—Boar, Myddelton House,
Waltham Cross.

Mars, Diana, and a winged lady who may be *Fame.* All have their arms raised, and perhaps were compelled to resign their divine functions in favour of holding torches, for their hands are closed round sockets. They have been badly used, and are now painted dark green. A Captain Hayes took them to Aislaby Hall about 1770, but the present owner is abroad while this is written, so no further information is available.

We can only connect John Michael Rysbrack (1693-1770) vaguely with lead statues. It is on record, however, that he modelled a big statue of Hercules, compiled from the Farnese Hercules, and from studies of pugilists and athletes of his own time.

Very possibly the *Hercules* at Shrewsbury (Fig. 307), in the Quarry Avenue, is a

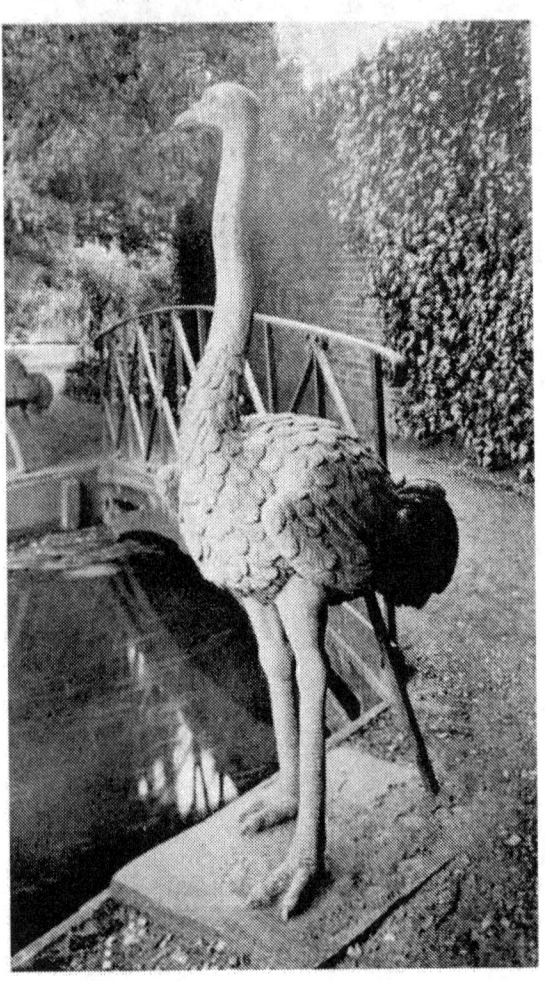

FIG. 316.—Ostrich, Myddelton House.

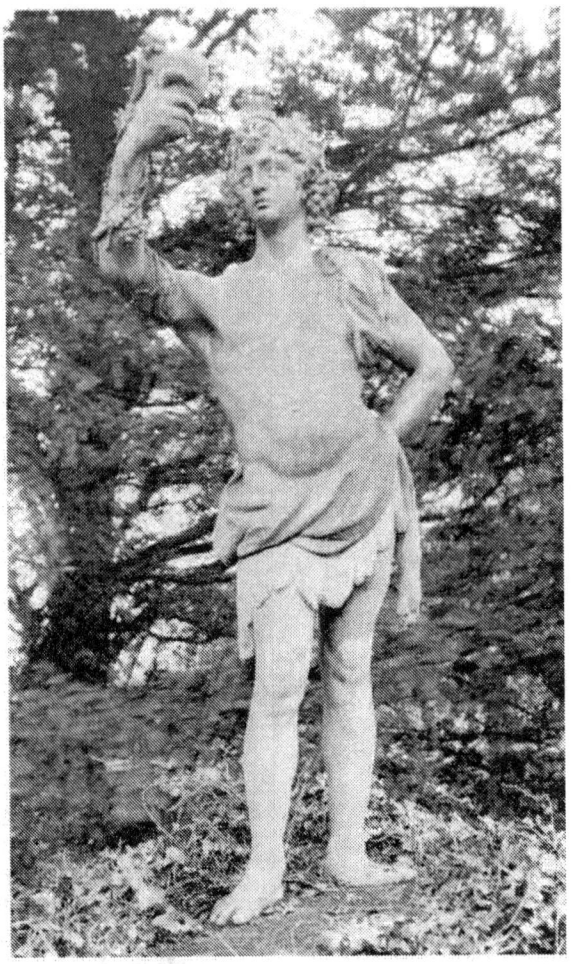

FIG. 317.—Bacchus, Enfield Old Park.

replica of Rysbrack's figure, an adaptation of the Farnese *Hercules*. The rains and airs of the Severn Valley have dealt very kindly with the lead, and have shaded the brawn and muscle of the god to the great enrichment of the modelling.

At Hardwick Hall, Derbyshire, there are six lead figures, but they are not native to the place. The gardens were laid out in the formal manner by the father of the last Duke of Devonshire, and the figures were then imported from Chatsworth. It has been suggested that they may be the work of C. G. Cibber. The records remain of his

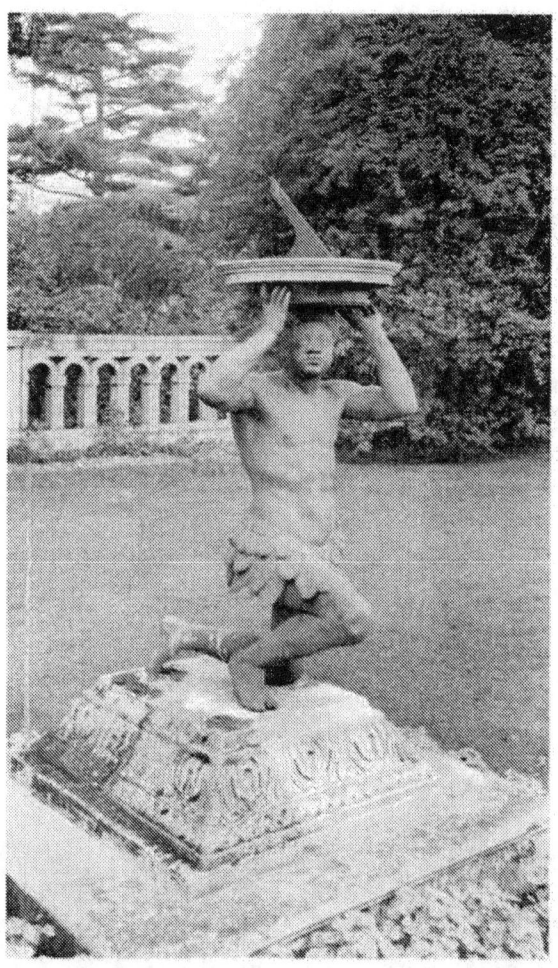

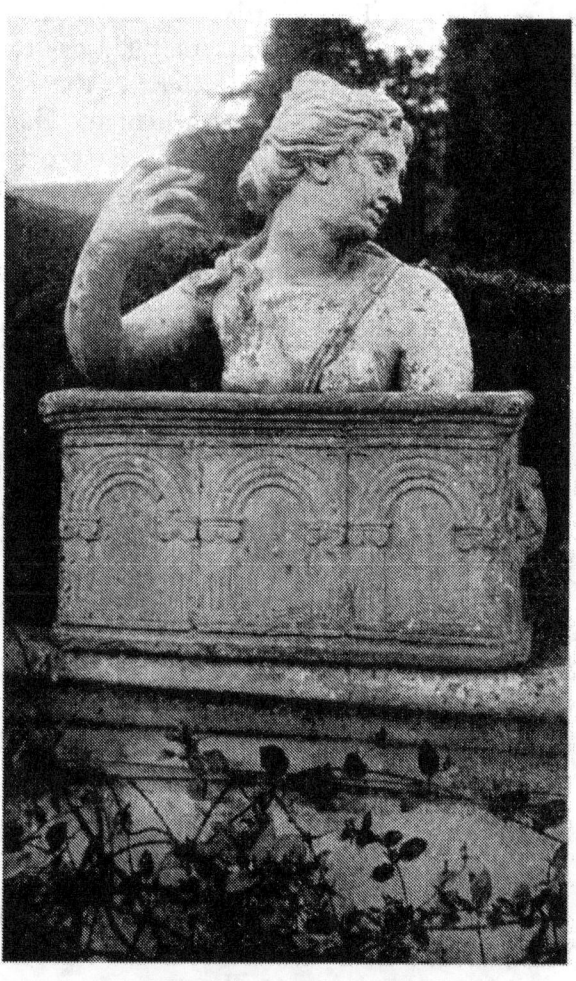

Fig. 318.—Kneeling Boy Slave, Enfield.

Fig. 319.—At Enfield Old Park.

employment by the first duke at Chatsworth, to adorn with statues and a fountain the lawn facing the south front. The lead figures now at Hardwick are, however, certainly later than Cibber, and it is probable that they stood by the south front, and were removed when the sixth duke replaced them by copies from the antique.

Of the six figures four are illustrated. The ladies have a solid Teutonic air, and while there is a certain cleverness in the draping of *Sculpture* (Fig. 309), there is a lady (not illustrated) with a violin whose clothing is an exercise in drapery instinct with the spirit of compromise.   It suggests the effort of an intelligent Papuan to absorb the researches of Professor Baldwin Brown into ancient Greek dress, and to apply the knowledge to native needs.   The goddesses who look after trumpets and painting (Figs. 311 and 310) are not very notable.   Of the youths, one is Bacchanalian with uplifted cup, and owing to the lead having given, is now leaning over in a way that befits a Bacchanal. The other is of somewhat lascivious aspect with a flute (Fig. 312).   It will be noted how cleverly the stability of the figure of this piping god is assured by making it lean against a tree trunk.   The Hardwick Hall figures are average examples of eighteenth-century type.   The ladies have a look of massive complacency, which would induce boredom in a gallery, but is not without merit in the restful atmosphere of a formal garden.

The leaden treasures at Glemham Hall are not confined to portrait statues.   While the head of the *Pan* (Fig. 314) lacks the subtle characterisation of the Castle Hill bust, the figure is a notable one, and it is unfortunate that the god has lost his pipes.   The tree trunk with its goat's skin is a thoroughly practical accessory as it helps to stiffen the figure. The hooded figure of *Winter* with arms akimbo, and lean thighs, is also admirable (Fig. 313).

At Godinton, Kent, is a charming pair of dancing figures, one at each end of the fish pond ; the boy has cymbals, the girl holds what apparently was once a branch in one hand, and in the other a bunch of flowers.   There is also a *Cupid* with sundial from the same pattern as the example illustrated in Fig. 301.

The lead fauna of gardens have no more notable representatives than the *Ostriches* (Fig. 316) and the *Boar* at Myddelton House, Waltham Cross.   Originally they all adorned Gough Park.   The birds stood on the top of the house, and the pair of boars (one has since been stolen) on the gate piers.   Mr John Ford, F.S.A., of Enfield Old Park, has happily got copies of the invoices, so we know the provenance of these delightful creatures.

To Captn. Goff.   Bot. of Jno. Nest, Sept. 21, 1724.   (*Note.*—" *Nest*" *is possibly John van Nost.*)

	£		
2 Estridges 6 ft. high	£20	0	0
2 Cocketresses	7	0	0
Carridg	0	14	0
	£27	14	0

paid Nov. 6, 1724.

Bot. of T. Maning.

		£		
1720 Aug. 23.	Neptune	£21	0	0
	Mercury and Fame	12	12	0
Nov. 17.	2 Boares	8	8	0
	2 large vases	25	0	0
	Waggon and Car	1	17	0
		£68	17	0

pd. Nov. 23, 1720

This " Goff " was Captain Gough of the Merchant Service of the East India Company and a director of the Company.   He was also father of Richard Gough, sometime director of the Society of Antiquaries, who wrote the earliest paper which dealt with lead fonts, published in *Archæologia* in 1789.   Doubtless his father's " cocketresses " (would that these charming creatures had not flown to limbo) stimulated his interest in leadwork; anyhow he is the father of its history. Perhaps his greatest monument is the persistence with which the mistakes he made in his paper have been copied and recopied in succeeding papers on the subject.

The *Boar*, shown in Fig. 315, was the Gough crest. The ostriches now stand on either side of a bridge over the New River, where it runs through the gardens of Mr Henry Bowles' house. I am told that these fine birds are not correctly modelled, as they should not have "flight feathers." Captain Gough must have had them made from sketches which his sea-faring acquaintances or he himself had secured, and either draughtsman or sculptor went wrong over the feathers. The skin of the legs is, however, well shown, and altogether they are notable work.

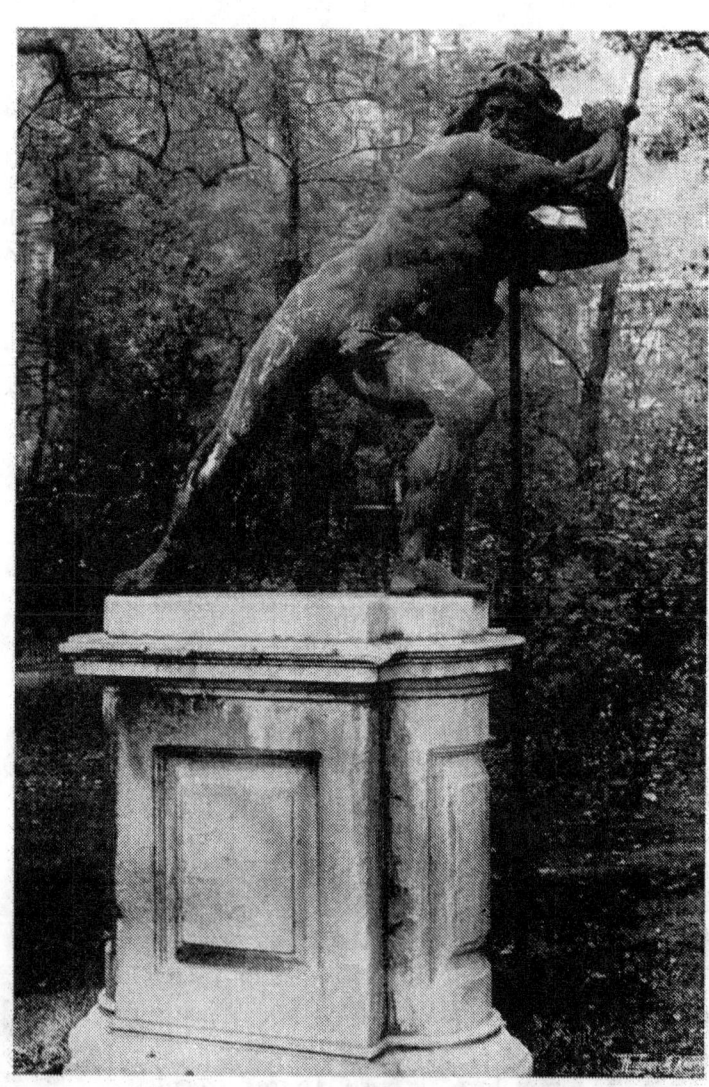

FIG. 320.—At Devonshire House, Piccadilly.

Not only is Mr John Ford the possessor of much leadwork, but of a collection of the *disjecta membra* of demolished historical buildings which may safely be called unique. The two carved stones which form the base for the *Kneeling Slave* of Fig. 318 once supported the chancel arch (one on either side) of St Mary Somerset in Lower Thames Street, the first of Wren's churches to fall to the destroyer.

The arcading in the background of the photograph came from the top of the tower of St Dionis Backchurch, also a Wren building, when it was destroyed in 1878 under the Union of Benefices Act. These two examples are given because they come into the leadwork picture, but they are merely representative of dozens equally interesting.

Of the *Kneeling Slave* himself it is to be noted that he is markedly younger in countenance than the elder African slave at Melbourne and elsewhere, and his history is known. He stood since about 1730 in the gardens of Bush Hill Park, and was bought originally by John Gore, who lived there and died in 1763, the last surviving director of

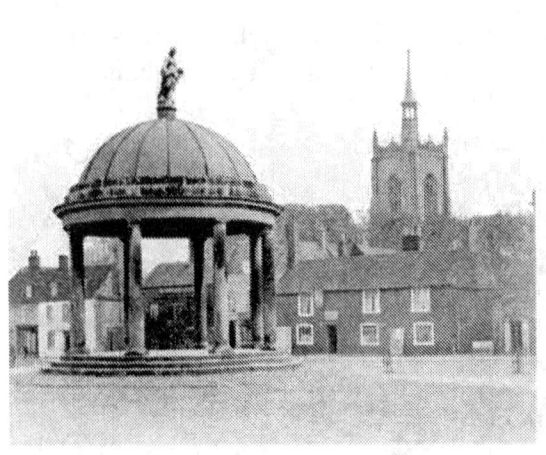

FIG. 321.—Butter Cross, Swaffham.
(The lead Spirelet in background is dealt with in earlier Chapter.)

the South Sea Company. High up on a parapet is a lead *Juno*. In the garden is a fine *Bacchus* (Fig. 317), and a dancing mounte-bank-like figure of very delicate modelling, which is German or Flemish, certainly not English.

The queer apparition of Fig. 319 is illus-trated rather for the arcaded jardinière than for the bust. The latter is all that remains of a complete statue, and in its mutilated state has found a resting place in the flower-pot, which from its arcading has an early font-like look. The top mouldings, however, betray it for a seventeenth or eighteenth century jardinière, but a pleasant one withal. There are also a *Shepherd* and *Shepherdess* in lead at Enfield Old Park, replicas of those of Figs. 305 and 306.

When Lord Burlington uttered his dictum against lead statues, on the ground that they tend to fall out of shape, and that arms became like "crooked billets," he doubtless had in mind such figures as that of Fig. 320. Despite that noble amateur's scorn, he filled the gardens of the Villa that he de-signed (not unaided) at Chiswick with lead statues, and this one was removed to Devon-shire House by the late Duke, when he dismantled the Villa. It is obvious that a material which needs to be stayed with iron rods is profoundly unsuited to a figure which does not stand well over its base. The Earl of Burlington had the sense to complain of the behaviour of unsuitable lead figures, but apparently not to choose those which were not liable to collapse.

At Devonshire House there are also a replica of the *Gladiator* at Burton Agnes (Fig. 280), and a youth bearing a lamb on his shoulder.

In Norfolk there are two delightful market crosses, at Swaffham and Bungay. Though not exactly alike they are similar, and consist of a circular colonnade with domed lead roof surmounted by a lead statue.

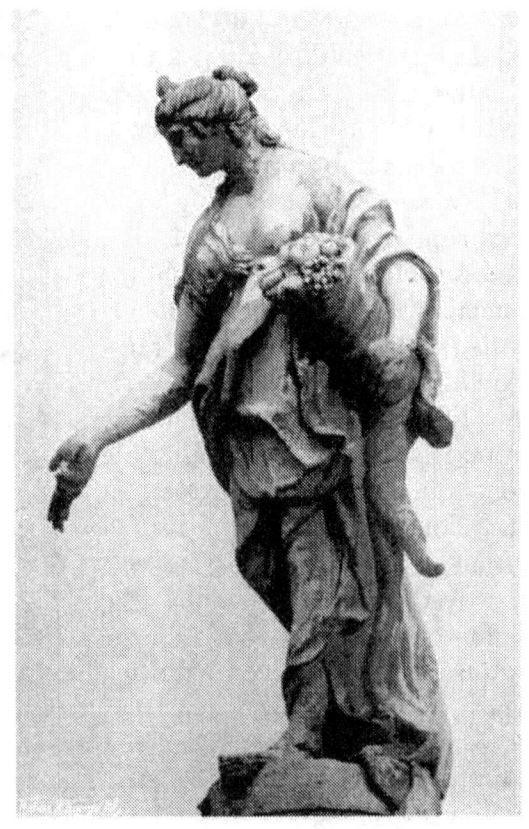

FIG. 322.—Ceres at Swaffham.

At Swaffham the figure is *Ceres* bearing the horn of plenty (Fig. 322). It is said to have been executed by a French artist, and cost £200, an amazingly big sum. The cross was built by the Earl of Orford in 1783. Butter was sold by the yard at markets held under the dome of this cross (so called doubtless because there is no cross). Let us mourn a decayed industry.

The similar cross at Bungay bears a lead figure of *Astræa*. It was set up in 1690, and

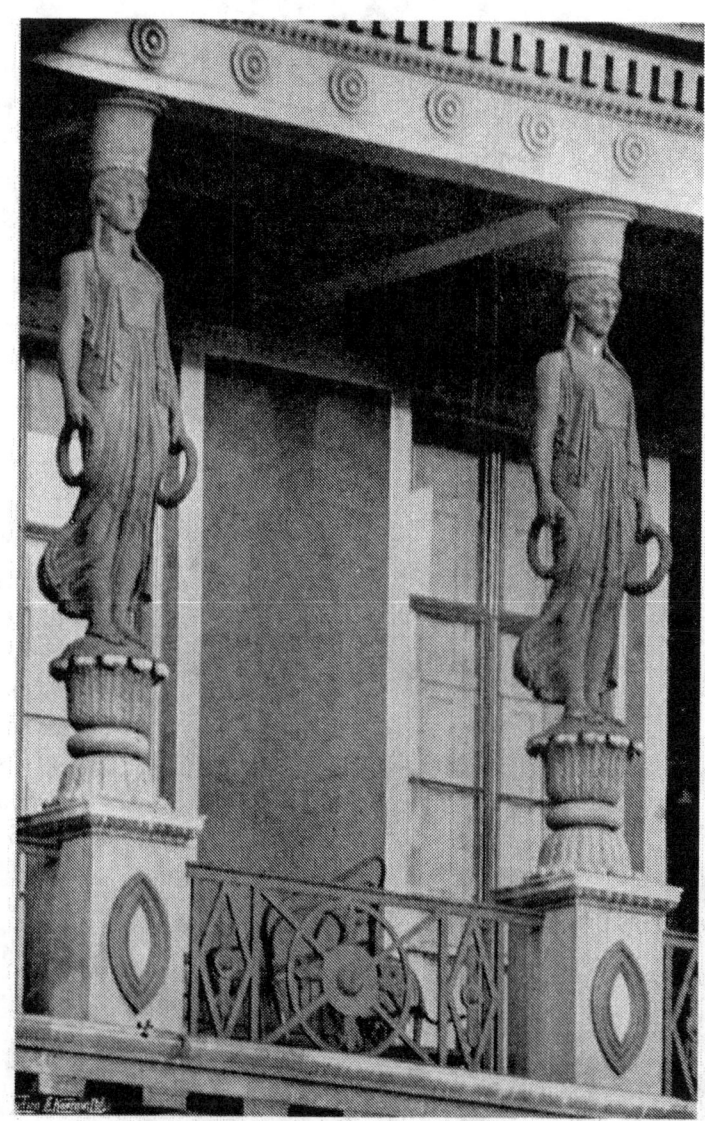

FIG. 323.—Lead Caryatides, Park Lane.

was also a butter cross. Amongst other pleasant uses to which it was put were as a cell for prisoners, a whipping post, and a place for the stocks. Under the dome a hook remains, from which hung a cage in which prisoners were exhibited.

Altogether *Astræa* has seen life during her 218 years on the dome.

The figure of *Charity* in lead is a not unusual ornament of almshouses and the like. At Great Yarmouth she appears at the Fishermen's Hospital, and bears an infant in her arms, while a young child clings to her knee. The hospital was built in 1702. A similar idea is expressed by a group on the pediment of the main front at Wimpole, where *Charity*, a girl, ministers the cup of cold water to *Poverty*, an old man.

Fig. 323 shows a comparatively modern example. On a balcony of a house in Park Lane are lead Caryatides, and very graceful they are with their windswept draperies.

They were erected about eighty years ago, and their great weight nearly pulled down the whole balcony. When repairs were being done, the figures were found to be full of large chips of white marble, obviously the waste product of some statuary's yard.

The last illustration of this chapter is not the latest in date, but a long chapter may be forgiven for disarranging a date, that it may carry the sting of a moral in its tail.

In 1903, Newcastle Street, W.C., was destroyed, and with it the workshop of Messrs Dent & Hellyer, a firm of plumbers established there in 1730. In a verandah of "Ye Olde Plumbers Shop" stood the lead figure of a *London Apprentice* (Fig. 324).

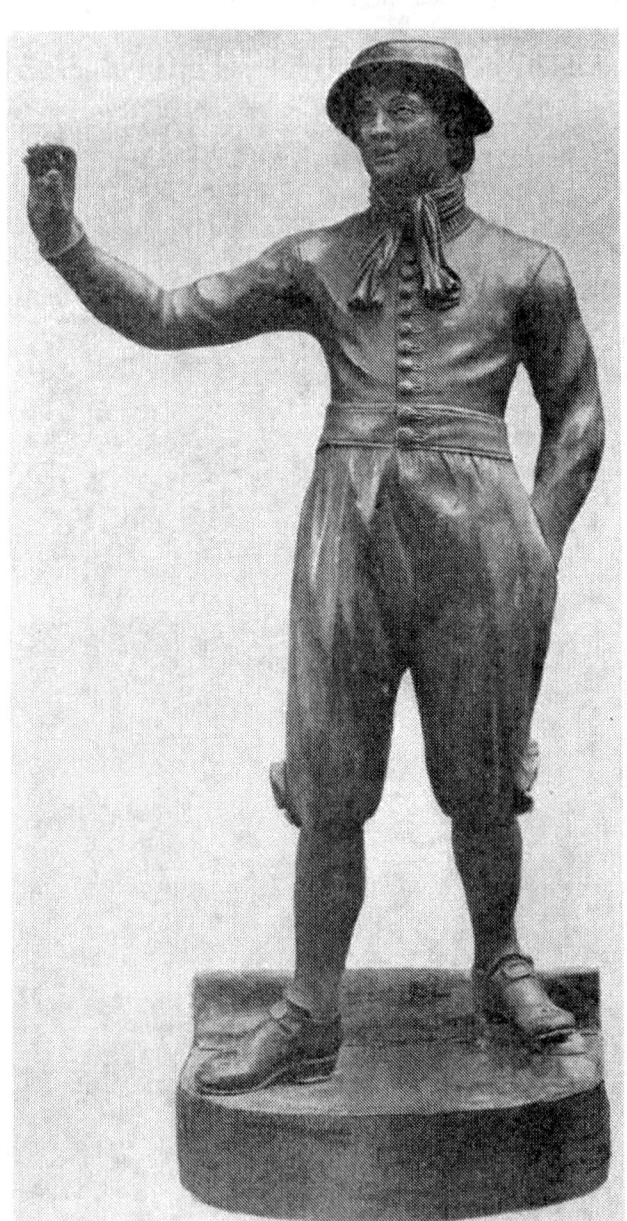

It is believed to have been modelled for Lancelott Burton, a predecessor, in 1769, of Mr S. Stephens Hellyer in the freedom of the Worshipful Company of Plumbers. Unsuccessful search was made at the old workshop for patterns of the *Apprentice*, and also of four other lead figures, now perished, that stood beside it. This suggests that the lead figure trade of the eighteenth century was confined to the statuaries of the Piccadilly lead yards and that the plumber proper confined himself, so far as decorative work was concerned, to cisterns and other domestic objects. Perhaps, however, the *Apprentice*, a lively and admirable figure, was cast in Lancelott Burton's shops and the mould forthwith destroyed. In 1906 the Plumbers' Company presented, in the hall of the Old Charterhouse, George Peale's pageant "The Masque of Lovely London" which had lain dormant since its first performance to Lord Mayor Wolstane Dixie in 1585. In the hall stood the leaden *Apprentice*, and the living apprentice in the pageant was clad like him as he spoke the plea—

"That lovely London may one day enjoy
The power that now lies dormant in the boy."

The Worshipful Company of Plumbers is to-day honourably distinguished by the zeal with which it fosters the practice of apprenticeship.

Thoughtful sociologists are agreed that apprenticeship must be added to the technical training in schools if right craftsmanship is to be restored. The leaden *Apprentice* stands therefore, not only as a fragment of London's history, but as one of the ideals in which are bound up the present aims and future hopes of the Art of English Leadwork.

FIG. 324.—The London Apprentice.

CHAPTER X.

VASES AND FLOWER-POTS.

Shenstone on Urns—Melbourne—Parham House—Hampton Court—Windsor—Wilton—Castle Hill.

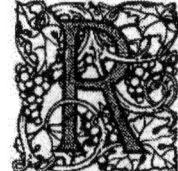EFERENCE was made in the last chapter to Shenstone's views about lead statues. Hear him on the question of vases: "Urns are more solemn if large and plain; more beautiful if less ornamented. Solemnity is perhaps their point, and the situation of them should still co-operate with it."

In Shenstone's famous garden at the Leasowes in Shropshire, there stood in the Lovers' Walk an urn, "inscribed to Miss Dolman," but it is not stated whether it, or the statues which are mentioned, were of lead.

It may be doubted whether the eighteenth century took very heartily to Mr Shenstone's claim for solemn urns, but some at least are a kind of tragic trappings in great gardens. At the Burlington Villa at Chiswick, one comes upon a charming vase in a shady walk near the big pool and garden house. It is solemn in the best manner. The great vase at Melbourne, Derbyshire (Fig. 325), is elaborately ornamented, but from its situation at the "crow's foot" in that fine garden may claim a deserved reputation for solemnity. Standing, as it does, where long grass walks meet, it pulls the design of the garden together in a notable fashion. It was cast in 1706 by John van Nost, who also supplied the lead figures. The cost of it does not appear, but in 1705 a Frenchman estimated that the carving on the stone pedestal would cost £6 exclusive of the stone. The lower part of the vase has four monkey-like creatures by way of supporters. Unfortunately, their support is more apparent than real, and has not prevented the vase from taking a marked list to one side. This is a technical fault that would have been avoided by a stout iron core in the stem. The upper part bears four heads, emblematical of the seasons. Spring, summer, and autumn range from girlish to womanly, and are wreathed with spring flowers, grapes and corn. Winter is a bearded, hooded man. The middle of the vase is covered with a delicately modelled masque of children playing and swinging, while in panels, above the swags that connect the seasons, are little scenes in the classical manner. The basket which surmounts all is rich with trophies of fruits, and altogether the composition is very handsome of its florid sort.

At Pain's Hill is a vase made from some of the same patterns, but smaller. The heads of the seasons are there, but no swags, and the basket is less plentifully supplied with fruits. On the top, however, sits a fox (!), and the same monkeys do duty at the base.

One of the finest of all garden vases is at Parham House, West Sussex (Fig. 326). This, with its flame top, is based in idea on the cinerary urn, and is a very sumptuous piece of modelling. It is free from the reproach of overloading which the Melbourne vase cannot fairly escape, and the relief is distinct without being insistent. The leaf work on the lid is particularly well done.

At Compton Place, Eastbourne, is a pair of handsome lead vases (Fig. 327) standing on the piers of the entrance gates. They are spoil from the Duke of Devonshire's dismantled villa at Chiswick, now given over to the unhappy uses of a private asylum.

At Myddelton House, near Waltham Cross, Mr Bowles has several lead vases. In Fig. 328 one of a graceful classical sort, with snake handles, is illustrated. There is a replica of this vase in Kew Gardens, and there are many more about. A pair was bought

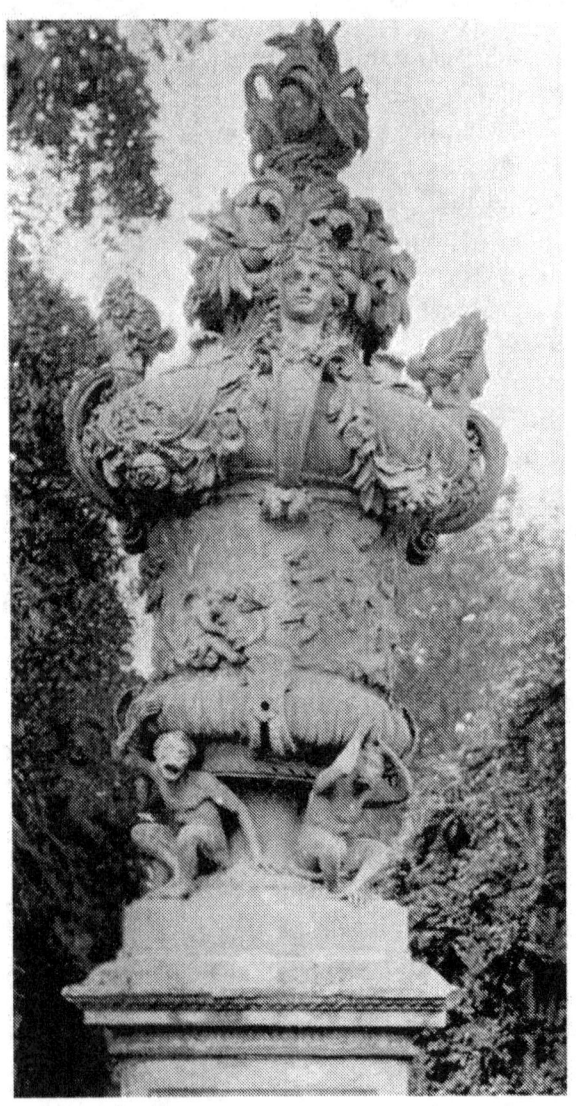

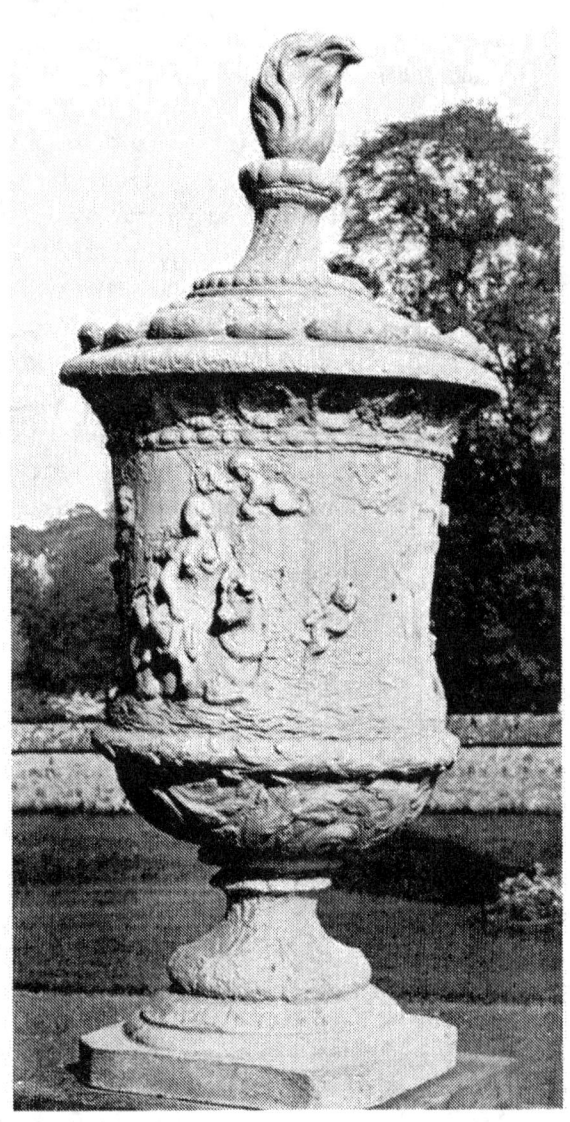

FIG. 325.—Melbourne, Derbyshire.   FIG. 326.—Parham House.

some little time ago on behalf of an exalted personage. The fact about this vase is, that it has been turned out in such considerable numbers in the last few years that it was worth while to make an iron casting pattern! It is a replica of a Greek vase of black marble in the Louvre. The original has swan handles, as have some of the modern replicas. The example illustrated has snake handles, which suit it well enough, but are merely the taste of the modern fashioner of "antiques."

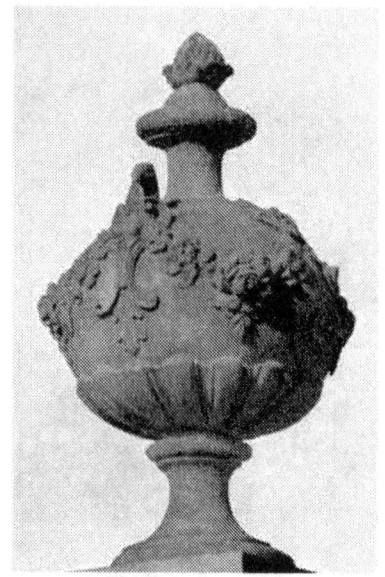
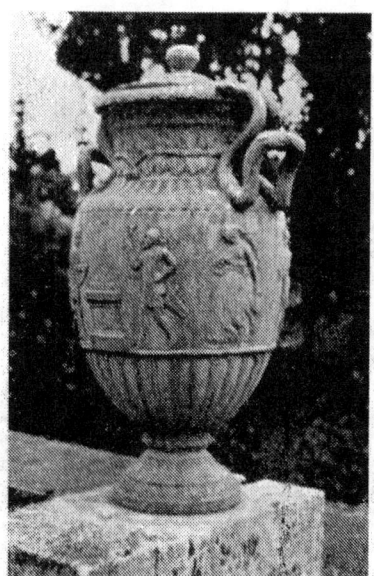
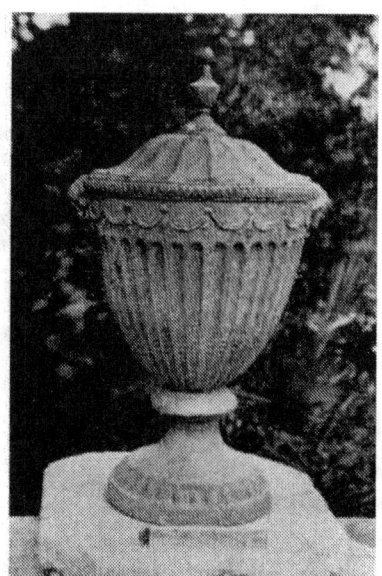

FIG. 327.—Compton Place, Eastbourne.   FIG. 328.—Myddelton House.   FIG. 329.—Myddelton House.

There are also at Myddelton House (Fig. 329) some delicately ornamented lead urns in the Adam manner. They accord very well with the formal balustrading on which they stand, and with the general air of trimness which is heightened by the orderly passing of the New River through the gardens. A similar vase, but with large swags, is also being turned out in large numbers in a London suburb.

Reference has already been made to Wren's use of vases on his lead steeples, as at St Edmund's, Lombard Street, and St Augustine's, Watling Street. Certainly at the former, and probably at the latter, these were of wood covered with lead, and not of cast lead made like the flower-pots.

At Hogarth's House, Chiswick, there used to stand on the gate posts a pair of lead vases, which are said to have been given to the artist by his friend Garrick. They are now to be seen in the dining-room by any one who takes advantage of Colonel Shipway's munificence in giving the house and its contents to the nation. Illustrations of them appear in the descriptive *brochure*, which can be bought at the house.

The vases of Fig. 330 at Temple Dinsley have boldly modelled mouldings and delicate reliefs, from which much evil paint has lately been removed. They are cast in four pieces and soldered together with a lapped joint, very neatly done.

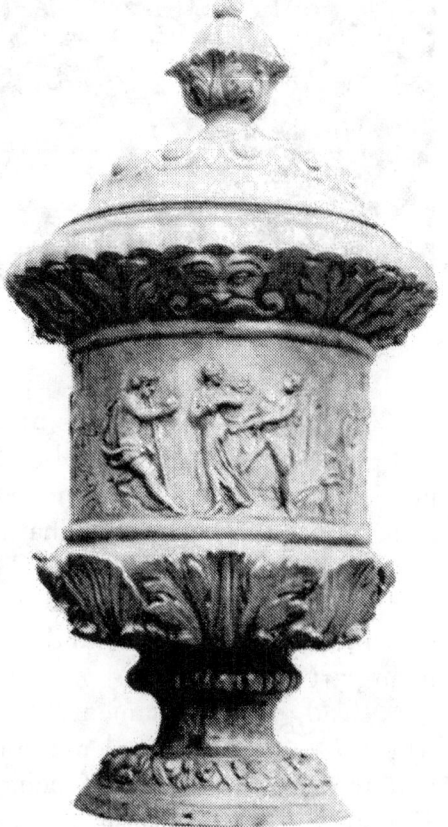

FIG. 330.—At Temple Dinsley.

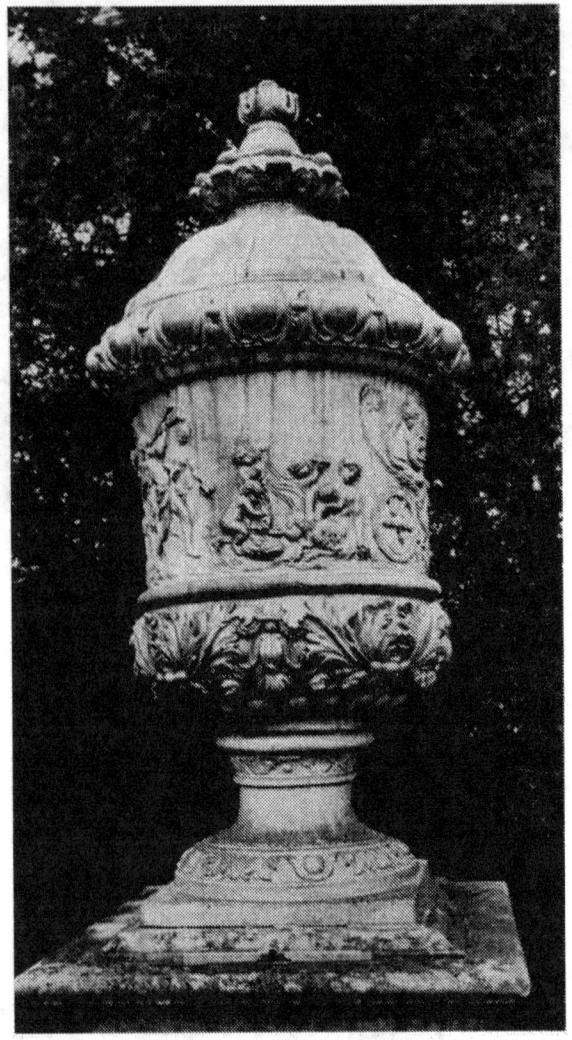

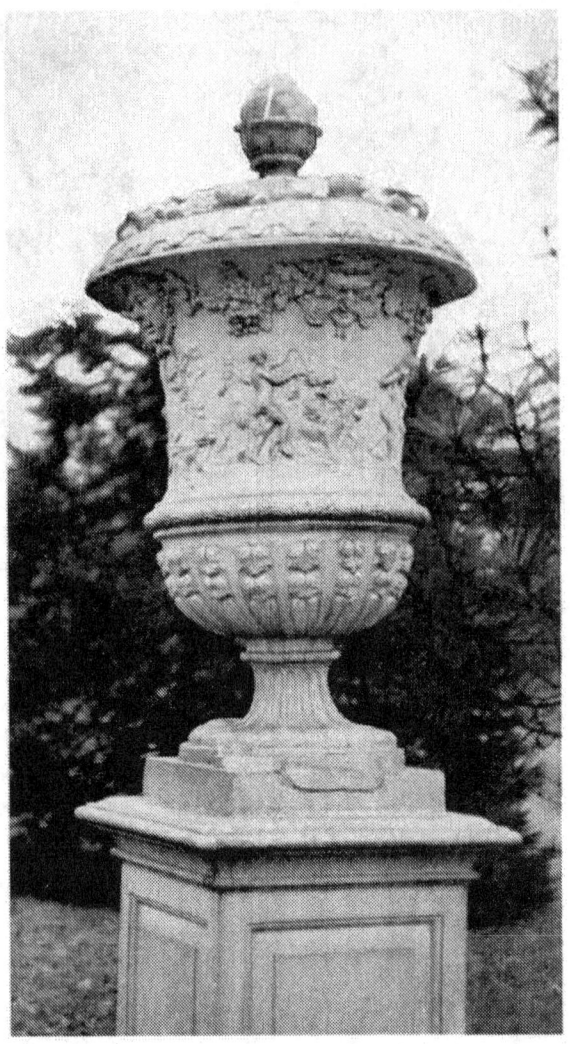

Fig. 331.—At Wrest Park.   Fig. 332.—At Wrest Park.

There are several vases at Wrest Park, but the point of a penknife judiciously used will prove more than one to be of cast iron. The author can, however, vouch for the two here illustrated, and both indeed confess their material to be lead, for they have taken a slight list to one side. That of Fig. 331 is one of a pair that flank the colonnade of the Bowling Green House. This delightful garden banqueting hall was built by the Duke of Kent in 1735, and doubtless the admirable vases are contemporary. More delicate in its modelling and, on the whole, less successful is the vase of Fig. 332, in the main part of the gardens.

Lead garden ornaments of the vase type naturally fall into two main classes, those which are urns of the solemn sort and make an appeal only to the eye, and those which add the practical value of being flower-pots. The variety of the latter is considerable. For sheer success both in proportion and ornament, the pair at Hampton Court (Fig. 333) are almost beyond criticism. As Mr Lethaby says, " The little sitting figures, slight as

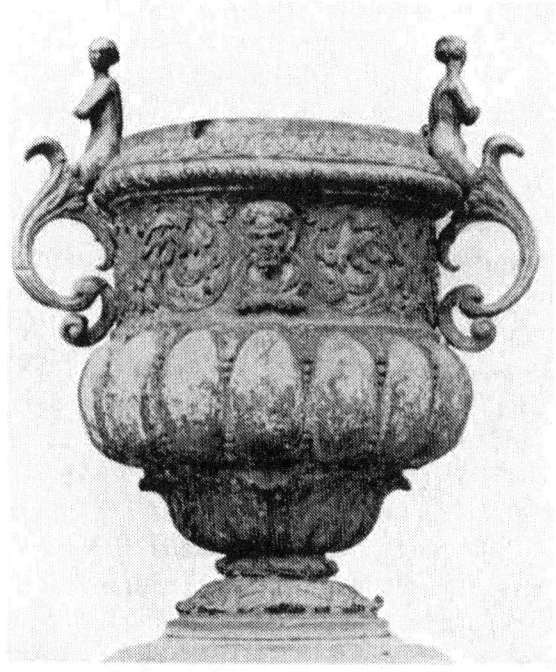

FIG. 333.—Hampton Court.

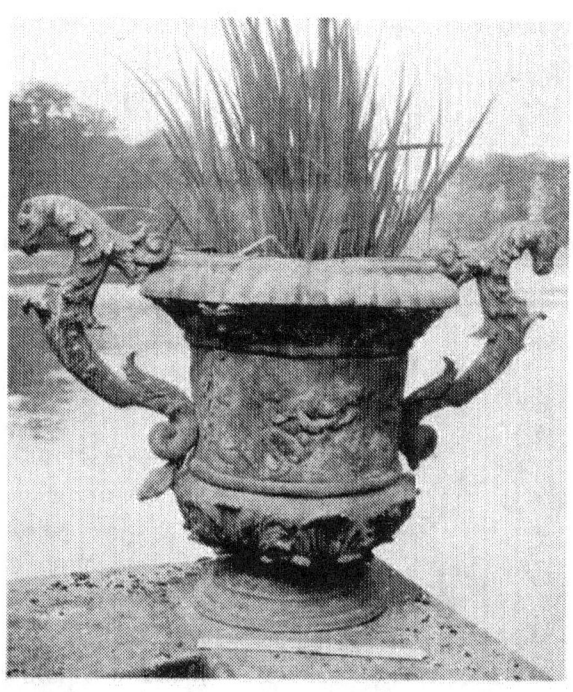

FIG. 334.—Studley Park.

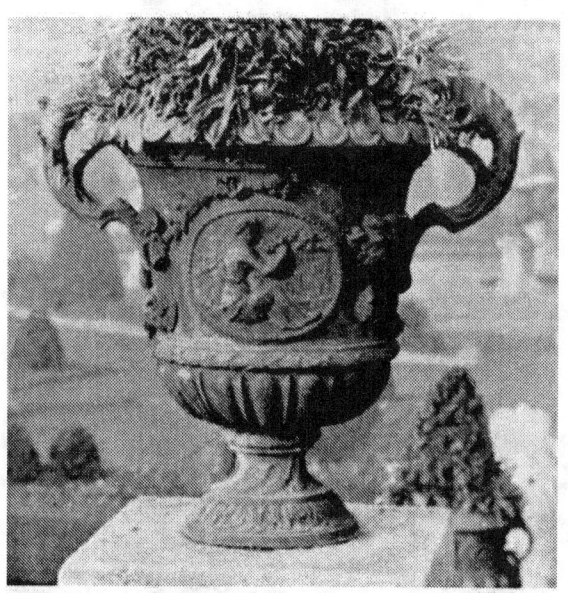

FIG. 335.—Windsor Castle.

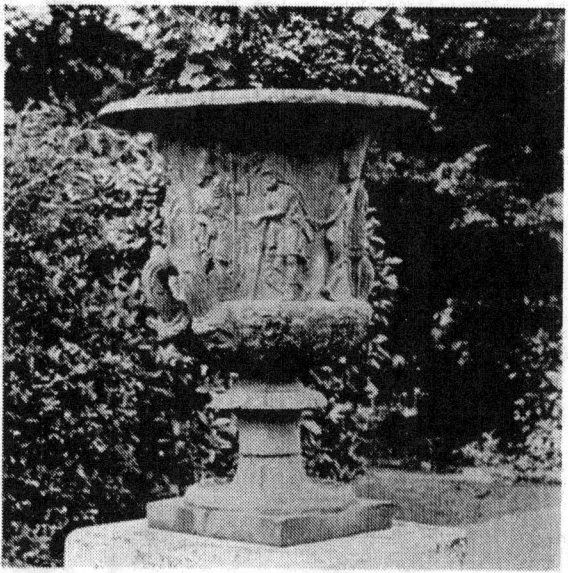

FIG. 336.—Charlton, Kent.

they are, are charming in their pose ; the folded arms and prettily arranged hair give us a suggestion of life which most of these things supposed to be in the classic taste lack."

A few old replicas exist, and also some modern copies, so well done that they would deceive in sale-rooms the very elect.   At Hampton Court these pots are sometimes the home of fuchsias, and the flowers nod in a charming fashion over the handles.   The

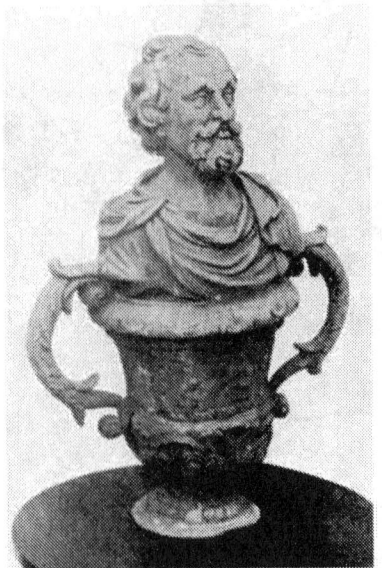

fuchsia is a wonderfully adaptable flower, and looks as appropriate in this refined and artificial atmosphere as it does when growing in great hedges in the wilds of Connemara.

At Studley Park, Ripon, there are four pots (Fig. 334) standing on a balustrade that overlooks the water.   The handles are of the arabesque griffin sort, and are common on pots of this shape.   The realms of classical myth have been ransacked to supply subjects for the low reliefs that decorate the bowls, and these reliefs are often continuous round the bowl, stopping only for the handles.   In some, however, as at Windsor (the photograph of Fig. 335 is reproduced by permission of H.M. the King), the classical figure or scene is enclosed in a little panel, rather in the Flaxman manner.   The base of the Windsor pot is rather small, and in this way not so practical as the Studley Park example, in which the stem element has been eliminated. The less stem there is to a pot of this sort the better, for lead vases are very apt to take a tottering pose.

FIG. 337.—Vase and Bust.

The examples so far dealt with have in common a general appropriateness to their material.   It would be impossible, indeed, to make some of them in anything but lead, the idea of bronze being rejected as unsuitable for English gardens.

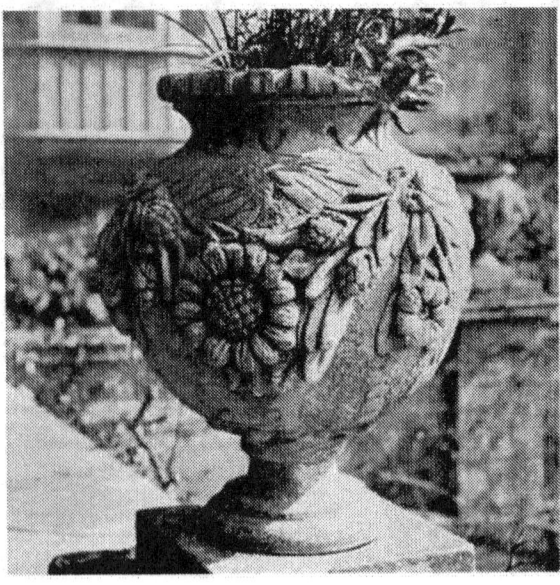

FIG. 338.—Wilton House, Wilts.

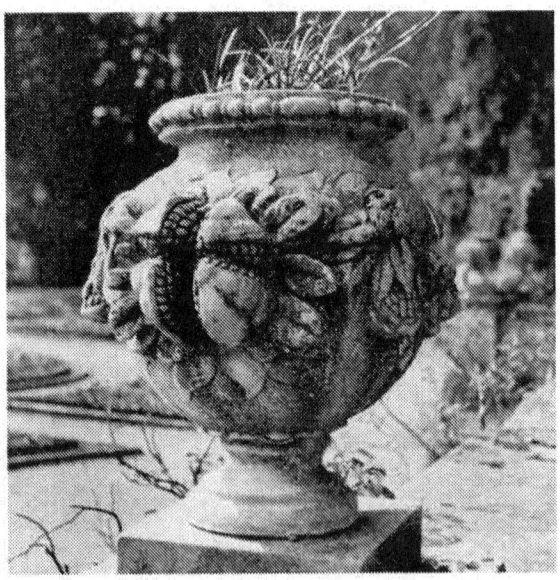

FIG. 339.—Wilton House, Wilts.

Of the Charlton House pot, shown in Fig. 336, less can be said.   It is obviously a terra-cotta design, and probably a simple copy of a terra-cotta vase.   The relief is very

Roman. There is a replica of this at Wootton Wawen Hall, and there are smaller vases of the same type, which seems most unsuitable for lead.

Fig. 337 shows a very queer hybrid of vase and bust. The vase is of a usual pattern, with acanthus handles and decoration round the base, and *amorini* in relief on the body of the bowl. It is in the possession of Mrs Frederick Leney, and was bought in 1794 by the grandfather of the last owner. How the bust came to be fixed in the pot, and what the mental attitude of the man who thought a bust a suitable alternative to a flowering plant, it is impossible to say. It is said that the bust represents Henri Quatre, but as the likeness is not striking and there is no royal emblem or badge to indicate that we have to do with a king, the attribution must be received with grave doubt. That it is a portrait bust, and French, is very likely, but in default of some evidence it would be unwise to be more definite. The total height of vase and bust is 26 inches. The

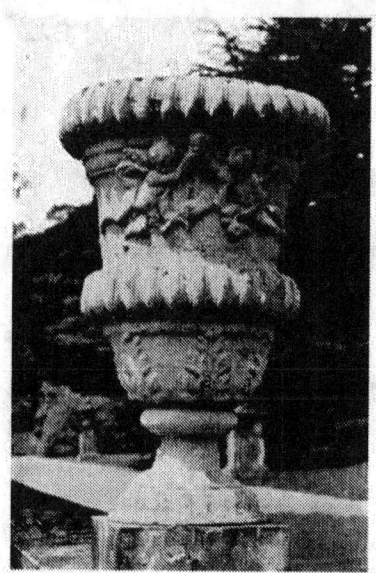

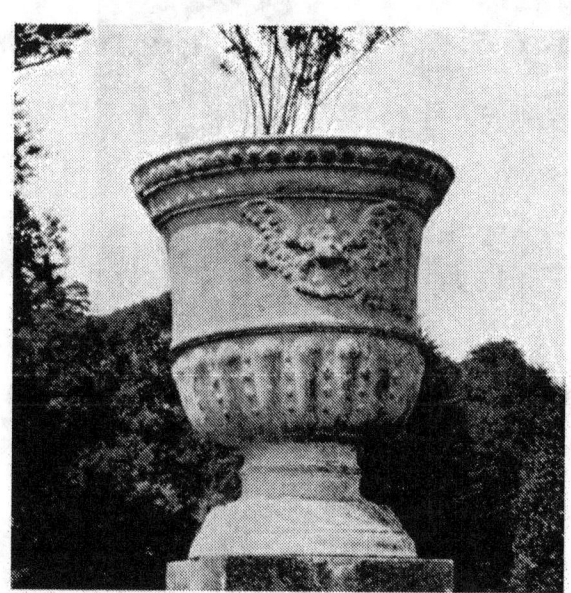

FIG. 340.—Castle Hill.   FIG. 341.—Castle Hill.

splendid gilt lead bust of Henri Quatre, now among the loan objects at the South Kensington Museum, is in every way infinitely finer.

At Wilton House, Wiltshire, is a series of flower-pots which are more of the vase than the pot type (Figs. 338 and 339). There are four patterns in all, varying in the flowers and fruits which form the swags. Very delightful they look, alternating with *amorini* on the piers of the balustrading which surrounds the Italian garden. From the fact that some of the *amorini* are cast from the same patterns as those at Melbourne, it is reasonable to guess that here we have more of van Nost's work.

At Castle Hill, Devonshire, there are lead flower-pots of two patterns. That of Fig. 341 stands well on a tall stone pedestal not far from the fine bust of Pan, and the mouldings are neat if not striking. The other (Fig. 340) is a fair example of the less attractive work of the eighteenth century. The mouldings are rather coarse, but the *amorini* cling to the bowl and support the coronet in a pleasant fashion.

The vase of Fig. 343 has a cherub which might have been cast from the same pattern that decorates the Castle Hill example, and the mouldings are simple and

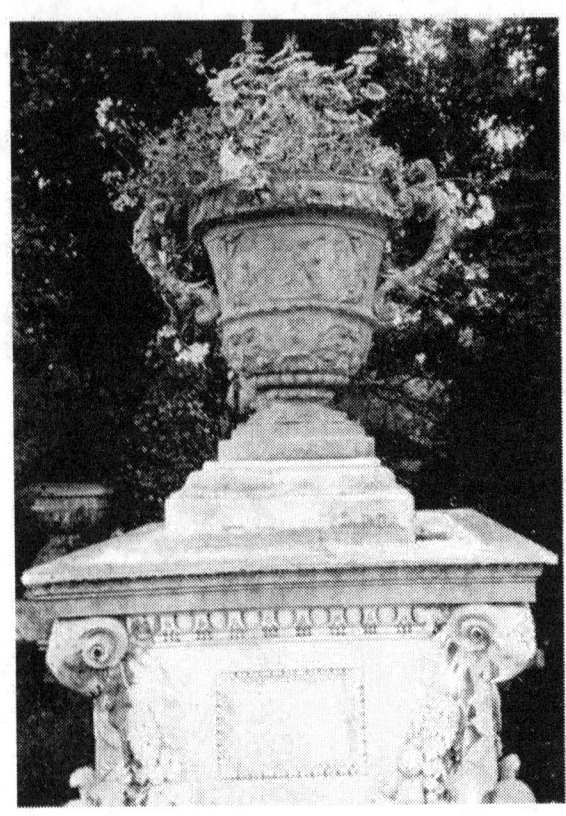

FIG. 342.—Enfield Old Park.

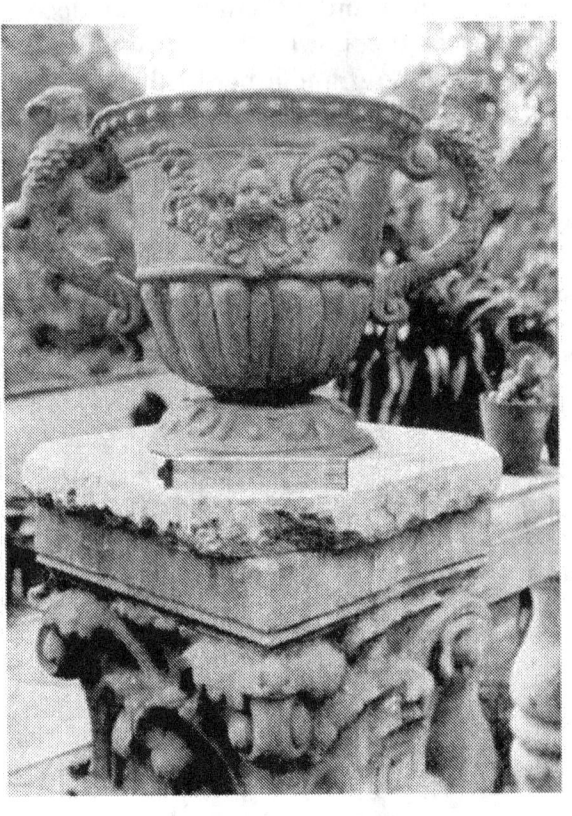

FIG. 343.—Myddelton House.

effective.   At Enfield Old Park Mr John Ford has a fine pot liberally decorated with acanthus ornament and figure reliefs (Fig. 342).

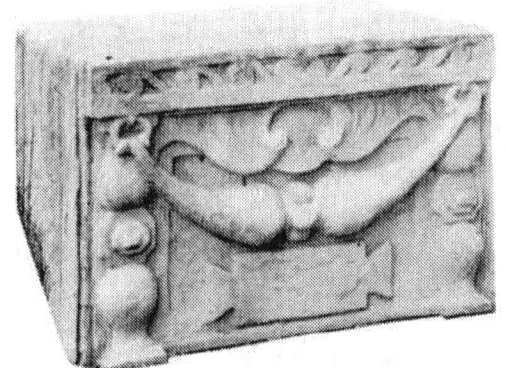

FIG. 344.—Lead Seat, Castle Hill.

At Drayton House, Northants, are many beautiful vases. One is an urn, rather in the Parham manner, but the majority are flower-pots with acanthus or griffin handles like those at Windsor and Studley Park. One, however, has lions' heads for handles, and in all the reliefs are unusually bold and elaborate.

At Penshurst is a vase that came from Old Leicester House in London. It is of the Studley type with acanthus handles terminating in horses' heads, and has a lid with pineapple top, which puts it in the urn category.

CHAPTER XI.

SEPULCHRAL LEADWORK.

Romano-British Coffins and Ossuaries—Mediæval Coffins and Heart Cases—Absolution Crosses—
Tomb Lettering.

EPULCHRAL leadwork is not a wildly attractive subject, but it has a peculiarly important place in the development of the decorative treatment of lead in England, because it is in coffins almost exclusively that we see Romano-British design. The subject cannot, therefore, be passed over, but notes on the various coffins found have been relegated to the Bibliography, and details are there given of the range of ornaments used. The example of Fig. 345 from the Maidstone Museum was found in 1869 at Milton-next-Sittingbourne, and is highly characteristic of Romano-British work. The cross-ornaments were made

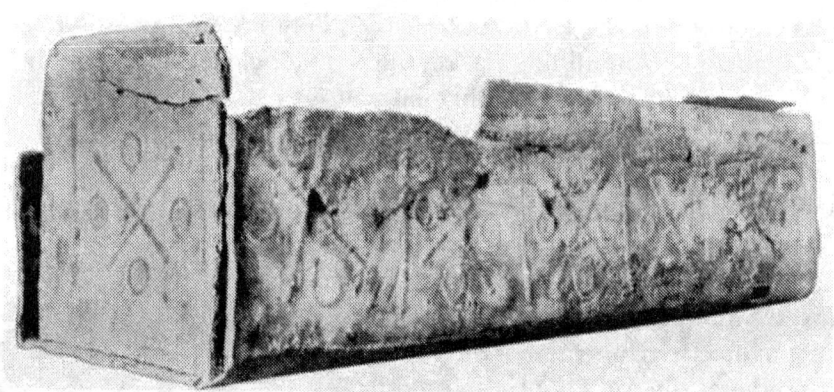

FIG. 345.—Romano-British Coffin, Maidstone Museum.

by pressing into the sand bed, before the lead sheet was cast, turned wooden rods of bead and reel design.

The same rod treatment, and also the rings, occur on Romano-British coffins at the British Museum, the latter now unfortunately in the basement, and inaccessible for inspection.

It is also seen on the Romano-British ossuaries at the British Museum (Figs. 346 and 347). Sol in his quadriga, on the example of Fig. 346, is the ancestor of the lively friezes of the Devonshire cisterns, just as the bead and real rod decoration led the way to the more sophisticated ornaments of the London cisterns. The ossuaries are technically admirable. The joints are burnt, not soldered, and the bead and reel rods, cast hollow to save metal, effectually brace the vessel.

A similar ossuary, but undecorated, is to be seen in Gundrada's Chapel. This brings us to the coffin of William de Warenne, at Southover Church, Lewes. It is one of the simplest of the mediæval types (Fig. 348), and in general treatment is more akin to the Roman coffins than to the examples with elaborate tracery that exist (but unhappily out of sight) at the Temple Church, London.

It is fortunate that careful drawings of the Temple coffins were made by Richardson, and these are reproduced in Figs. 349 to 352. The character of the ornament is very like that of the Long Wittenham and Warborough fonts (*q.v.*), and Richardson attributes the work to the beginning of the thirteenth century.

The burial of the coffins, and the sanctity of the fonts, have preserved to us these very beautiful and characteristic studies in the decorative possibilities of leadwork, and there is little doubt that in the more ordinary

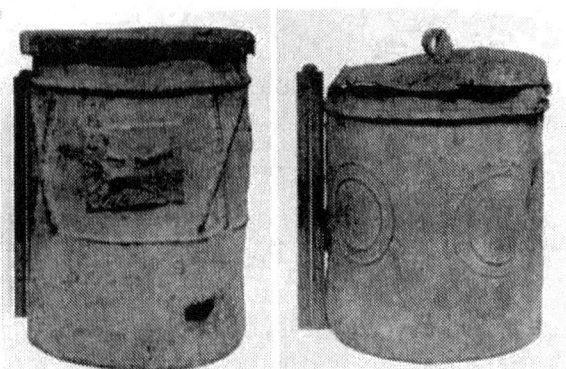

FIGS. 346 AND 347.—Ossuaries at Bristol Museum.

plumbing works the craftsman indulged a like fancy, but its products have disappeared. It will be noted that while the treatment of the Temple coffins is far in advance of the Romano-British, the rope moulding is retained to enclose some of the tracery panels, and for crosses, &c. In most cases the pattern was doubtless a piece of rope pressed into the sand.

The coffin of Fig. 349 is so much more elaborate than the others that it doubtless held the remains of an important personage. Coffins of this shape that followed the head outline are less usual than the box form. Probably the Temple coffins are among the earliest of mediæval times, as the Romano-British practice died out, and did not come in again until about the middle of the twelfth century. From then until late in the seventeenth century lead coffins were largely used, and were buried either with an outer wood or stone coffin or without.

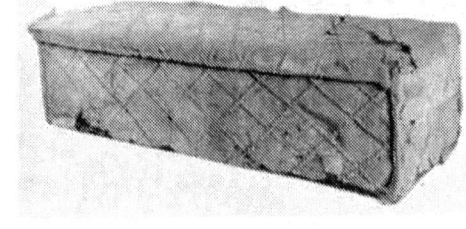

FIG. 348.—Coffin of William de Warenne, Lewes.

Reference has already been made to the Reliquary at Folkestone (Fig. 124). Akin to such objects are the heart caskets now illustrated. In the *Trésor* of Rouen Cathedral is preserved the plain box which held the heart of Richard Cœur-de-Lion. It is the inner of two cases, the outer being undecorated and much damaged. The lid of the inner box (Fig. 353) is engraved "✠ HIC IACET COR RICARDI REGIS ANGLORVM." The heart itself was found "withered to the semblance of a faded leaf," and was reburied in the choir in a new casket. Originally the lead cases were enclosed in

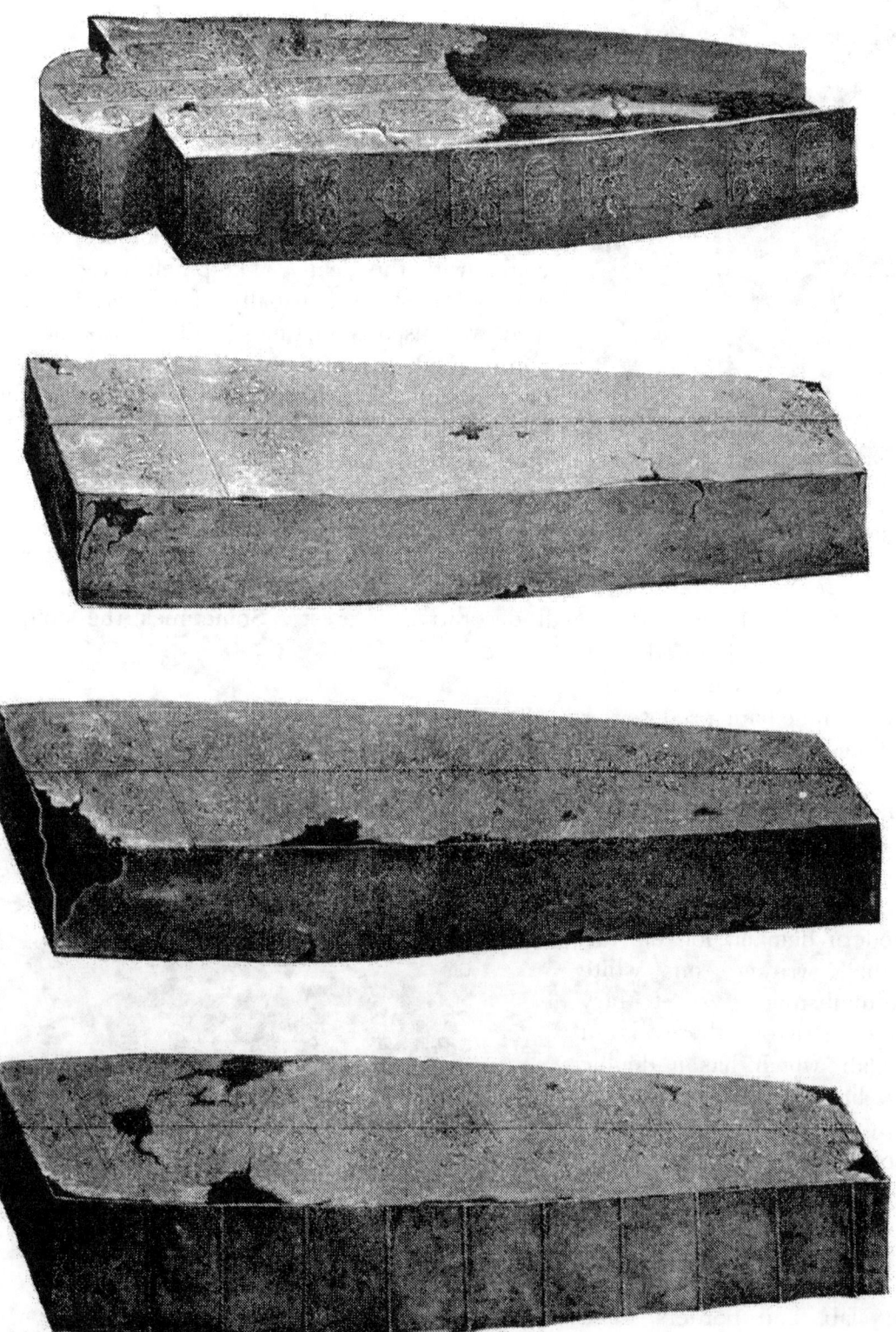

Figs. 349 to 352.—Lead Coffins Found at the Temple Church, London, and Re-buried.

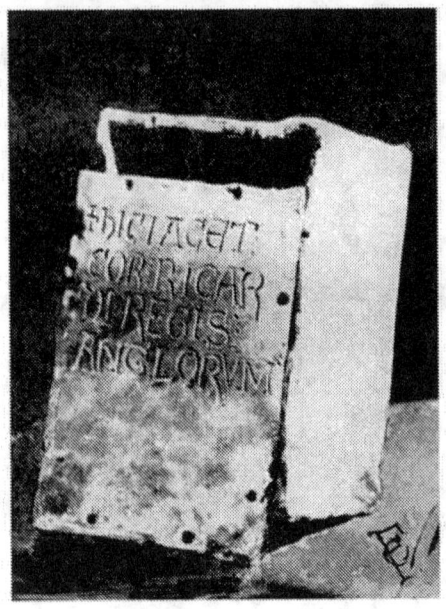

FIG. 353.—Heart Casket of Richard I. at Rouen.

a sumptuous gold and silver casket, which was sold in 1250 to raise money for the ransom of St Louis.

A later but very interesting example is that of Fig. 354. On the lid is a spear-head enclosed by a garter, and engraved on the bowl are the words: "Here lith the Harte of Sir Henrye Sydney. Anno Domini 1586."

Lead was largely used for objects enclosed in coffins with the dead. The paten and chalice buried with a priest were usually of pewter, not lead, but lead was used sometimes. The absolution crosses laid on the breast of the deceased were very frequently of lead, and the Bibliography gives many references. One is said to have been found in King Arthur's grave, and Mr Lethaby reproduces Camden's drawing of it and its inscription. A judgment as to its authenticity may well be left to experts in the Arthurian legend. Another found at Southampton commemorates one Udelina, and is engraved with the "Ave Maria." These objects have small decorative interest. Sometimes the coffin plates were of lead and lettered. A good example is that of Theobald, the immediate predecessor of St Thomas à Becket as Archbishop of Canterbury.

One of the most decorative but rather rare uses of lead was as a filling for incised inscriptions, a use revived to meet the modern demand for an imperishable writing on white marble tombstones. At St Mary Redcliffe, Bristol, there is a tomb slab which has a double border line, and between the lines a Latin inscription in common form, which seems to commemorate (fixed pews prevent a full reading) Johannes Blecker and Ricardus Coke. A cross extends the whole length of the slab, and borders, text, and cross are incised in the stone, and filled flush with lead.

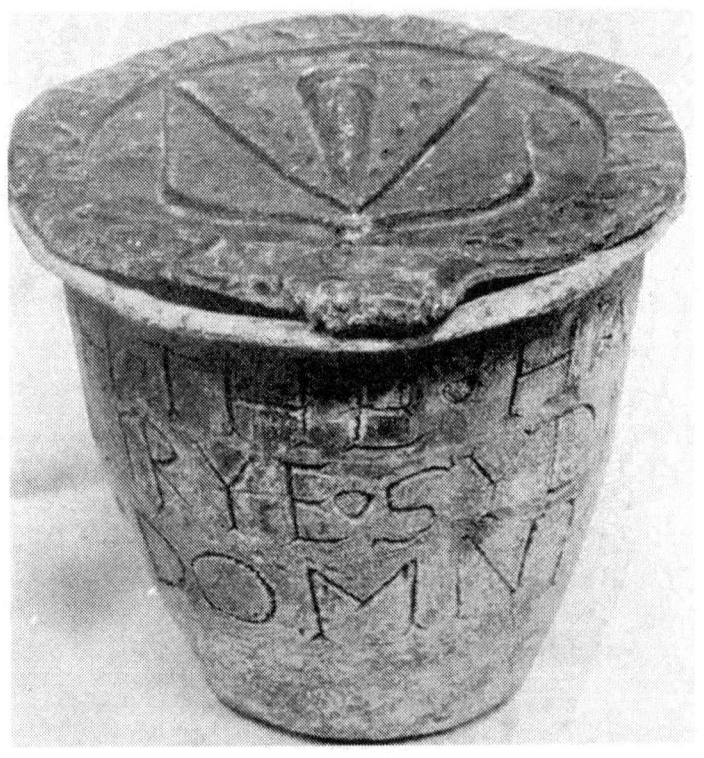

FIG. 354.—Heart Casket of Sir Henry Sidney, British Museum.

There is also an eighteenth-century inscription to one Lucas Stritch, incised, and without lead filling.

Lead grave slabs were used too in the eighteenth century. There is one at Wilmington, 22 inches by 15 inches, dated 1757, to the memory of one Thomas Ade and his family. It has a long inscription, and is a plain casting with raised letters.

Brass as a material for mural memorial tablets was sometimes set aside for lead. In the family pew at Dorney Church near Windsor, are the plates which have been described as memorial tablets. They are, however, coffin plates taken from a vault, and bear dates 1768 and 1774. Mr Lethaby mentions a lead wall tablet to Lady Corbett in Burford Church, Salop, dated 1516, but there are difficulties attached to getting a photograph of it.

As this chapter goes to press Mr Philip M. Johnston, F.S.A., reports a very notable find of three mediæval lead coffins at Tortington Priory, Sussex. The ornaments include various floral and star-shaped devices within a diamond lattice-frame, a cross in rope moulding, and a variant of the Greek honeysuckle. The latter is a singularly interesting ornament, as will be seen by Fig. 354A, while the four-leaved pattern of Fig. 354B compares in beauty with the decoration of the best fonts of the same period. Two of the coffins will find a home in the museum of the Sussex Archæological Society at Lewes, and the lid of one, it is hoped, in the British Museum. Mr Johnston is to be congratulated on a material addition to our knowledge of late twelfth-century leadwork.

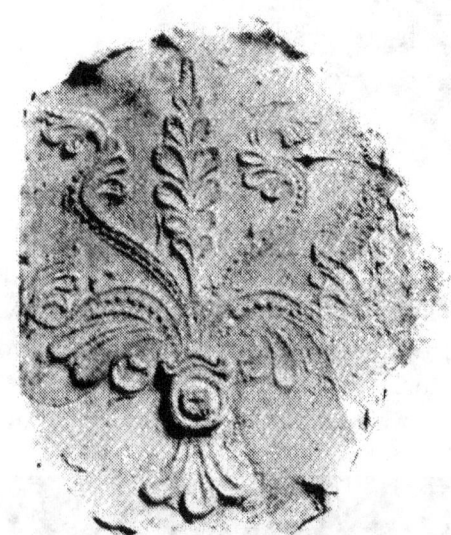

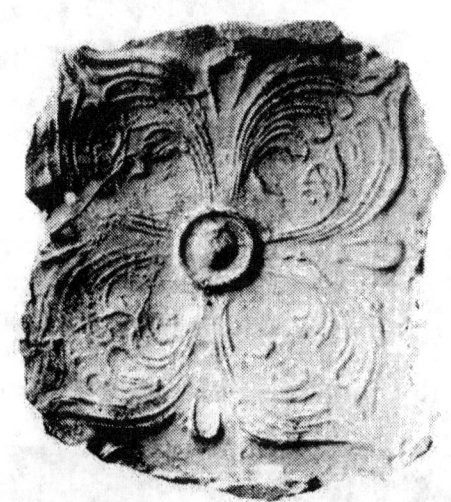

FIG. 354A.—Honeysuckle Ornament. FIG. 354B.—Four-leaved Ornament.

ORNAMENTS FROM LEAD COFFIN FOUND AT TORTINGTON PRIORY, SUSSEX.

CHAPTER XII.

VARIOUS OBJECTS AND DECORATIVE APPLICATIONS OF LEAD.

Roman Pigs and Pipes—Pilgrims' Signs—Papal Bullæ—Ornaments on Woodwork—Charms—Tobacco Boxes—Ventilating Quarries.

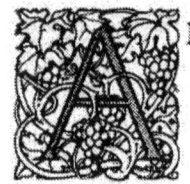 N "omnibus" chapter is not a very satisfactory way of providing a place for odd items which are difficult of classification, but it is perhaps a better device than to smuggle them into the introductory chapter as is sometimes done. In this book, moreover, there has been a steady purpose to emphasise those uses of lead which are practical and capable of more extended revival. With one or two exceptions, the objects dealt with in this chapter belong solely to history.

In the pig of lead found at Chester (Fig. 355) we have lead in its simplest form as a

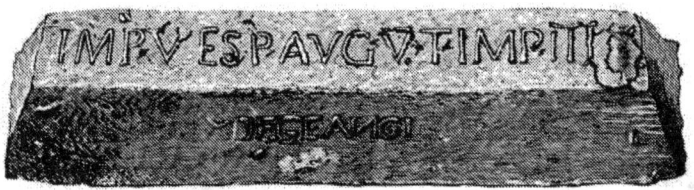

FIG. 355.—Roman Pig.

manufactured article. This example was a stray from a consignment of pigs paid to the Roman occupiers of Chester by the Deceangi, a Flintshire tribe that busied itself with lead mining. It bears, as do most of the Roman pigs, the name of the reigning emperor.

The pipe shown in Fig. 356 is particularly interesting, as the inscription tells a long story. Roughly translated, it runs, "These pipes were laid when Vespasian and Titus

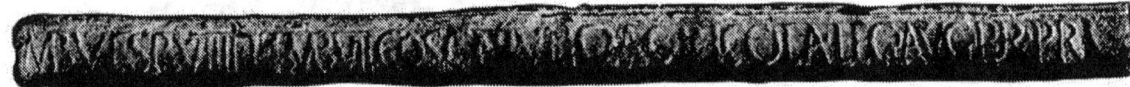

FIG. 356.—Roman Inscribed Water-Pipe.

were Consuls for the eighth and ninth times respectively, and when Cnæus Julius Agricola governed the Province of Britain." The date is A.D. 79, and the pipe is of interest as showing that the elaborate water supply of Rome found its imitators in the Roman colonies in Britain. Of this there is further proof among the Silchester finds, which include a flanged pipe about 16 inches long and 2 inches in diameter, and fragments of sheet lead with edges snipped to a rough fringe.

The jointing of the Chester pipes is of two kinds, both shown in Fig. 357. The upper was formed by pouring molten lead into a mould of earth round the ends to be joined; the lower has the surface comparatively smooth, and appears to have been made like a modern wiped joint. The Silchester pipe referred to above has a keeled longitudinal seam. Other pipes have a longitudinal butt joint, which was probably soldered, but the solder has perished.

There are no decorated lead objects at Silchester, but several steelyard weights with iron eyes cast in. Mr Lethaby has figured a Roman jewelled lead cup in the British

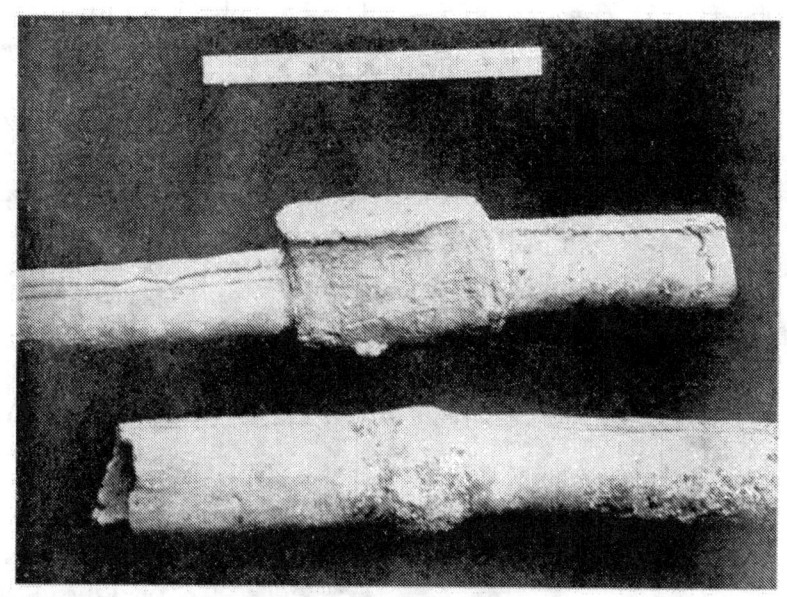

FIG. 357.—Roman Methods of Jointing.

Museum, but it was probably made abroad. In general decorative effort seems to have been reserved for the sepulchral objects described in the last chapter.

When we come to mediæval times, the wealth of small objects is almost bewildering. The most interesting of these are the Pilgrims' Tokens.

Erasmus in his "Pilgrimage" represents one of his interlocutors as meeting a pilgrim and addressing him thus: "Thou art . . . laden on every side with images of tin and lead." The custodians of shrines did a thriving trade in these small memorials of pilgrimages, which most commonly took the form of round, oval, square, or lozenge shaped plaques having either a loop for sewing to the dress or pins for use as brooches. These signacula represented an infinite variety of subjects, of which a good idea can be formed by reference to the catalogue of the London Guildhall Museum. Most of the Guildhall tokens have been found in the Thames. An enormous quantity has also been dredged from the Seine.

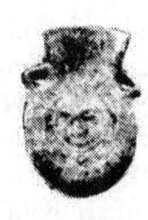
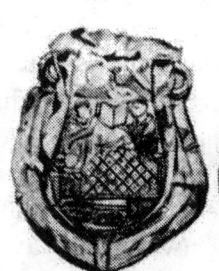
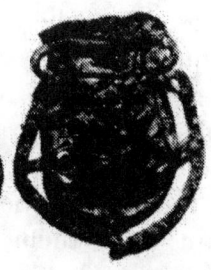

FIG. 358.—Small Ampulla, York Museum.

FIG. 359.—Drawing of Reverse of the Canterbury Ampulla, York Museum.

FIG. 360.—Canterbury Ampulla, York Museum.

The ampullæ sold at Canterbury were among the most popular. They have been variously said to have held a solution (one would suppose dilute) in water of the blood of St Thomas à Becket, dust gathered round the saint's shrine, or oil from the lamps burning there. Whatever they held, they are in effect little leaden bottles 3¼ inches long, and were hung round the neck. On one side (Fig. 360) is a bishop in robes with mitre and staff. On the narrow fascia round the ampulla is the legend. "*Optimus egrorum medicus fit Toma bonorum*"—The best physician for the good invalid is Thomas. On the reverse (Fig. 359) is a representation of the rite of extreme unction, which is being administered to the sick man by two priests. Fig. 358 also shows a small ampulla. Fig. 361 shows five examples from a private collection, including a *St Edward the Confessor*, a *Virgin and Child*, and *a Crucifixion*. Other common forms are a *W crowned* for St Mary of Walsingham, *scallops* for St James, and a *T* for St Thomas à Becket. The legend on the Canterbury ampulla indicates the popular belief in the curative properties of some at least of the tokens. Sufferers from ague would put their trust in Sir John Schorne, a saint of high repute in that connection. On an emergency (doubtless in the intervals of curing ague) he conjured the devil into a boot, and is represented on his token with the enemy thus conveniently restrained. Other signs were the *Vernicle*, or likeness of Our Lord, and the *Head of St John Baptist*.

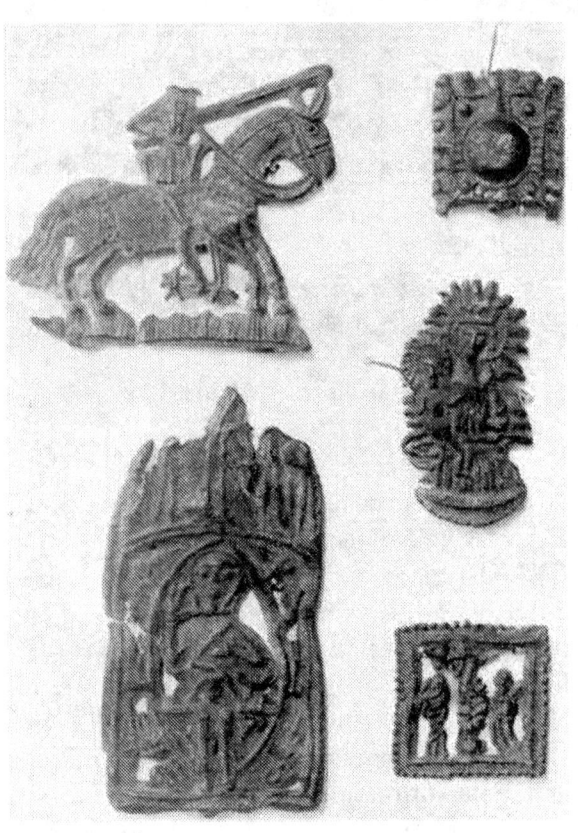

Fig. 361.—Pilgrims' Tokens (actual size).

A curious classical parallel to these mediæval objects is to be found in the lead figurines of the sixth century B.C., found at Sparta on the site of the sanctuary of Artemis Orthia. The types represented include heraldic animals, goddesses, and warriors. They were cast from moulds on one side only, and from their rough technique it would seem that the same methods were employed as for the mediæval signacula. Their purpose was votive, and save for the fact that the Spartan offered them at the shrine, whereas the mediæval Englishman took them away by way of remembrance, the separation of about twenty centuries means but a small difference in intention and execution.

A considerable number of the mediæval stone moulds in which the tokens were cast remain. Shrines were not responsible, however, for all these tokens. They were used in abbeys as vouchers for attendance in choir, like the timekeeper's brass numbers in a modern factory. Lead medals, too, were struck for the Festivals of Fools in the Middle Ages, and mock coinage was struck in lead by the Boy Bishops, who were elected to commemorate the Murder of the Innocents. Altogether the output of small decorative lead objects in mediæval times was great, and collectors have sought them eagerly. Demand creates supply, and about 1857 two ingenious workmen named O'Flanagan, also known to fame as Billy and Charley, conceived the brilliant idea of forging them in great numbers, and "discovering" them during excavations. Archæologists either believed or

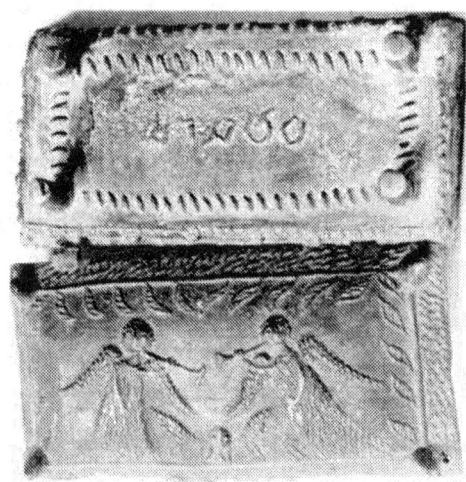

FIG. 362.—Top and Bottom of the Box.

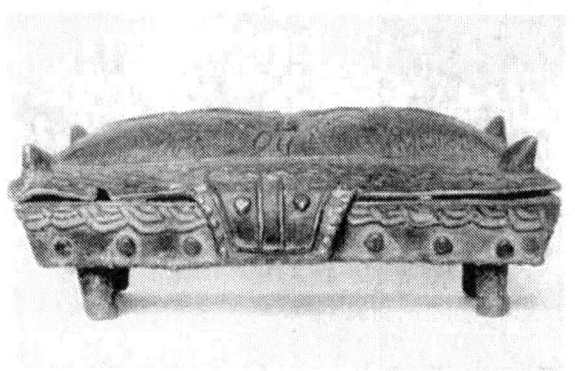

FIG. 363.—The Box with the Lid on.

disbelieved in the discoveries, and many hard words were said, and legal proceedings even were taken. It was sufficiently proved that the output of Billy and Charley ran into many thousands, and at the Guildhall Museum the so-called "Dock" forgeries are set apart and frankly labelled. The mock tomb of Figs. 362 and 363, consisting of a box with four feet and a lid, is obviously a forgery of this period, and probably the most ambitious that was achieved. Other examples are spear-heads, daggers, seals and rings. Many are decorated with dates of the eleventh century in Arabic numerals!

Papal seals or bullæ, whence the document itself got the name of bull, form an important series of small lead objects, of considerable historical interest. In 1878 Pope Leo XIII. ordained that papers of minor importance should have wax seals, lead being reserved for the more solemn documents. The earliest bulla in the British Museum is one of John V. (685-686), and from his pontificate until thirty years ago, every papal document had its lead seal appended. When the communication was a pleasant one, it was attached by threads of red and yellow silk; if in *forma rigorosa* the thread was of hemp, a grim suggestion.

Fig. 364 shows a series of four bullæ found in Sussex. The obverses bear the name of the Pope, and the reverses conventional heads of St Peter and St Paul with the labels over them, SPA (for Sanctus PAulus), and SPE (for Sanctus PEtrus). Three of the popes figure in the " Divina

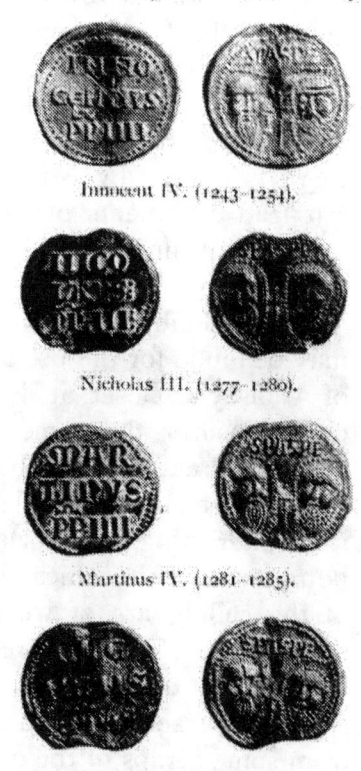

Innocent IV. (1243-1254).

Nicholas III. (1277-1280).

Martinus IV. (1281-1285).

Clement V. (1305-1314).

FIG. 364.—Papal Bullæ found in Sussex.

Commedia." Nicholas III. was in Hell amongst the simonists; Clement V., who exiled Dante, was "licked by ruddier flames," while Martin IV. had the easy fate of fasting in Purgatory to purge his sin of gluttony.

There are lead impressions of seals in various museums, which are apt to mislead. They (or some of them) have the appearance of antiquities. Figs. 365 and 367 show

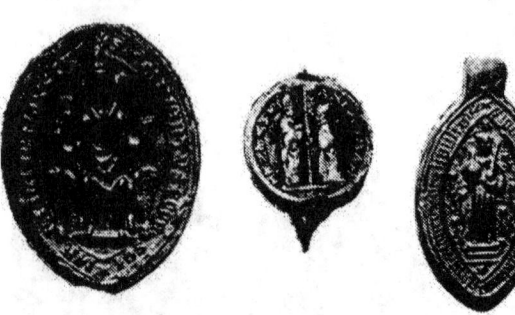

FIG. 365.   FIG. 366.   FIG. 367.
(In York Museum.)

examples at York which have been taken for pilgrims' signs, &c. They are simply modern casts of conventual seals. The little medallion of Fig. 366 is probably foreign, and was apparently used as a seal on a cord like the many examples of lead seals used by cloth and other merchants for sealing bales of cloth in bygone days. The Post Office of to-day uses similar seals, but does not waste ornament on them.

Lead has been used for every sort of unlikely purpose, for things as diverse as tickets for eighteenth-century dances, and the book cover of an Anglo-Saxon manuscript of Alfric's homilies. It may be hoped that no enthusiastic leadworker will regard either of these as suitable precedents.

Among its less usual architectural uses may be mentioned its substitution for wood carving in the ornamentation of rood screens and the like. At Worsted, Norfolk, the screen panels have figures painted on a gesso ground, and the bands of ornament beneath the figures and the spandrels above them are (or were, for the church was *restored* a few years ago) of lead painted and gilt.

In Mr Francis Bond's book on "Screens" there is a note by Mr W. Davidson on the gilt lead ornaments of the Ranworth screen and the Burlingham pulpit. The Ranworth ornament is "a close imitation of a star-fish."

It is doubtful whether much justification may be found for the use of lead on the ceiling of Wolsey's Closet at Hampton Court. It clearly usurps the place of plaster, and for no visible reason. Doubtless the work is by an Italian hand, and while its richness makes it an interesting study (see Fig. 368) it must

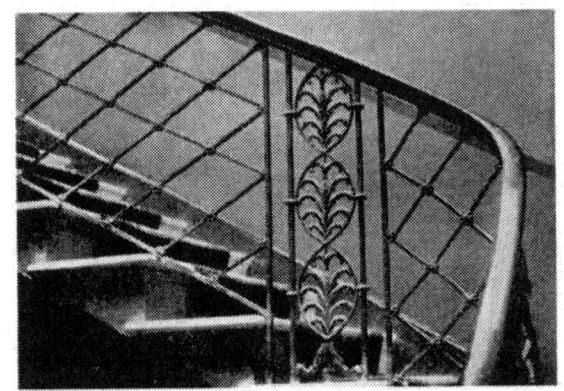

FIG. 369.

be regarded as technically a freak, and need not here be discussed at length. The ribs of the ceiling are of wood and the panels of *papier maché*, but the leaves at the intersections are of lead, as are also the letters of Wolsey's motto on the frieze.

Round the beautiful painted chest in the parvise of Newport Church, Essex, runs a gilt lead traceried band of exquisite delicacy. The existing work is a careful restoration from some scraps of the original, which are to be seen at the South Kensington Museum.

The use of lead for such purposes as the decoration of furniture, is open to some question, but in the case of the Newport chest the end fully justifies the means, for the same effect of delicate richness could not have been obtained by the woodcarver.

Mr Harold Brakspear, F.S.A., has drawn attention to (and has figured in *Archæologia*,

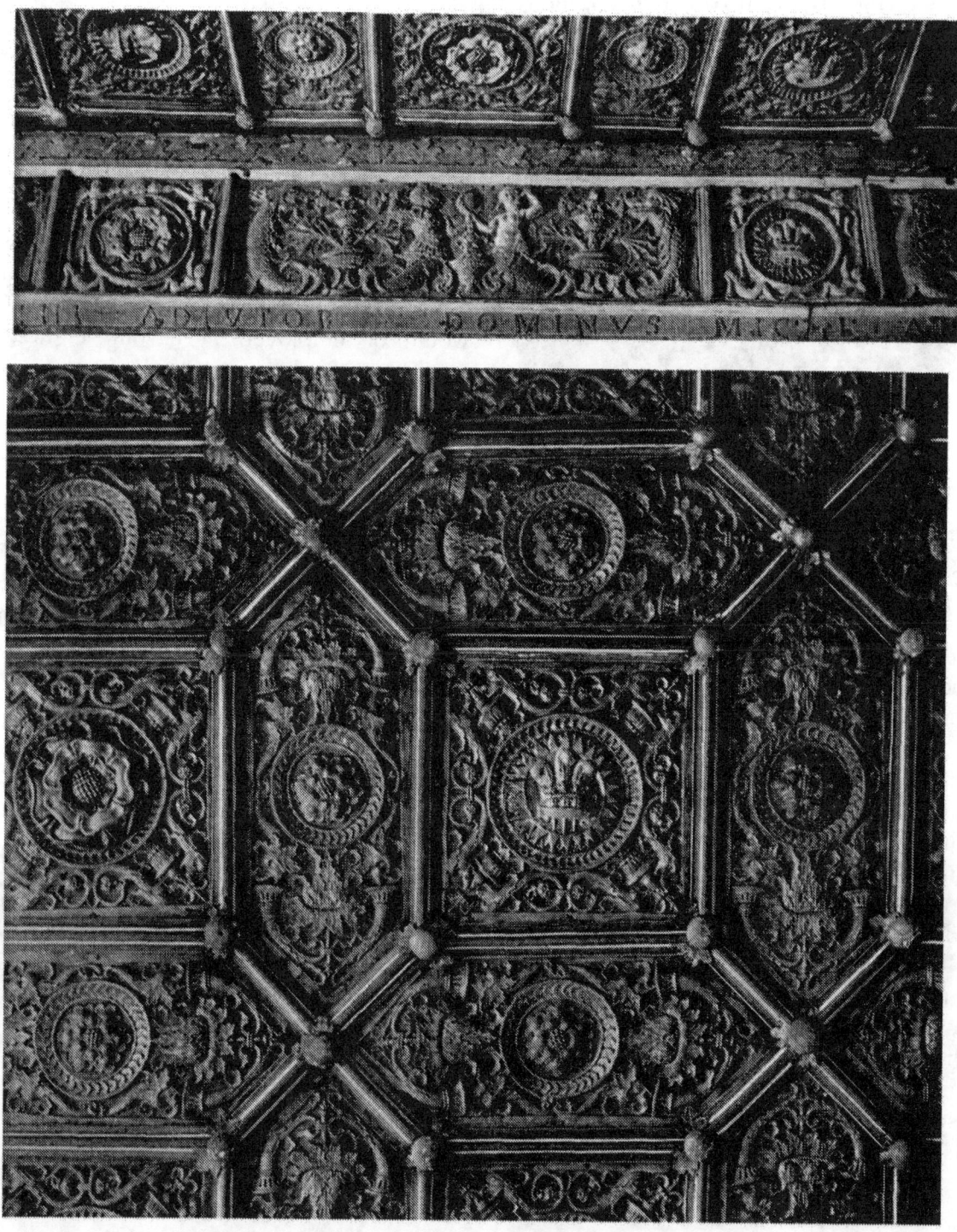

FIG. 368.—Ceiling, with Lead Enrichments, Hampton Court.

vol. lx., part 2) some little lead panels of fifteenth-century open tracery, found at Stanley Abbey, similar in form to those of Fig. 373. He points out that though they are generally supposed to be ventilators, the fact that rivets were found attaching a small piece of sheet iron to which the leadwork was originally fixed, goes against this supposition. Obviously rivets and sheet iron have nothing to do with lead glazing, and it seems likely that we have here a case of lead tracery being used to decorate an iron box or other object of domestic use, and that so far it is analogous to the decoration of the Newport chest.

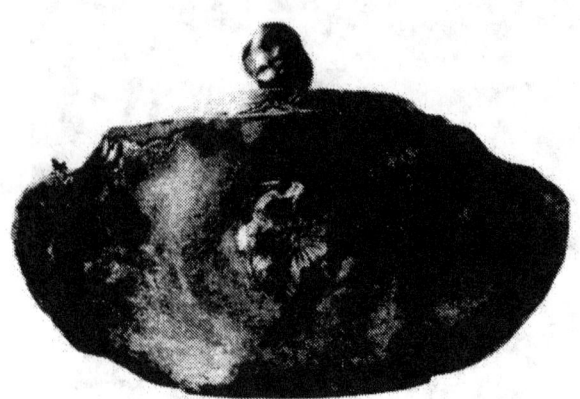

Fig. 370.—Tobacco Box, Maidstone Museum.

Cognate in character, though widely separate in date, is the inlaying of the west doors of St Pancras Church by Inwood with lead mouldings. In this case, however, lead is simply a cheap substitute for wood. Robert Adam used lead for the enrichments of mantelpieces and the like, as *carton pierre* would be employed. In some eighteenth-century mantelpieces panels in low relief depicting some conventional classical scene were sometimes cast in lead. Doubtless the patterns used for garden vases thus served a double purpose.

There is something to be said for the eighteenth-century practice of making the ornaments of wrought-iron staircase railings in lead. Fig. 369 shows a scholarly example of this, but the lead is here stiffened by tin or antimony into an alloy of considerable hardness. Pure lead would obviously have been too soft. Here lead takes the place of bronze or brass for cheapness' sake. Speaking generally it seems fair to employ lead for modelled enrichments where a large number are required of the same design, as for example the gilt stars that were so freely used on Gothic ceilings, and parts of the pendants of the ceiling of Hampton Court Chapel. It is, however, difficult to find a suitable commentary on the restorer of a church near Oxford, who finished off a rood screen with a cresting cast in lead from an old wooden model, and grained it oak colour!

As lead is the metal associated with Saturn, an often unfriendly planet, the purveyors of magic and spells did not neglect it when the agreeable business of curse-making

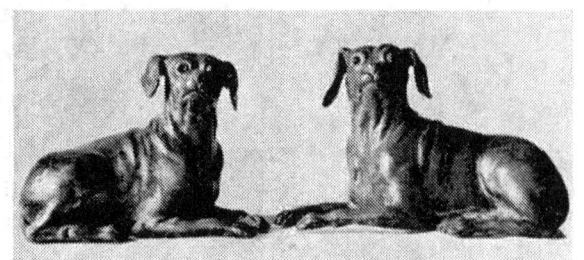

Fig. 371.—Lead Dogs.

was afoot. Some years ago an engraved lead tablet of Romano-British date was discovered at Bath. It is doubtful whether it records a curse on nine guests who were suspected of stealing a tablecloth, or a statement that one Quintus received 500,000 lbs. of copper coin for washing a lady named Vilbia. If the latter, their hydropathics seem to have cost them more. A lead disc inscribed with the symbols of Saturn has been found in a Cornish garden, deposited there for magical purposes.

Mr W. Paley Baildon, F.S.A., in a highly entertaining paper has illustrated and described a lead plate engraved with eighty-one squares on one side, and, on the other, " That Nothinge maye prosper Nor goe forwarde that [Raf *crased*] Raufe Scrope take in hand," and underneath this pious wish are the names " Hasmodai, Schedbarschemoth, and Schartatan, with three astrological symbols. These pleasant names belong to the spirits of the moon, who are thus invoked against the unhappy Scrope.

For coinage lead, owing both to its softness and the ease of forgery, is obviously unsuitable, but owing to the small supply of royal coinage at various periods local issues of lead tokens were made to supply the pressing need of currency. They were used chiefly in the sixteenth and seventeenth centuries, and in Ireland largely at the end of the eighteenth.

The British Museum contains many examples of foreign medallions in lead, exquisitely and delicately modelled. Many of these were doubtless struck or cast to test the perfections of die or model, and though in original intention fugitive, have survived by accident.

The distinctive colour and texture of lead make it more appropriate for some subjects, even if finely detailed, than bronze, and the admirable condition of the many remaining small lead medallions and delicate reliefs is sufficient answer to the objection that they have undue liability to damage.

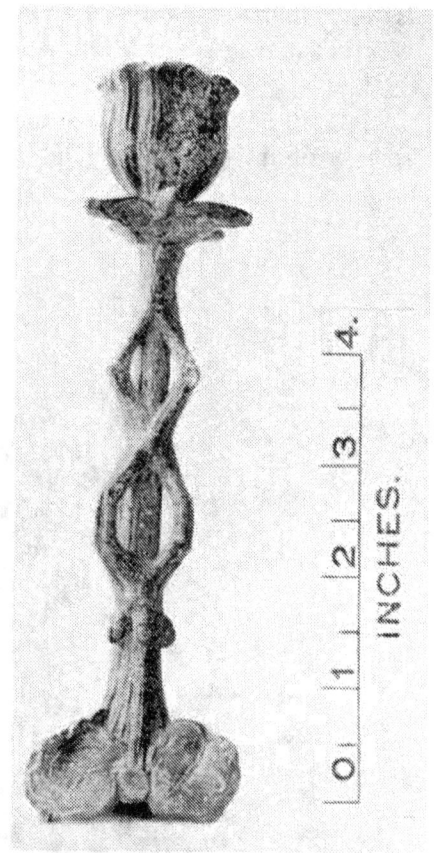

Fig. 372.—Lead Candlestick, Maidstone Museum.

Lead was used considerably in the eighteenth and early part of the nineteenth century for tobacco boxes. A common form is a square box on small feet with hunting scenes in low relief on the sides. In the Maidstone Museum is a lead box (Fig. 370), said to have been dug up at Tel-el-Kebir in 1882 by a soldier, who found it full of wheat. There is a rosette on each side, and the handle of the lid is a negro head. The soldier was probably a relation of " Billy " or " Charley " aforementioned. Negrohead is an historic brand of tobacco, and if the pot was found at Tel-el-Kebir, it was certainly taken there from England. The finding of wheat in it was an artistic touch, worthy of the land of mummy wheat. Tobacco stoppers of quite elaborate patterns were also made of lead as early as the seventeenth century.

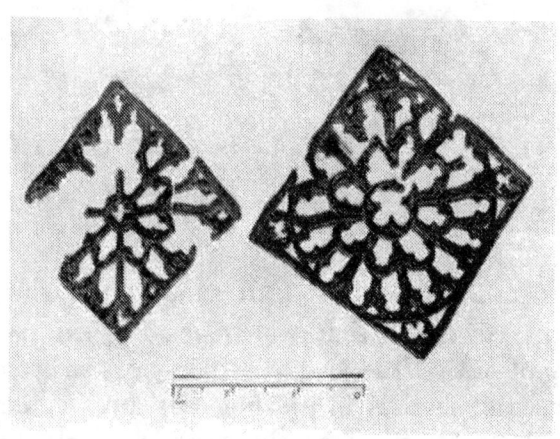

Fig. 373.—Quarries, York Museum.

The delightful dogs of Fig. 371 take us

further back. They are probably of Queen Anne's time, and well represent the spaniel type, that was popular then. They are in the possession of Colonel G. B. Croft Lyons, F.S.A.

It must be admitted that for most domestic objects lead is unsuited. Pewter, by reason of its fine texture and hardness, is in every way more suitable for such things as

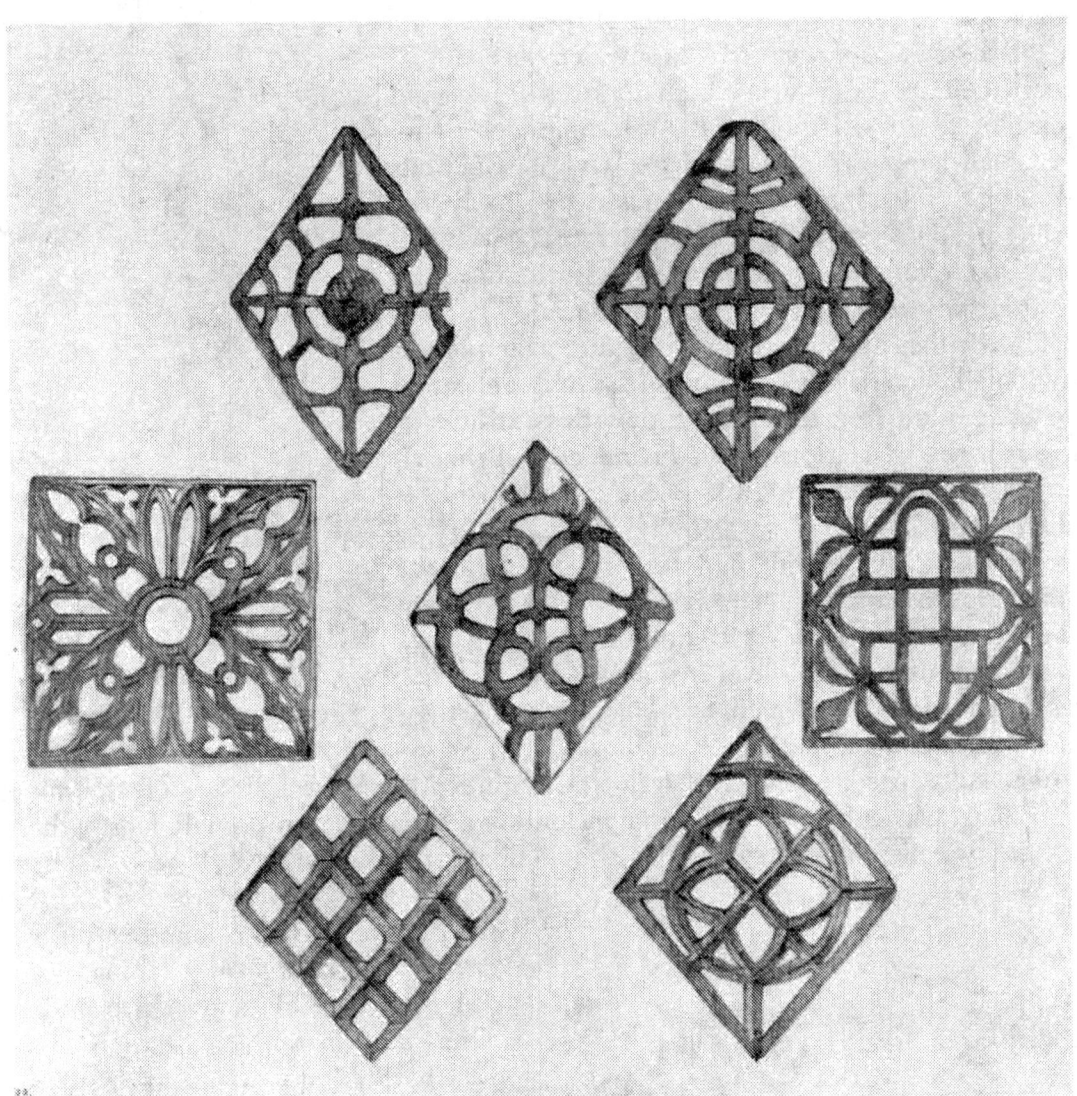

FIG. 374.—Lead Ventilating Quarries.

candlesticks. There is, however, in the Maidstone Museum a lead candlestick which is shown in Fig. 372. The commonest kind of pewter is that which has a great proportion of lead, and this candlestick is probably of such bad pewter rather than of good lead. Among the most important of all the uses of lead is in glazing, but any detailed study of this belongs more properly to a history of glass, as the lead is clearly the subordinate material. There is one class of objects, however, lead ventilating quarries, which perhaps may here be described, as their beauty depends wholly on the modelling of the lead itself.   There are two examples in the York Museum (Fig. 373), and Fig. 374 shows a series got together by Mr J. Starkie Gardner, F.S.A.   The square example with Gothic tracery is particularly delightful.   At South Kensington is one that bears the name of the plumber who made it.   There are many at Hampton Court. They are used, one or two in each window, in place of glass quarries, as air inlets, and are perhaps the only contrivance for ventilating which is not markedly ugly.

The glazing of fanlights over eighteenth-century front doors was frequently done with leading of delightful outlines, and with rosettes and other enrichments.   Illustrations of these are omitted, as they belong rather to the history of leaded glazing, which is another story.   In the early days of fire insurance, when one's house needed to be labelled to secure the kindly attentions of the firemen, the labels were frequently of lead. The author has a very pleasant example in a Royal Exchange tablet, which was coloured and gilt.   There is a good collection at the London Guildhall, including signs of the Hand-in-Hand, the London, and the Sun Offices.   Parish boundary marks were often cast in lead.   The City of London made lead shields-of-arms as ownership marks, and at the Guildhall is a well-modelled lion, with " M C 1693 " beneath, the mark of Morden College.   The device vulgarly known as the Southwark Arms, which is the ownership mark of the Bridge House estates, was frequently cast in lead.

It is hoped that the Bibliography of this volume will not be altogether neglected. The notes give references to many odd uses of lead which are not of enough importance to be incorporated in the main text.

CHAPTER XIII.

MODERN LEADWORK.

Fonts—Rain-water Heads—Cisterns—The larger architectural Uses—Figures on Buildings and in Gardens—Fountains—Vases—Clock-faces—Sundials—Gasfitting—Inscription.

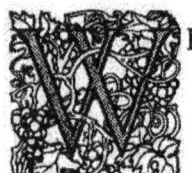

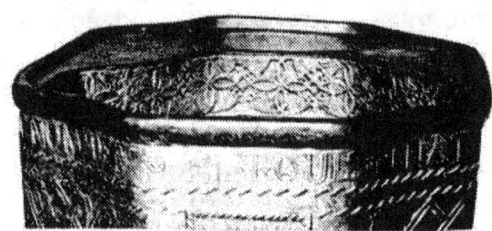

FIG. 375.—Font at Edinburgh.
(Showing Decoration inside Bowl.)

WHEN the late Mr J. Lewis André wrote in 1888 a paper on English Ornamental Leadwork, he said : "I am compelled to come to the conclusion that most of the applications of ornament to leadwork belong to bygone times, and that a revival at the present day is hardly to be expected." Twenty years have gone by, and happily Mr André is proved to have been no prophet. The revival is real, active, and increasing. Its products will now be illustrated in the same order, roughly, as in the chapters dealing with the old work.

Fonts.

Among modern fonts there seem to be none that rival, or indeed endeavour to imitate the splendid figure treatment of Norman times, when apostles and saints sat beneath elaborate arcading. The font of Fig. 376 is, however, very fully treated, and has much unpretentious charm. The relief is soft and flat, and the symbolism interesting. The fish in the wide middle band are the common symbol of Christianity, and their natural swimming motion suggests the living waters of baptism. On the upper band appear four panels which represent the elements, a symbol which seems natural rather than spiritual, and the lowest band is made up of lilies, also a symbol of baptism.

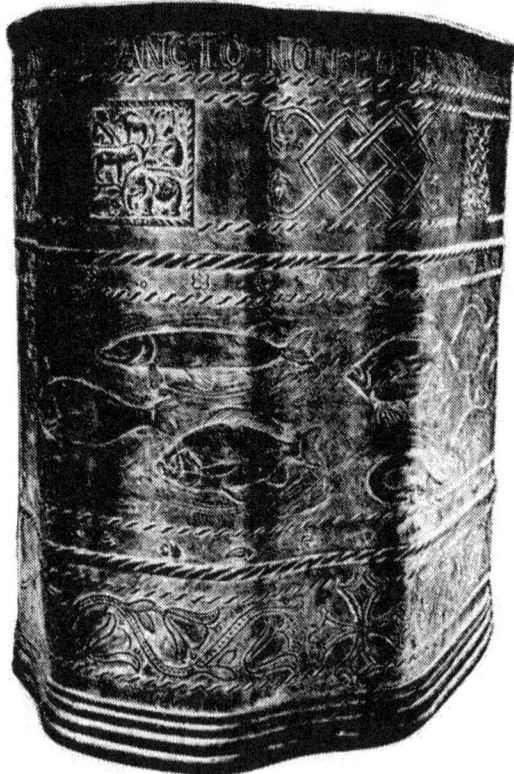

FIG. 376.—Font at Edinburgh.

The inscription round the top reads :—

"NISI QUIS RENATUS FUERIT EX AQUA ET SPIRITU SANCTO NON POTEST INTROIRE IN REGNUM DEI."

One of the most interesting features of this font is its practical arrangement. Reference to the illustration (Fig. 375) will show that there is a small basin provided at one side.

The main part of the font is filled with water which is blessed by the archbishop

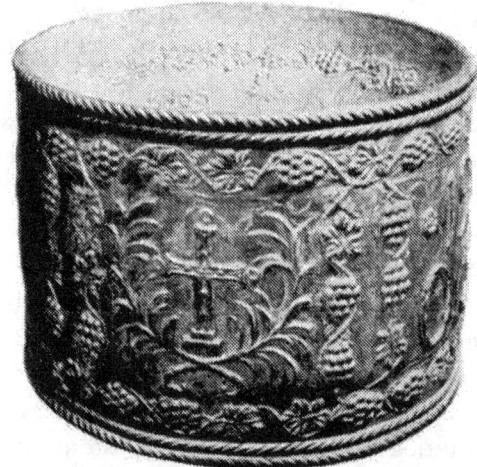

FIG. 377. St Alban's, Leicester.

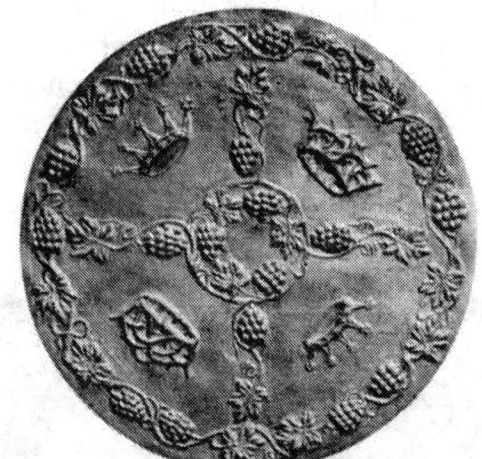

FIG. 378.—St Alban's, Leicester.
(Bottom of Bowl.)

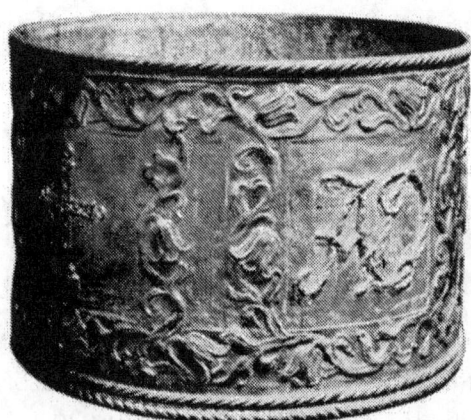

FIG. 379.—Font with Lily Decoration.

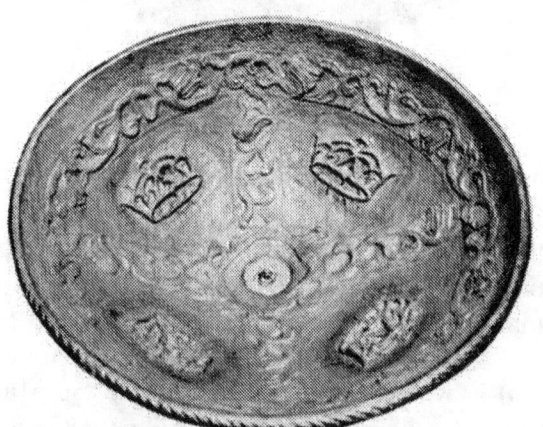

FIG. 380.—Saucer Top of Font.

once every year. The infant to be baptized is held over the small basin, from which the water used in the rite runs to earth. The font is an unusual but interesting shape on plan. The addition of the small oval basin indicated an octagon with two cardinal faces longer than the others. By making the cardinal faces rather convex, and the diagonal faces a little concave, a vague cruciform suggestion is given, and the outlines take on the easy flowing feeling that is so appropriate to the nature of the material. The font is

3 feet 6 inches high, and stands on a stone plinth, which hollows as it meets the floor to allow room for the toes of the officiating priest—a very practical thought.

The font was made by Mr Bankart for Mr R. S. Lorimer, R.S.A., for a Roman Catholic church in Edinburgh, and its whole treatment is original without being strained or precious.

The fonts of Fig. 377 and Fig. 379 are also by Mr Bankart. The former is at St Alban's Church, Leicester, and was made for Mr Howard Thompson, architect. An interesting feature is the decoration of the bottom of the bowl. It is a fresh and good idea to mitigate the usual bareness of the inside by ornament, and the crown of thorns and the crown celestial are added as emblematic of the difficulties and rewards of the Christian life entered by the gate of baptism. The vine is less appropriate, as being identified with the other of the two great sacraments, and, however pleasant a treatment decoratively, is a confusing emblem on a font.

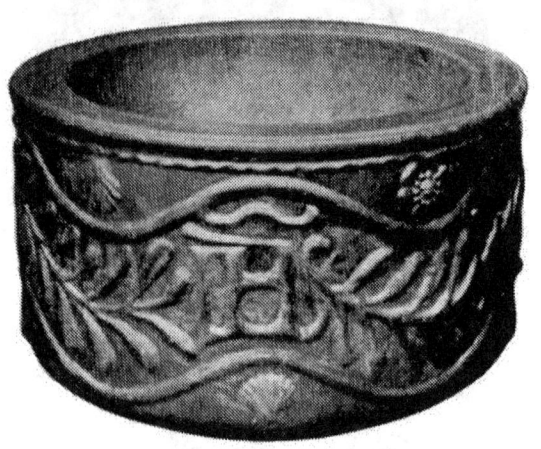

FIG. 381.—Font at Brithdir.

In the example shown in Fig. 379 the lily is again used as on the Edinburgh font, and though the A.D. and the date are a somewhat aggressive size, the design is more satisfying than that of Fig. 377. A most interesting feature of both these smaller bowls is in the saucer-shaped top, which is shown placed on the bowl in the case of Fig. 377, and separately in Fig. 380. With bowls of considerable water capacity, such as these, there is a practical difficulty in filling them, and this is often overcome in an odious way by the placing in the font of a small jug and basin, as though the font were a kind of spiritual lavatory. The saucer top is a practical way out of the difficulty, as it holds but little water. Dr Yeatman-Biggs, Bishop of Worcester, was consulted as to the liturgical propriety of the saucer, and he agreed to its use, provided that it were made readily removable.

The rubric of the Church of England provides, "if the child may well endure it, the priest *shall dip it* in the water discreetly and warily," and this use obtains in a few parishes. Were the saucer top fixed to the bowl this would be impossible; by its being made loose the font is suitable for both immersion and sprinkling.

Mr Arthur Grove modelled the font shown in Fig. 381 to the design of Mr H. Wilson, and it was cast by Mr Dodds for St Mark's Church, Brithdir, Wales. The decoration is of that soft and simple kind so entirely suitable to leadwork, and the broad horizontal margin round the top of the bowl emphasises a heavy material. It is a most admirable thing.

Rain-water Pipe-heads.

The revived interest in the use of lead for pipe-heads and gutters has had to struggle with some rather evil influences.

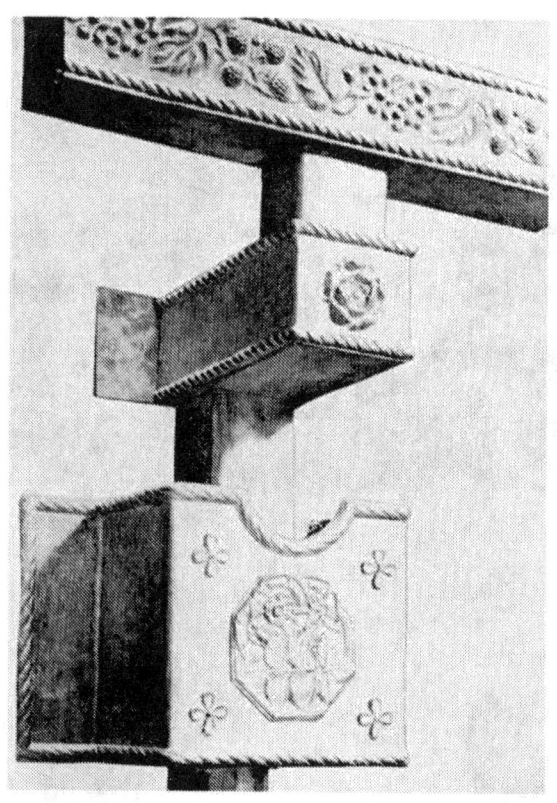

Fig. 382.—Intermediate Head instead of Swan-neck.

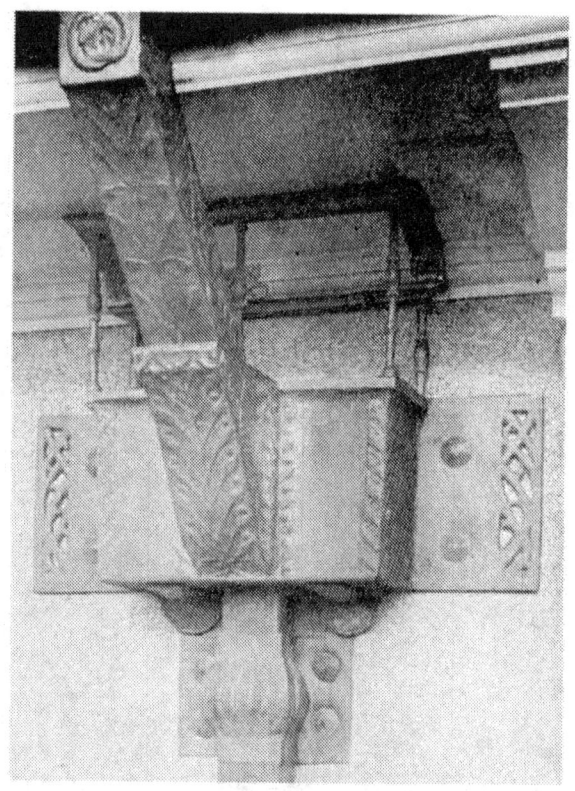

Fig. 383.—Welbeck Abbey.

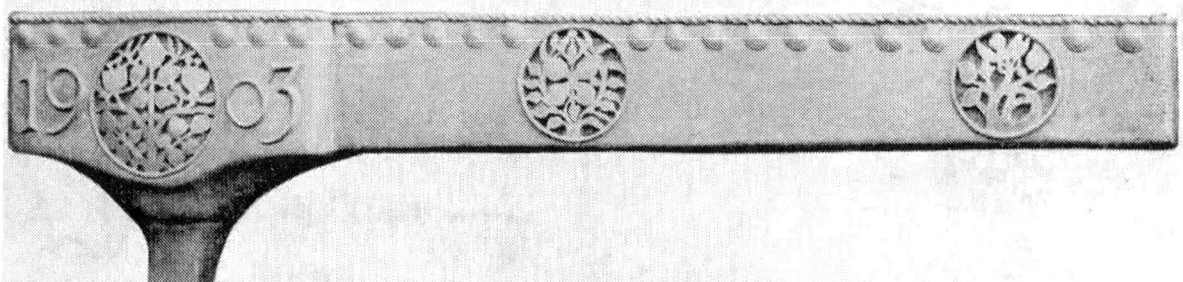

Fig. 384.—Designed by Mr Arthur Grove.

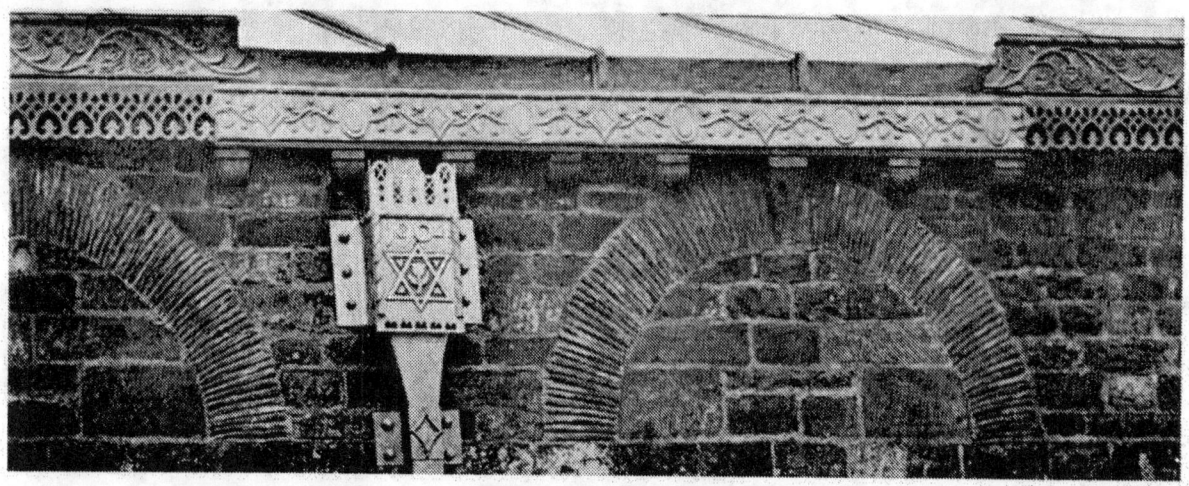

Fig. 385.—Charwelton Church.

P

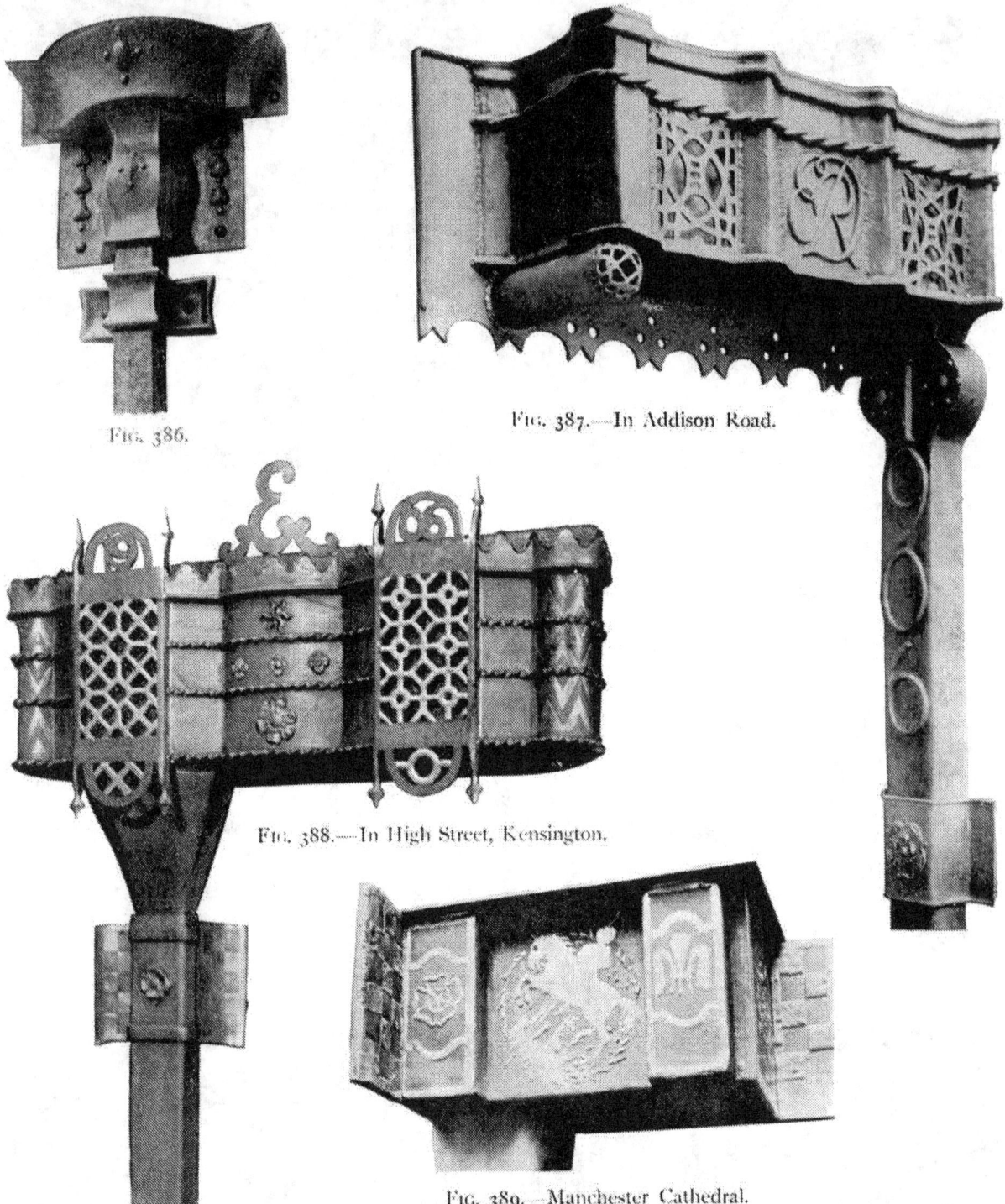

FIG. 386.

FIG. 387.—In Addison Road.

FIG. 388.—In High Street, Kensington.

FIG. 389.—Manchester Cathedral.

Since the end of the eighteenth century, when the traditional treatments of lead died out, cast iron has held almost undisputed sway. It is true that the conditions of modern building usually put lead pipes and heads out of the question on the simple score of cost. Moreover, cast iron, if reasonably heavy, is a quite satisfactory material; it only becomes ridiculous when historical leadwork is used as a slavish basis for its design. There is, happily, a growing perception that cast iron has a character of its own, and that it can be treated to look like itself. When, however, lead as a decorative material was rediscovered, the ideas of leadwork design were quite incoherent. Some astonishing results followed, notably the transfer to leadwork of the sense of sharpness, which is proper to iron, but distressingly comic in lead. The happy mean in leadwork is to secure easy, gracious lines without degenerating into amorphous sloppiness.

One of the difficulties involved in the use of the eaves gutter is the swan-neck from the gutter to the pipe-head. It is a practical necessity, but generally an ugly one. Two efforts to get away from the ordinary type are illustrated. Mr Bankart, in the example of Fig. 382, has effected a rather cumbersome alternative by interposing between the gutter and the pipe-head an intermediate head of large projection. The result is not

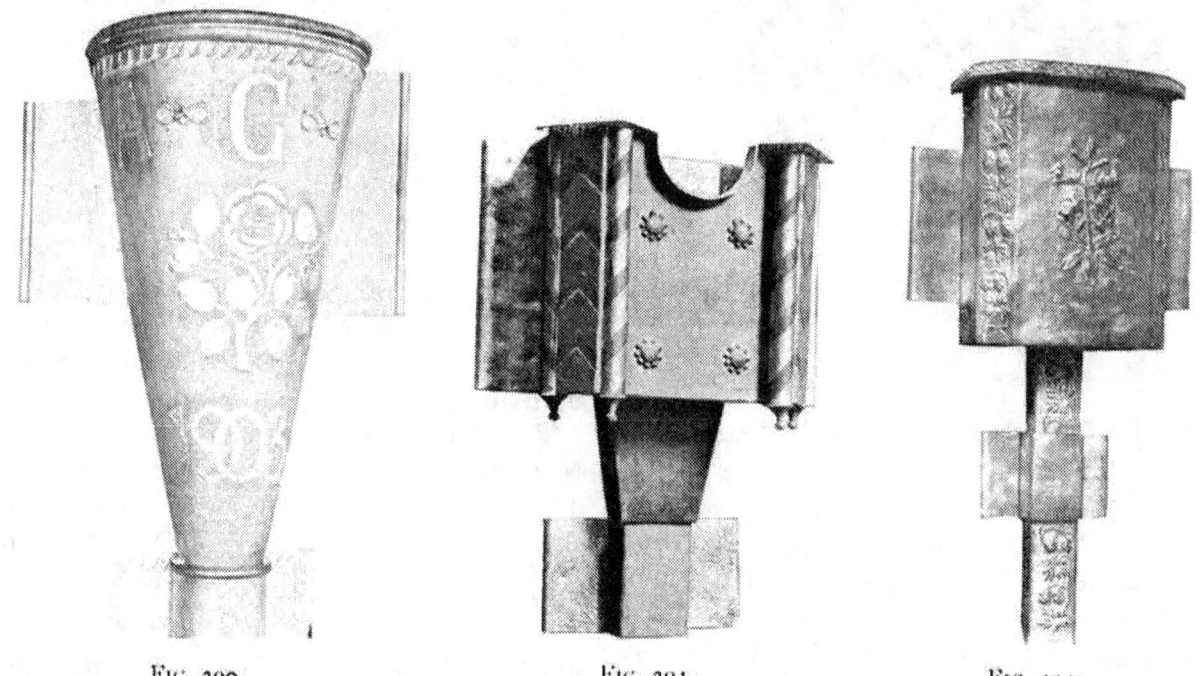

Fig. 390.   Fig. 391.   Fig. 392.

in any way so successful as a method adopted in 1895 by Mr H. Wilson at Welbeck Abbey (Fig. 383). Here the swan-neck is recognised as a practical need, and, so recognised, has been decoratively treated. This treatment is as original as it is successful, and gives an idea which may well be repeated, viz., of regarding the swan-neck and head as two parts of a whole. The projecting lip on the front of the head not only prevents an awkward break in the line of the swan-neck, but pulls the two parts together in a very happy way. The least usual feature is the little superstructure of slim lead balusters. It is simply ornamental, as it does not suspend the head, which is supported beneath by stout iron staples, and does not seem a very useful addition. The decorative treatment of the head is admirable, both in the soft modelling on the projecting lip and swan-neck done by Mr H. W. Finch, and in the simple piercing of the ears.

The head of Fig. 384, designed by Mr Arthur Grove, is a successful translation, as

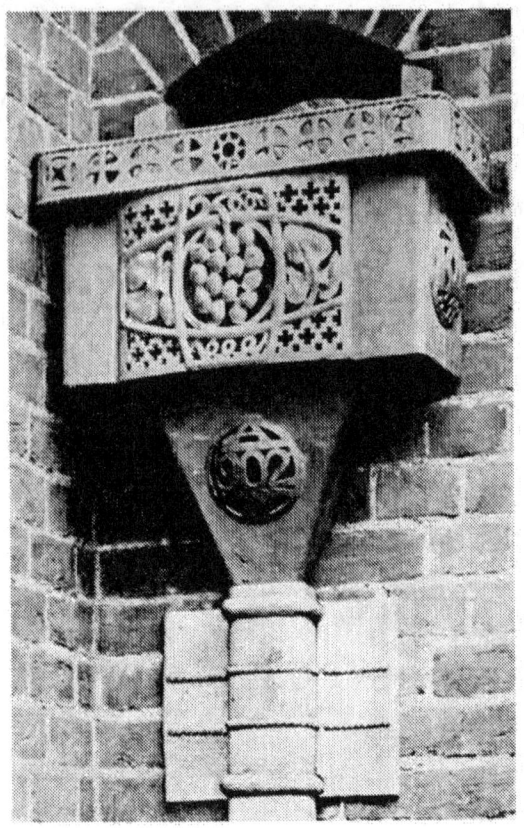

FIG. 393.—Designed by Mr F. W. Troup.

to treatment, of the pierced heads which we find at Knole and Haddon Hall, but it is entirely modern in feeling. The little shell-form ornaments beneath the rope-moulding give an agreeable spottiness, and the increased projection of the left-hand end and its funnel outlet preserve the character of pipe-head. Long heads are apt to degenerate into simple gutters, and so lose their character.

At Charwelton Church, the late Mr Christopher Carter designed an admirable system of water leadwork (Fig. 385). The parapet gutter guides all the water from the low-pitched roof to the break over the trough gutter, which in turn discharges into a funnel-shaped pipe-head. The stone corbels on which the trough rests give an easy sense of stability. The pierced valance which hangs from the lead parapet is in pleasing alignment with the trough, and reverts (no doubt unconsciously) to an early Aberdeen use of such decorative lead valances. The arrangement is altogether well conceived, and the ornament thoroughly suited to the material, and yet modern in feeling.

The two heads of Figs. 386 and 387 tend more to the feeling of historical leadwork. Mr F. S. Chesterton would seem to have studied the Knole heads in deciding on a turreted type, as Mr Lutyens has done in some of his leadwork. In one detail Mr Chesterton is delightfully archaic, but with entire success. Hardened students of leadwork may

be excused if they get a shade weary at times of rope mouldings. The horizontal bands in this case are of lead strip, twisted and soldered on. In this they recall a Romano-British coffin at York, a far cry for a precedent. The head of Fig. 387 is on the coloured house in Addison Road, designed by Mr Halsey Ricardo, and is vigorously coloured and gilt. The shaped outline of the back continued below the box of the head is unusual. The ears of the old heads have generally square outlines. The shaping, however, is a legitimate opportunity for variety.

Messrs Wimperis & Best have

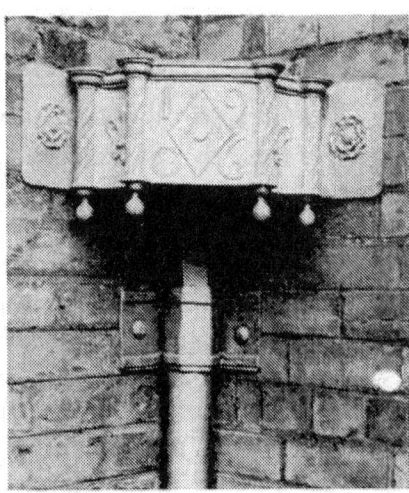

FIG. 394.—Horsley Hall.

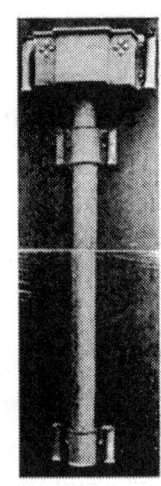

FIG. 395.

succeeded (in the head of Fig. 388) in a design showing some originality of form without any ill-treatment of the material, by no means an easy task. The moulding of the top is gay without being trivial. This head is from the works of Messrs Singer of Frome.

The majority of such modern pipe-heads as are designed and made on right lines, are built up of cast sheet metal. Messrs Singer use both this method, which is simple plumbing, and also box patterns such as are employed by ironfounders. There is much to be said for the latter method, particularly where several heads are to be made of one design and size, but it is an objection that the surface of the lead is always a sand surface. The method of building up from cast sheets gives the alternatives of using either the sand surface or the cooling surface.

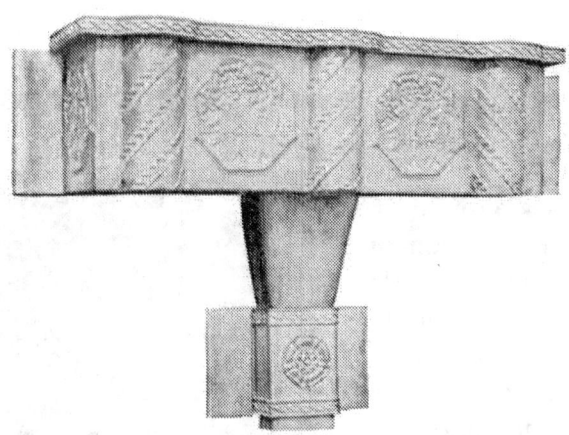

FIG. 396.—By Mr Bankart.

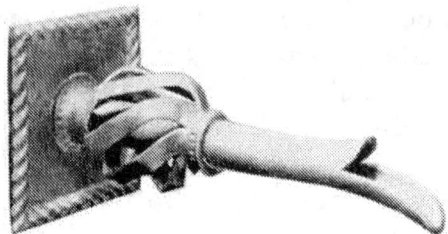

FIG. 397.—Piscina Outlet.

Furthermore, with box patterns there is more temptation to depart from a natural treatment of the metal, and indeed entirely to forget it.

Of the many heads made by Mr Bankart, illustrated in Figs. 389 to 392 and 396, it may be said that they show originality, while they preserve the right traditional feeling. Fig. 389 is one of a series fixed at Manchester Cathedral. The lily, St George and the Dragon, and the fleur-de-lis are the chief tinned ornaments, and are appropriate enough, for the cathedral is dedicated to the Blessed Virgin, St George, and St Denys. The St George ornament needs special comment. It is almost pictorial, and though there is ample historical authority for masks and small figures in cast relief, I know of no similar use of tinning

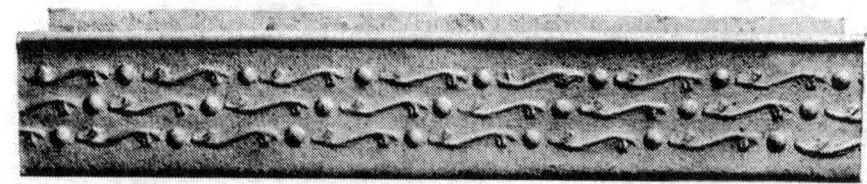

FIG. 398.—Lead Gutter.

for figure work. The treatment is, however, purely conventional, and seems perfectly justified. The long plain funnel of Fig. 390 is a happy example of the pipe-head reduced to its simplest and most practical form. The floral ornament redeems it from baldness, and the head is a pleasant change from the sometimes distorted and troubled outlines which derive from wild searches after originality. The character of the flower ornament is sound. Some of Mr Bankart's early work showed an undue delicacy in its surface ornament, and suggested embroidery rather than leadwork, but his later work is masculine and unaffected. The head of Fig. 392 is good, but the "embroidery" criticism may be levelled against it to a small extent. The surface decoration of the pipe is attractive.

The barber's pole and chevron decorations of the head of Fig. 391 are done in bright tinning, and the design generally is simple and appropriate. It is based on the turreted fancies of the seventeenth century, but with enough difference to make the feeling frankly modern. The shaping of the top edge gives it an architectural character, yet without affectation.

The early seventeenth century inspired the example of Fig. 396, and the decoration is simple and appropriate.

The head of Fig. 393, designed by Mr F. W. Troup, and made by the late Mr Dodds, has good simple outlines, and the pierced ornament is unaffected and pleasant.

Messrs George Wragge Ltd. have carried out many important pipe-heads to the designs of various architects. The example of Fig. 394 was made for the restoration of Horsley Hall, Hexham, to the design of the architect, Mr G. H. Kitchen. It is a sober thing, in strict subordination, as heads should always be, to its architectural surroundings. The head of Fig. 395, also made by Messrs Wragge, is one of the simple sort welcome on any building, and markedly

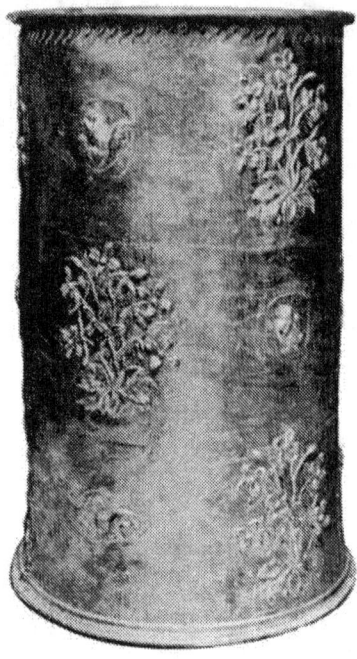

FIG. 400.—A Garden Tank, by Mr Bankart.

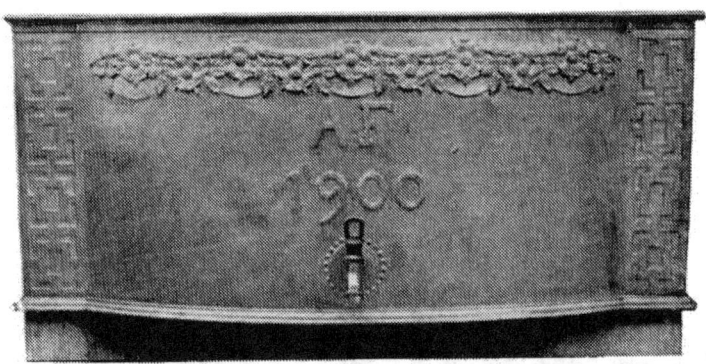

FIG. 399.—Designed by Mr Ernest Newton.

better than a head full of design, unless the design is restrained and appropriate. The gutter of Fig. 398, made by Messrs Henry Hope & Co., has decoration of an excellent simplicity.

Earlier than pipe-heads were gargoyles, and on Hardwick Hall is an example, which has been copied by Mr Bankart for another purpose (Fig. 397). It is fixed on an external church wall to discharge water from a piscina into an earth drain, an open-air arrangement which seems open to some liturgical objection.

The same treatment of bulging and piercing appears on the stem of a pewter sepulchral chalice of the thirteenth century, which is in the possession of the Society of Antiquaries.

Cisterns.

Leaving rain-water heads for cisterns, one welcomes the many admirable things which have been done for the beautifying of formal gardens. Figs. 402 and 403 show examples based on the traditional lines of dividing the surface into small compartments, and putting

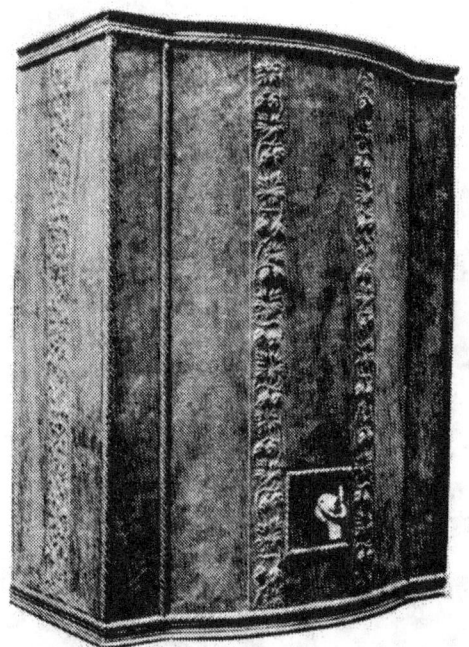

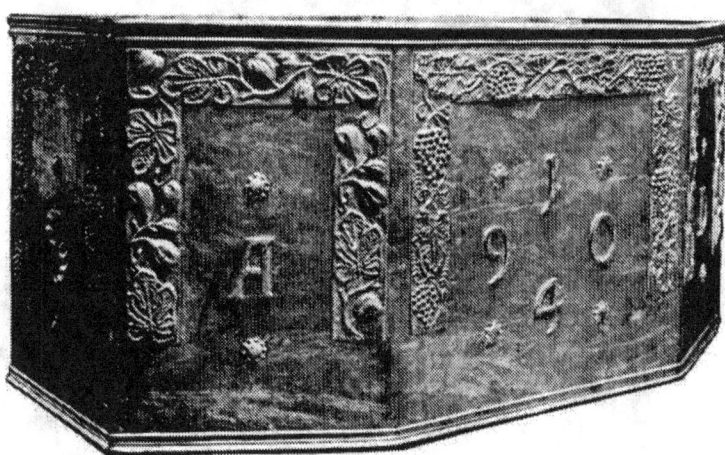

FIG. 404.—Made by Mr Bankart.

FIG. 401.—A Garden Tank.

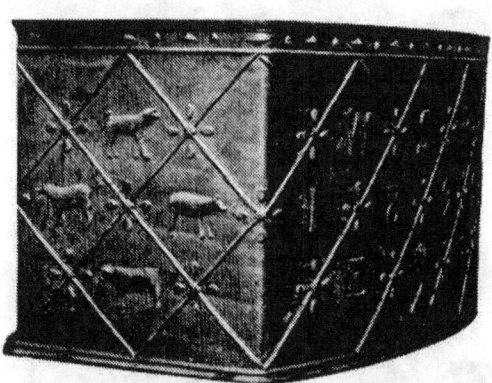

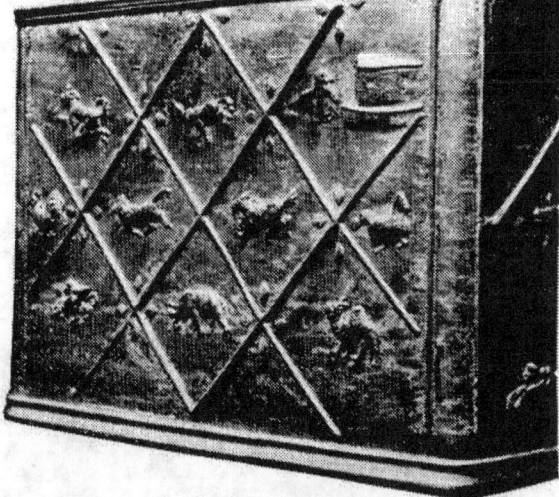

FIG. 402.   "Noah's Ark" Cisterns.   FIG. 403.

a little ornament in each. They are decorated with the same subject, Noah's Ark, and show the widely differing treatments which can be employed with propriety in such work.

In Fig. 402 the models are of the simplest. The wooden creatures of the child's Noah's Ark were impressed in the sand, and show the grain of the wood quite unaffectedly. In Fig. 403 the animals, Noah, and his ark are freshly and vivaciously modelled, and the camel swings after the hasty elephant in most convincing fashion. The donkey is peculiarly delightful, and the creatures altogether are very engaging.

Decorative humour is ordinarily a dangerous trade, but here it is successful.

Both these cisterns were made by Mr Dodds, as also that of Fig. 399, a dignified design by Mr Ernest Newton. In the old cisterns the varieties of shape were few. They

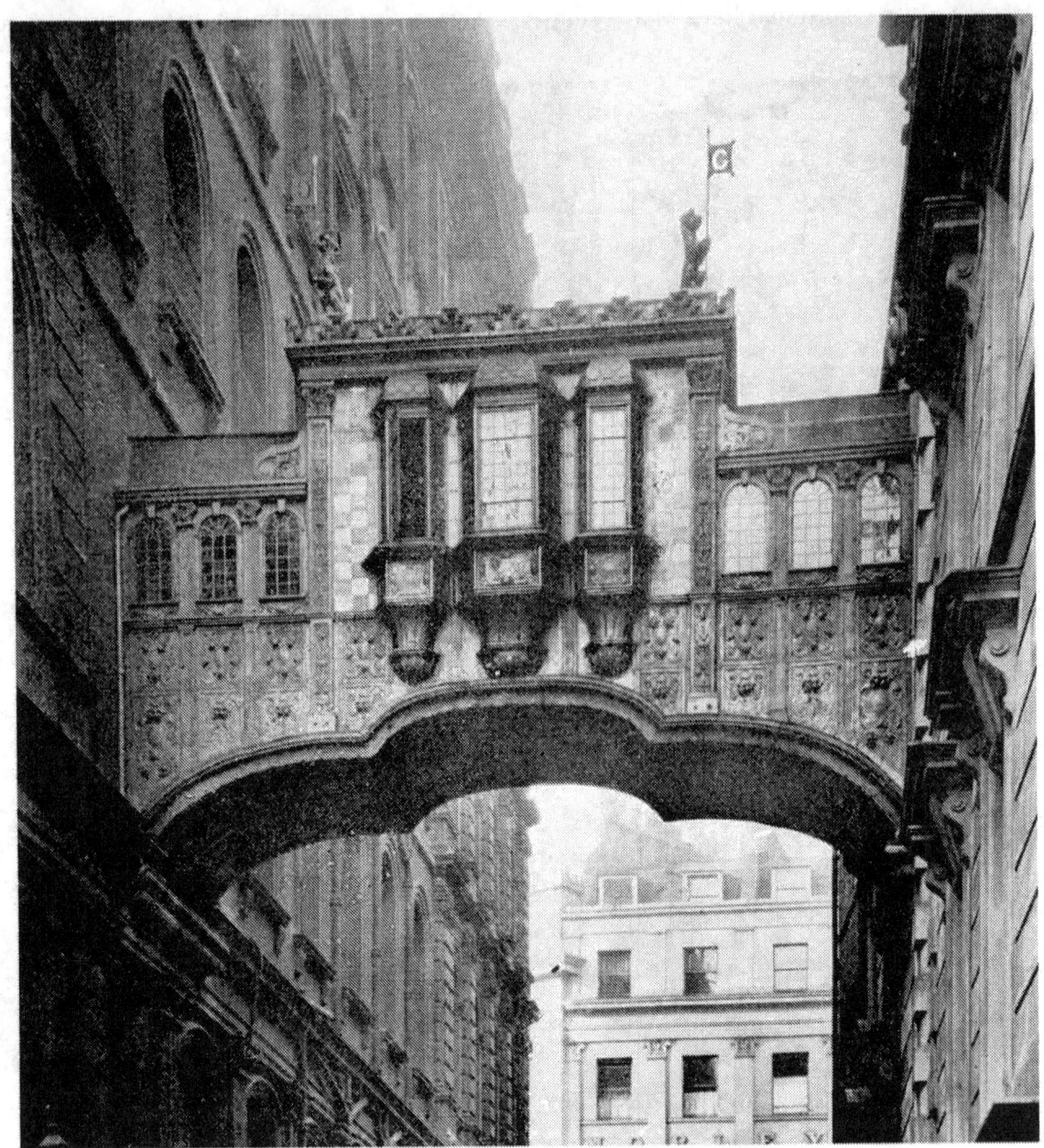

FIG. 405.—Leaded Bridge by Mr J. Starkie Gardner.

were circular and segmental, rectangular or regularly polygonal. Irregular plans add interest, however, and a moderate divergence from the more obvious shapes is a safe departure from traditional methods. The frieze of the cistern of Fig. 399 is pleasantly formal, but has a slight sense of sharpness not quite satisfactory.

The disposition of the bands of ornament on the tank of Fig. 401 is unusual and attractive. The height of the tub made originally by Mr Bankart for his own garden (Fig. 400) is a notable feature. There is no old cistern of anything like these proportions; that at Lincoln Cathedral is the nearest to it. The bunches of flowers and the little creatures—a newly-hatched chicken, a squirrel, &c.—are appropriate garden decoration. The informality of the thing is a feature that one likes, as a change, in a craft which usually relies for safety on a stiff conventionality.

Larger Constructional Uses.

When one turns to spires there is little to record. Many modern leaded spires have been built, and some spirelets of a very elaborate character, e.g., by Street on the Law Courts, but traditional methods have been closely followed in most cases. The spires of Gothic style have generally been built without large spirelights, the absence of which is characteristic of the

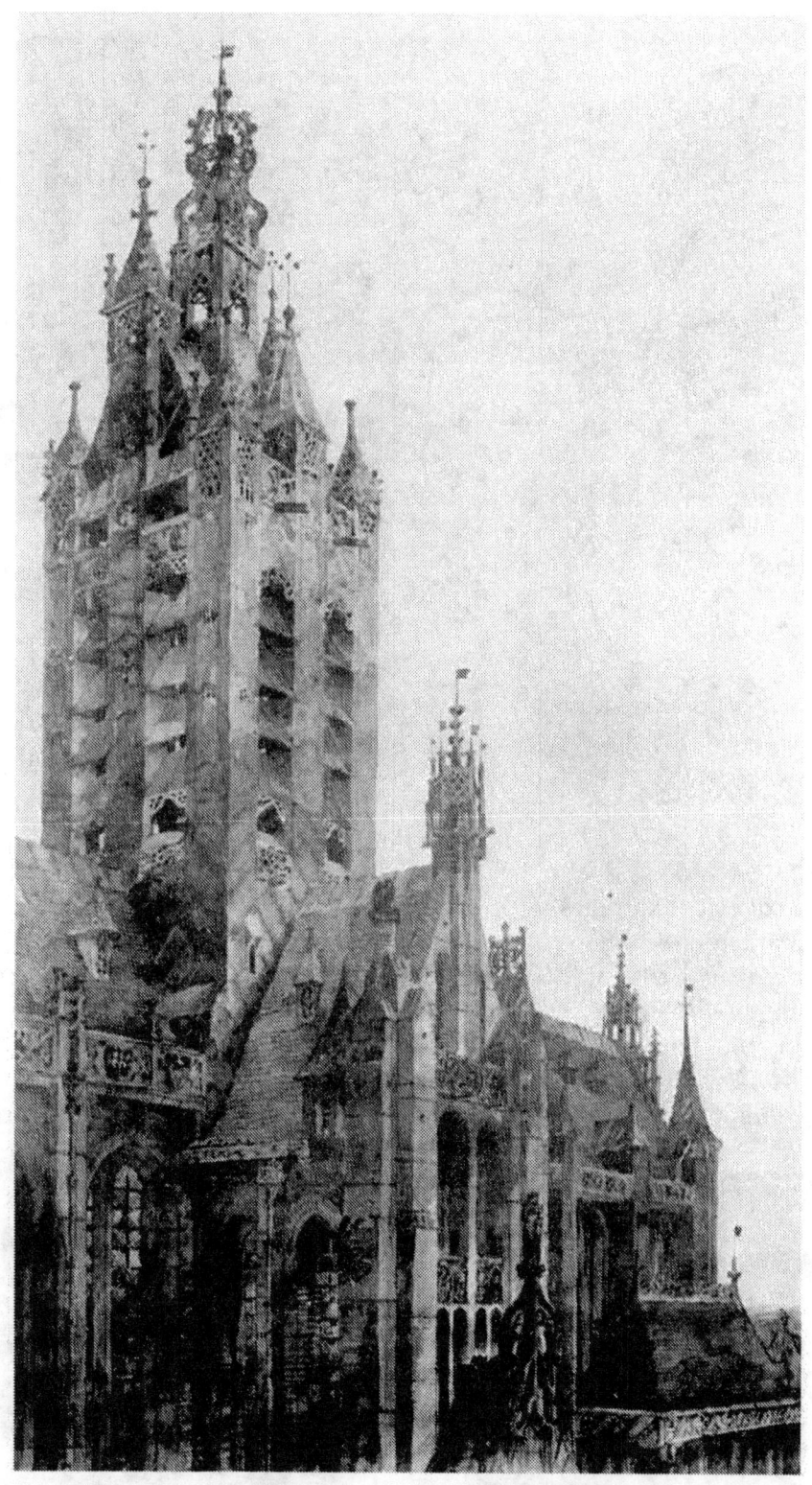

FIG. 406.—Die Bleiern Kirche, Strelsau. (Sir Charles Nicholson, *inv. et del.*)

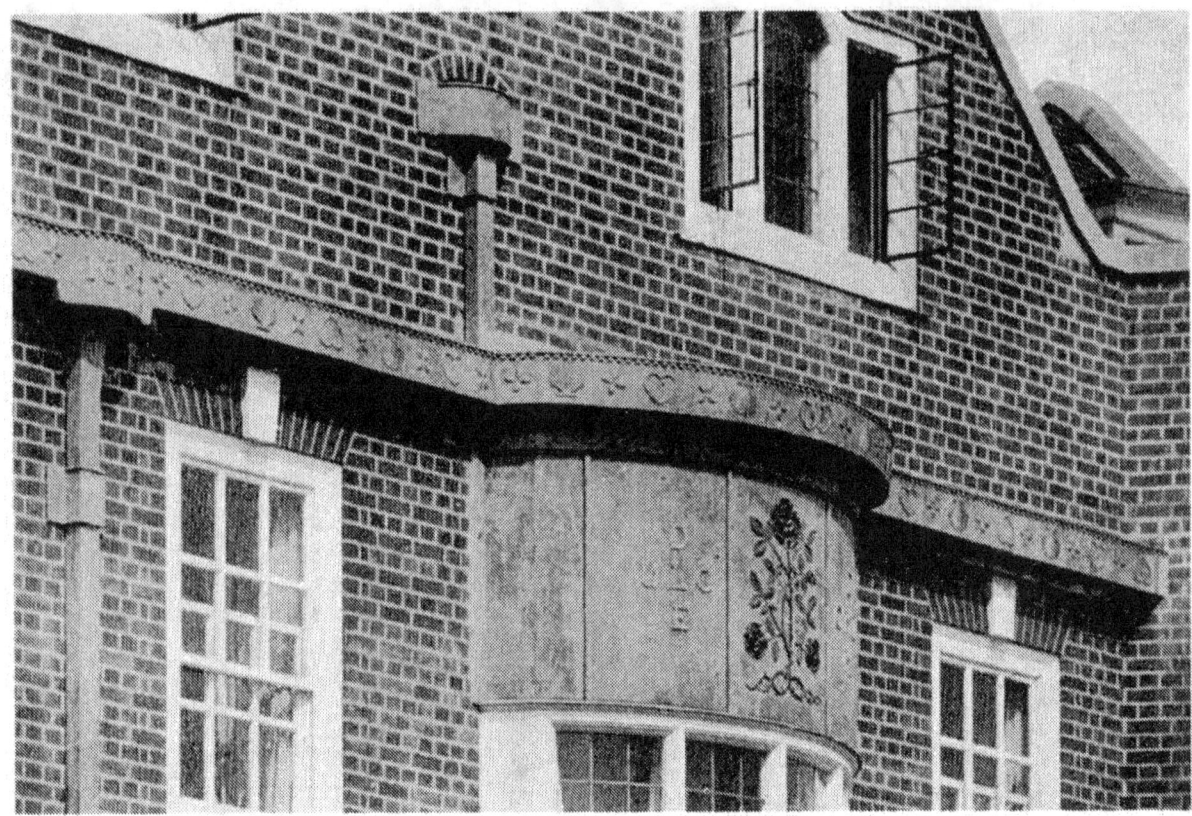

FIG. 407.—Redcourt, Haslemere.

mediæval examples. It was, perhaps, Sir Gilbert Scott's failure to grasp this outstanding character of the great early leaded spires that accounts for the unloveliness of the leaded spire he built on St Nicholas, Lynn. It consists of a lower, straight-sided, octagonal stage, with great mullioned windows on four faces and broaches on the other four, and for the upper stage, an ordinary octagonal spire. The broach is one of the earliest, as the big spirelight is one of the latest features in the development of leaded spires, and the attempt to merge conflicting traditions breeds a sense of anachronism as well as ugliness.

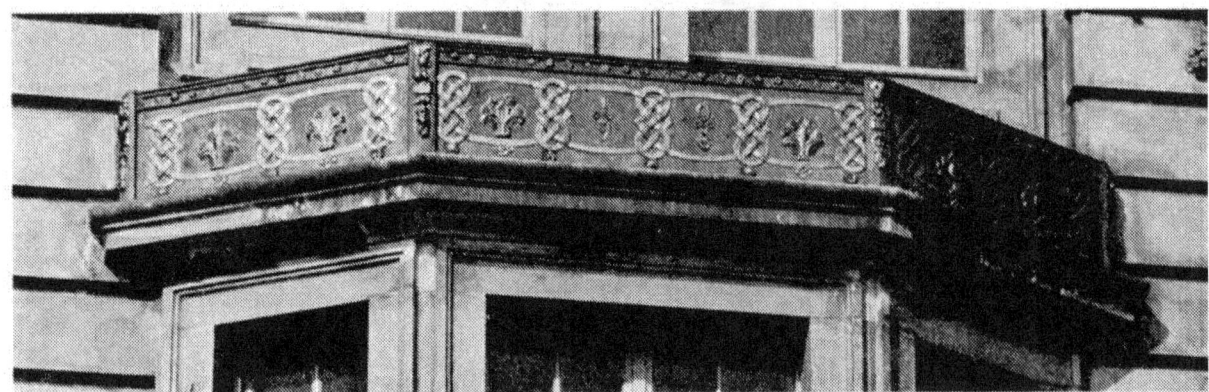

FIG. 408.—Insurance Building, Pall Mall.

Something by way of constructive suggestion for the future may perhaps be made. Mr Lethaby when dealing with lead as a roofing material points out that metal architecture was in early days the architecture of the poets. That is hardly its character to-day. It is unquestionable, however, that much thought has been given to the use of iron construction, if haply it might be made as beautiful as it is often useful. Critics of architecture have laid down with dogmatic impressiveness that, concealed in the womb of time, there must be an adequate steel architecture which shall be æsthetically satisfying, but its arrival lingers.

The illustrations of Chapter V. show how beautiful lead spires can be and are. They certainly held a high place in the affections of the mediæval architect. The lead gave him no trouble; he gained infinite variety of surface by different arrangements of the rolls; he outlined great cartoons on the faces of his spires (as at Châlons-sur-Marne), and blazoned them with gold and colours; he wanted the metal-cased architecture of the poets, and he got it; his difficulty was that he could not keep it. His timber framing was in danger of fire from above and fire from below. Lightning conductors have minimised if they have not rendered impossible the former disaster, but there is always the danger to a timber spire from fire arising in the belfry stage or in the body of the church.

There is, however, a sound alternative. Spires can be built in steel and sheathed in lead, and will defy the flames. Here there is room for effort, and the possibility

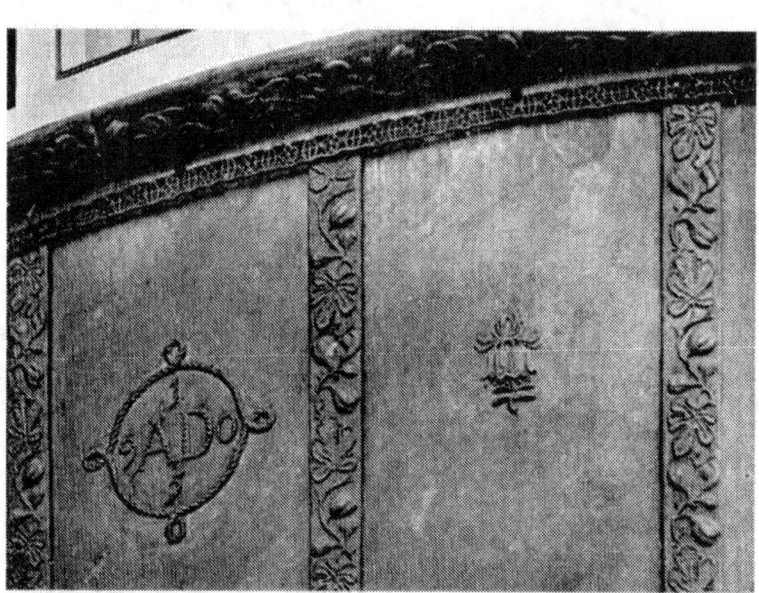

FIG. 409.—Sandroyd School, Cobham.

of notable achievement. The construction should present no difficulties. The spire has but to carry itself. Here is a field, not unimportant even if it be small, where steelwork may come into its own; may come faithfully and gracefully; may be the metal bones of a metal architecture. It preserves the initial idea of a spire that it is a glorified roof; and the lead surface gives opportunities for colour treatment that a stone spire cannot give. Had the mediæval architect found the material to his hand, it seems reasonable to suppose that we should be pointing to-day to his leaded steel spires as notable examples of the Gothic spirit. Fig. 406 shows a design for a leaded steel tower which Sir Charles Nicholson has done to illustrate this suggestion, and it will not be attributed to the author's friendship if this Bleiern Kirche is described as being instinct with the poetry and mystery which are the characteristics of great architecture. It may be hoped that some ecclesiastical Mæcenas will be found, for whom can be materialised this dream church encrowned with lead. So far it has only been built in Strelsau, and its date is February 1906. Strelsau is little visited by architectural tourists, but when it is visited the natives speak of the Prisoner of Zenda.

People have gibed, and justly, at the papering of steel skeletons with stone, of which the Tower Bridge is one of the most dismal examples. Had the bridge been treated as was the little leaded bridge over Northumberland Street, Strand (Fig. 405), what a magnifi-cent and honest structure it would have been ! Mr Starkie Gardner, who built this bridge connecting the Grand Hotel with its annexe, for Mr William Woodward, has pleaded the merits of this admirable fireproof construction for streets of shops. The fronts could then be almost entirely of lead and glass, but so sane and practical a method of building presupposes a drastic modification of the building by-laws. The large flat surfaces which are the natural outcome of ferro-concrete construction also lend themselves to decora-tive treatment with lead panelling.

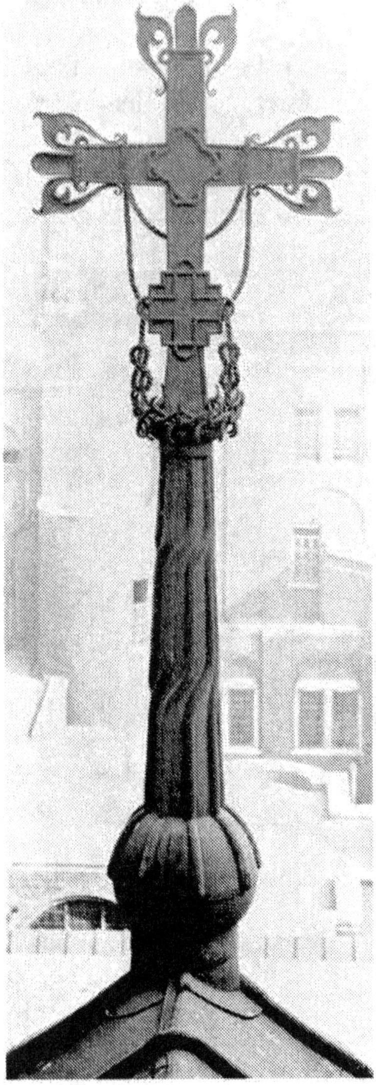

One modern use of lead for covering buildings has so little root in the past that it may be regarded almost as an invention, viz., the sheeting of brickwork.

Mr Ernest Newton has been active in this, and his happy example has been somewhat widely followed.

The charm and value of Mr Newton's handling of the lead sheeting at Martin's Bank, Bromley, and at Red-court, Haslemere (Fig. 407), are greatly increased by the skill with which he has brought this unusual treatment into relation with the normal uses of lead for gutters, heads, and down-pipes. Particularly is this the case at Haslemere, where the sheeting of the circular bay beneath the gutter has an effect entirely natural and even inevitable.

The decorations on the gutters are of that simple unaffected sort which accords best with any extensive use of lead.

One is ordinarily a little tired of heart-shaped orna-ment, but it should be remembered that Mr Newton was employing it before the dreary vagaries of New Art had made this natural outline wearisome. The heart outline was, moreover, consistently favoured by plumbers in the seventeenth and eighteenth centuries, and may be regarded as traditional in leadwork. The work was done by Messrs Wenham & Waters.

FIG. 410.—At Westminster Cathedral.

The main ornament on the Haslemere bay has been vigorously coloured. Mr Newton has employed the quite straightforward medium of oil paint, and has therein departed from the older method of transparent colours. The objection to oil paint is that it veils the texture of the metal. Perhaps a better way is to have transparent colours, such as madders, ground in a wax medium and painted direct on the lead, the whole being afterwards treated with parchment size. Brilliance is increased if the lead be tinned or gilt before the colour is applied, and initial gilding will add to the effect, even if the colour to be used is solid—*e.g.*, vermilion. For any

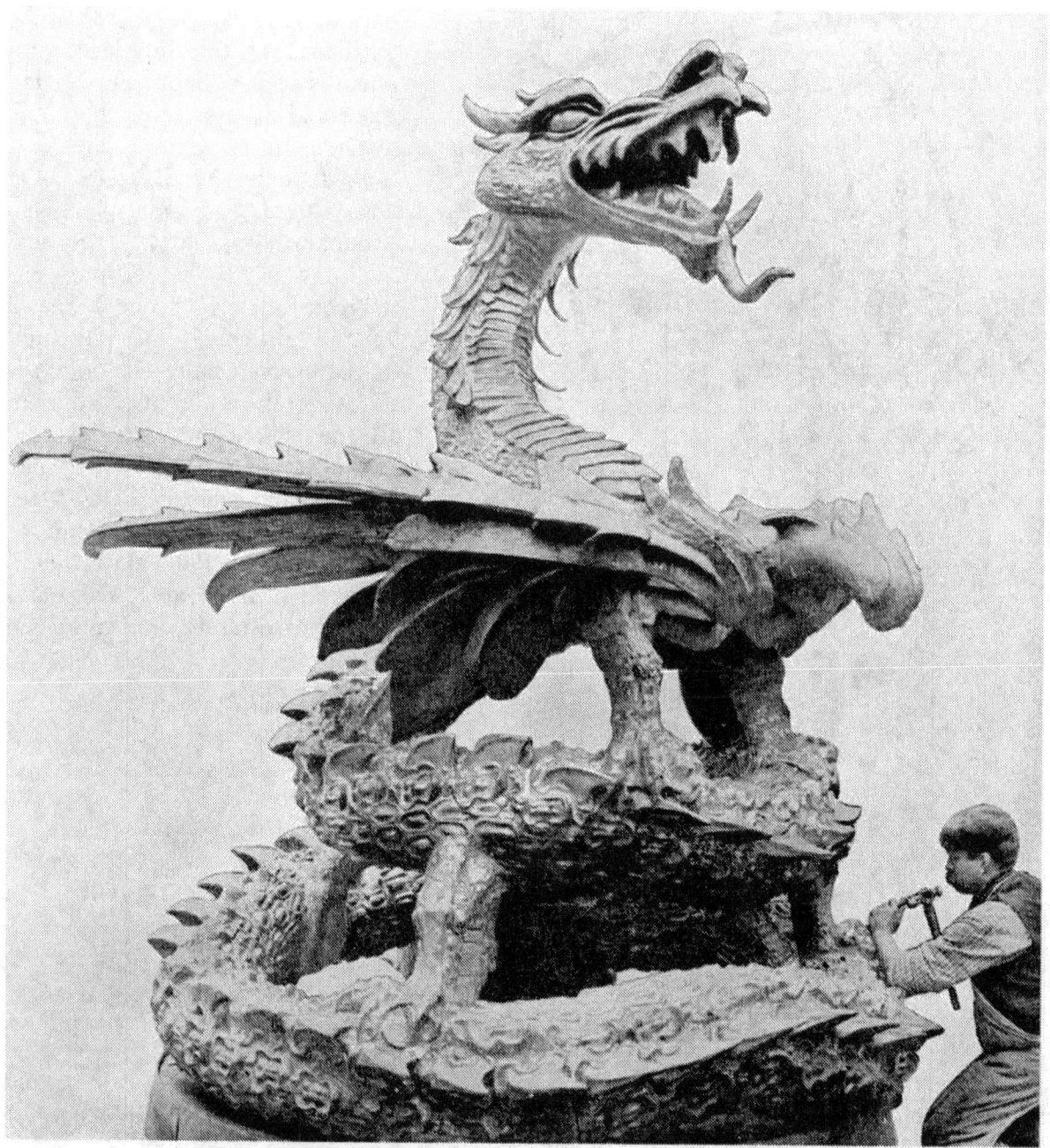

FIG. 411.—The Dragon of Wales, Cardiff Law Courts.

colour treatment except gilding, which is always satisfactory, a reasonably clean country air is needful; in a smoky town the colour, however applied, will mock the effort in a few months.

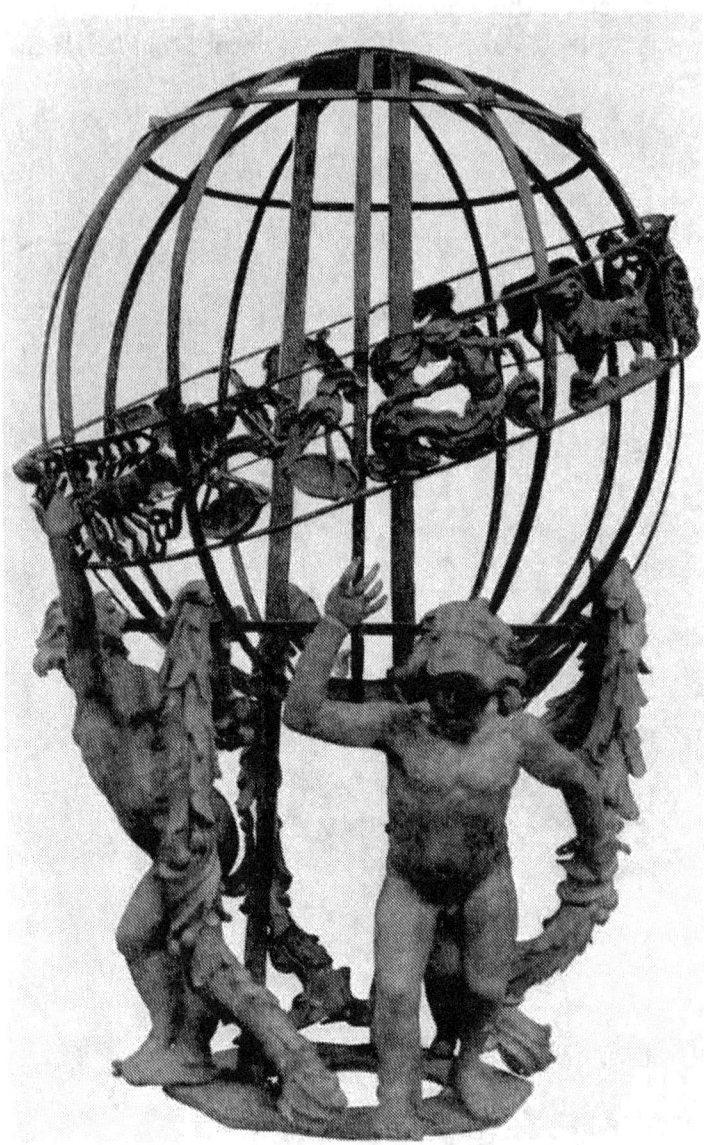

FIG. 412.—*Putti* Upholding Globe.

Mr Guy Dawber has heavily gilt the delightful leaded parapets to the bays of his Insurance Building in Pall Mall (Fig. 408), and the brilliance of the interlaced ornament is of very happy effect. Here the lead is fixed on a concrete backing 4 in. thick. This work was done by Mr Bankart, as was also that at Sandroyd School, Cobham (Fig. 409). An added delicacy is given by the slight pierced valance on the other side of the gutter. This piercing is taken up on a more elaborate scale for the rain-water head adjoining. In the ordinary way the restrained use of ornament, such as the latter example indicates, is the best treatment, but the general richness of detail of the Pall Mall building demanded a greater elaboration, and the result is eminently satis-

factory. For such work the milled sheet lead of commerce is a hopeless, textureless, pasty material to be avoided. Cast sheet should always be used.

Amongst the larger exterior uses of lead may be mentioned some of the late Mr Bentley's work. He was an enthusiast in leadwork, and as far back as the sixties built the little chapel of the Convent of the Nuns of the Perpetual Adoration at Taunton. The flèche is surmounted by a leaden figure of an angel in the manner of the great French roof-builders, but the flèche itself is shingled instead of

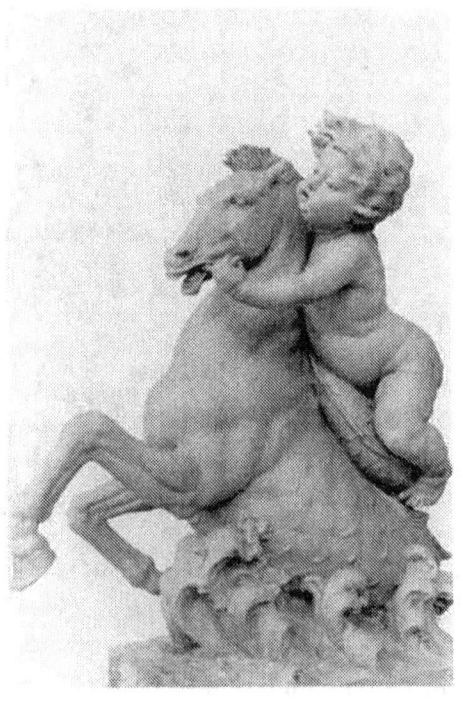

FIG. 413.

being leaded. The pipe-heads and roof-work at Westminster Cathedral, executed by Messrs Matthew Hall & Co., are full of interest. The dome of the campanile is a most refined piece of leadwork design, and the headcross on the choir roof (Fig. 410) repays study.

There is a lead spire-let on the church at Watford which Mr Bentley designed, slender, and in delightful contrast to the massive flinty tower.

Figures.

When we turn to lead figures, their principal use

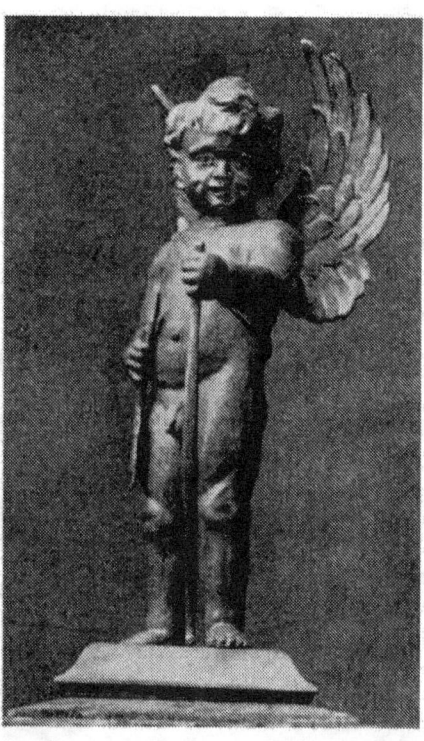

FIG. 414.—Finial on Summer House.

in modern work has been in gardens, but the biggest decorative work in cast lead ever done in this country is the great dragon on the New Law Courts at Cardiff. It is 8 feet high and weighs 4 tons. The model was made in clay by Mr H. C. Fehr for Messrs Lanchester & Rickards, and the plaster cast of this model was used by Messrs Singer of Frome as a pattern for reproduction in lead. It was cast in ten pieces and soldered together. It is a lively piece of modelling and a bold essay in massive heraldry. It seems, however, rather too lively

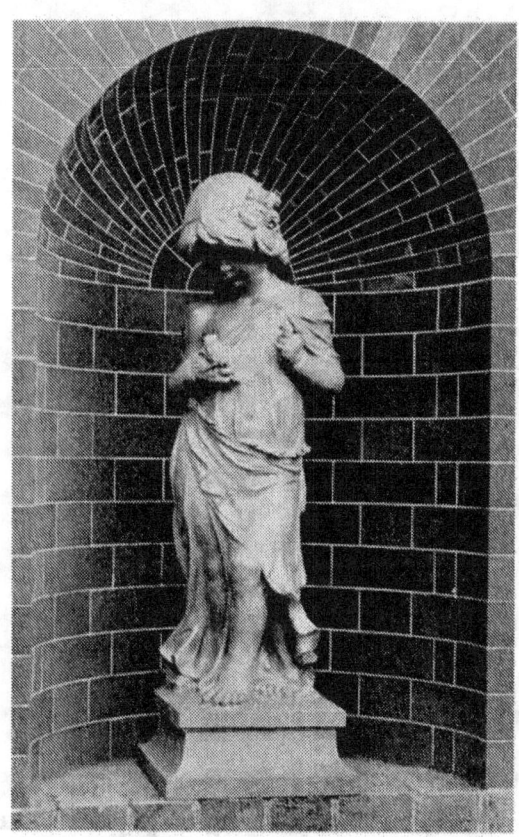

FIG. 415.—At Barnet Court.

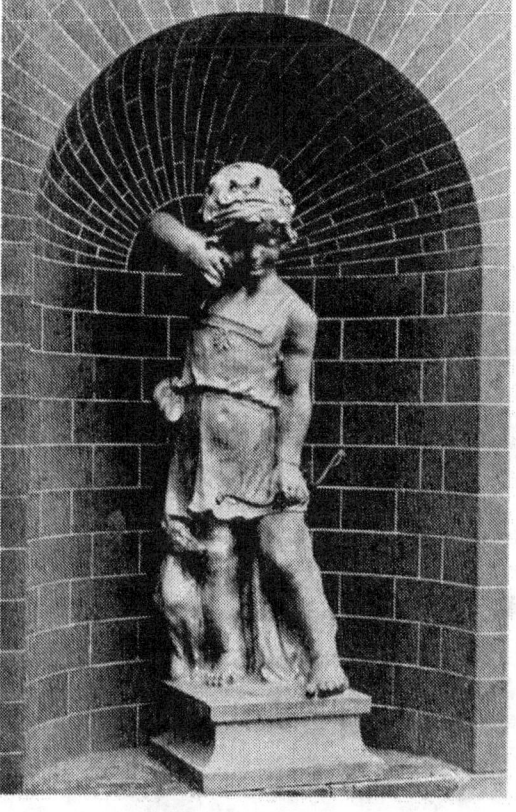

FIG. 416.—At Barnet Court.

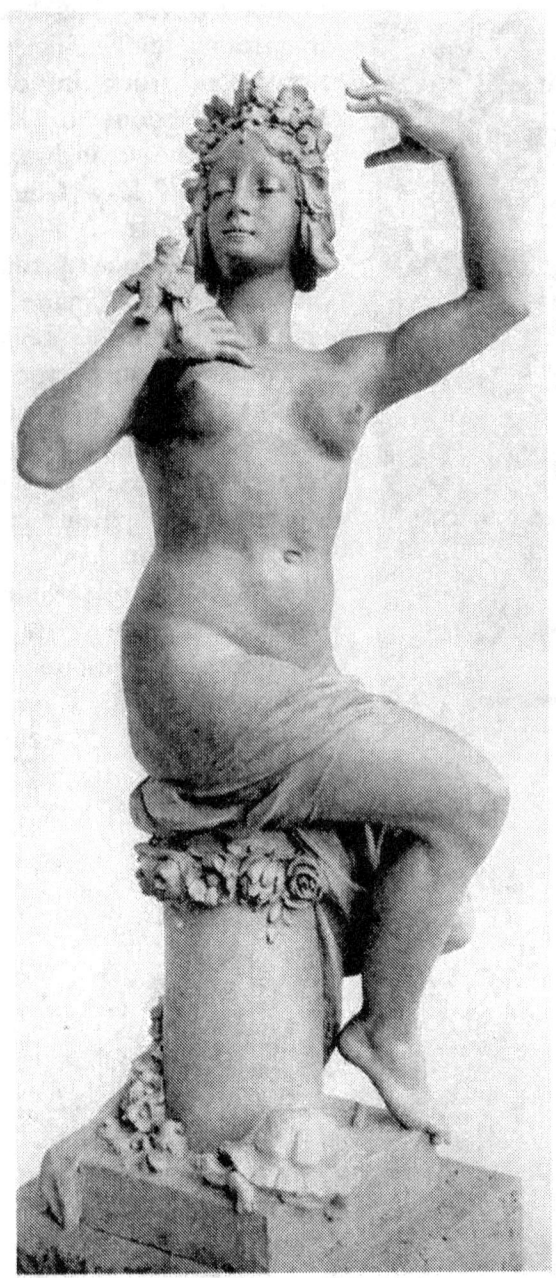
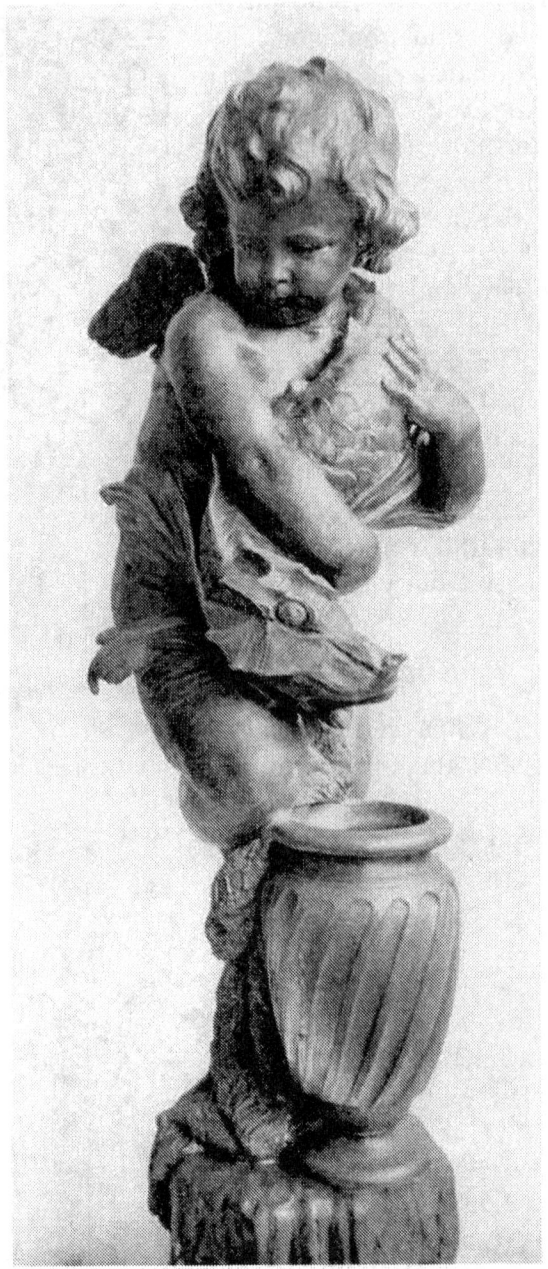

FIG. 417.   By the Bromsgrove Guild.   FIG. 418.

for so grave and admirable a building, and one could wish that the national aspirations of
the Principality had been satisfied by some less disturbing presentment of the Dragon of
Wales. As to the fitness of casting such a detail in lead, there is, however, no doubt.
The character of the subject forbids stone, bronze would be a wastefully costly material
for work so far removed from close view, and the architects are to be congratulated on
reviving a good tradition by employing lead.

A trio of *putti* upholding a burden is an old enough, but always attractive device.
The group shown in Fig. 412 has strong characteristics. It was designed and executed

by the Bromsgrove Guild from rough sketch suggestions made by Mr J. J. Burnet, architect. A pleasant feature of the scheme is the encircling of the openwork globe by a band decorated with the signs of the zodiac. These, and indeed all the details, are freshly and agreeably modelled, and with the softness appropriate to leadwork.

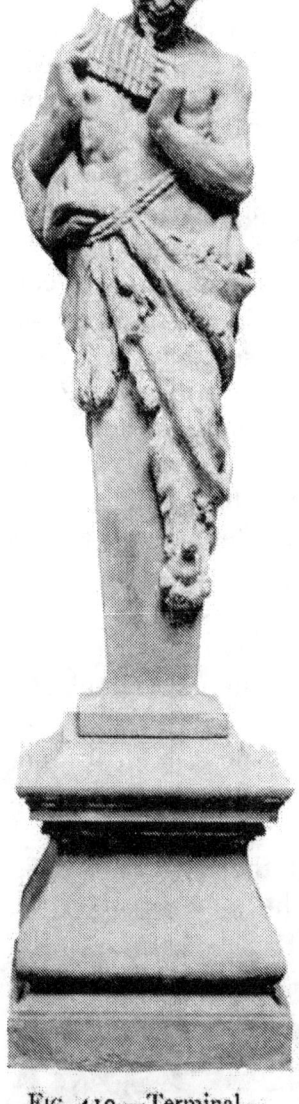

FIG. 419.—Terminal— "Pan" for Ardross Castle.

The Bromsgrove Guild was also employed for the two delightful figures at Barnet Court (Mr Arnold Mitchell, architect) shown in Figs. 415 and 416, and for the angel for a lych-gate (Mr W. E. Webb, architect) of Fig. 420.

The little people at Barnet Court are tenderly done. The sportsman with his acute hound is evidently bent on very moderate bloodshed, while his little sister is actively concerned for the comfort of her frog. They are both admirable, and look the better for being in their brick niches.

The British climate is more appropriate for draped figures, such as those at Barnet Court, than for the nude, like the Bromsgrove Guild's statue, shown in Fig. 417. It may be doubted whether the posed arm is a wise feature in a lead statue, as it is apt to become the "crooked billet" of Lord Burlington's criticism, but the figure is a charming conception, and on a sunny day would be an exquisite touch of life in a garden. One can imagine it posed in the midst of an ornamental water, surrounded by some such watery figures as the boy riding the sea-horse (Fig. 413). This is a peculiarly happy piece of modelling, also by the Bromsgrove Guild. It is as impossible as it is unwise to make rules, but in a general way it may be said that nude figures for the garden are better used in connection with ornamental waters. These Bromsgrove figures seem to owe something to French in-

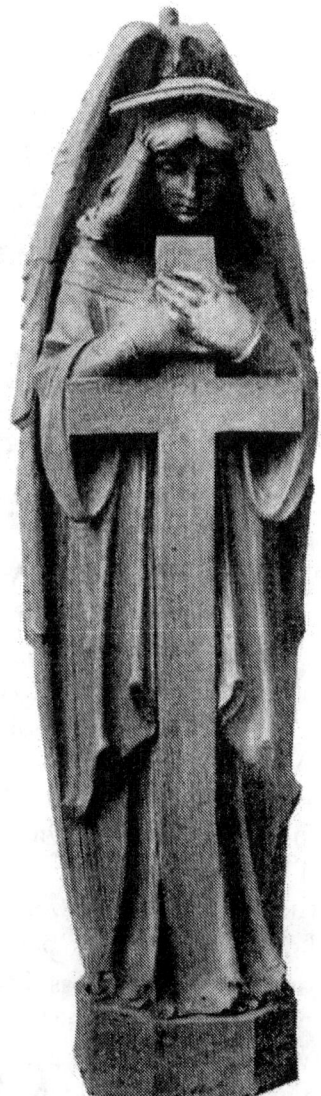

FIG. 420.—Angel on Lych Gate.

fluence, and a very proper influence it is, when it is remembered how much the idea of formal gardens owes to the great French gardeners of the past.

The cupid of the heavy legs (Fig. 414) is a pleasant archer, though he looks rather middle-aged. He serves as a finial on a reed-thatched summer-house at

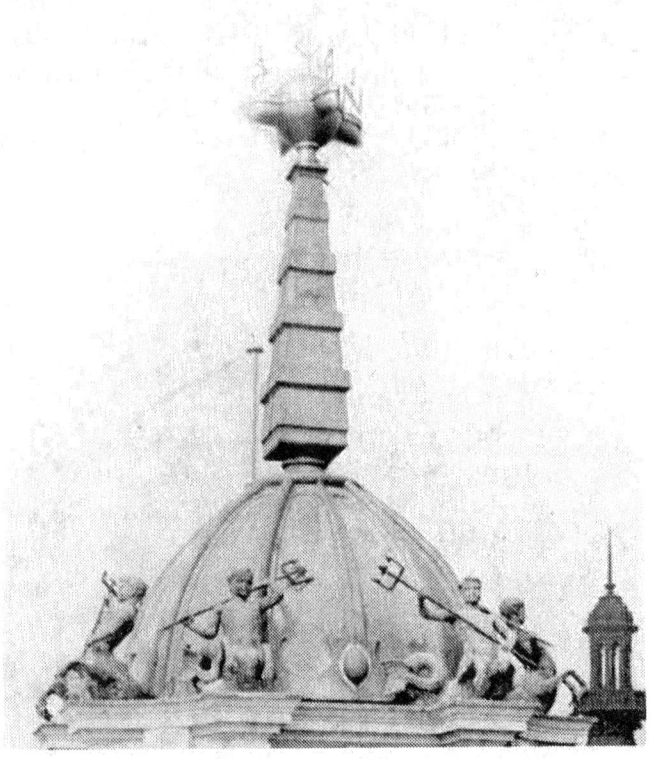

FIG. 421.—Hamburg-America Steamship Offices.

Kinfauns Castle, Perth, and was made by Mr Charles Henshaw of Edinburgh for Mr F. W. Deas.

When all is said, there is no figure more absolutely appropriate to the garden than *Pan*, and the terminal figure at Ardross Castle (Fig. 419) is a worthy successor to the *Pan* at Glemham Hall, if it lacks the fine dignity of the Castle Hill bust. It is a far cry from the Piping God to the Lady of the Lych Gate (Fig. 420), which is hardly so successful as the garden figures from the Bromsgrove studios. Perhaps it is a fad to cavil at lady-like angels, but if the unseen ministers are to be represented as markedly of one sex or the other, there seems more justification for a male tendency. It must be admitted, though, that the artist in this case is on the side of the big battalions, as the modellers and limners of angels are, for artistic purposes, almost universally feminist. Figures of this type are peculiarly suited to lead, as there are no outstretched arms to run the risk of damage or collapse.

Mr Arthur T. Bolton has made very effective use of leadwork at the new Hamburg-America Steamship Offices in Pall Mall (Fig. 421).

For the covering of the dome and obelisk sheet-lead, cast in sand, 7 to 8 lbs. per foot, has been used, and this part of the work has been done by Messrs Dent & Hellyer. The smaller gussets between the main ribs are in one piece, and in the larger gussets there is a central welt uniting two sheets. The welt is recessed at the back of the big boss, which is of beech, with the lead sheet beaten over it. The joint between the dome and the boss is wiped. The base of the obelisk is a large collar wrought in one piece. This required very careful work in contracting the lead to form the neck between the circular flange bossed over the ribs and the square base of the obelisk. There is one vertical seam only to the obelisk, and the raised bands cover the horizontal joints. The vane is in cast bronze. The Tritons were modelled by Mr W.

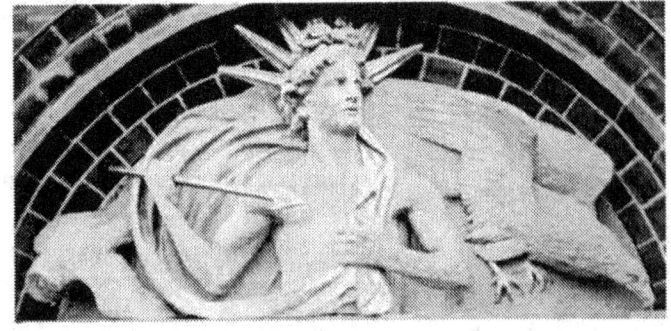

FIG. 422.—Ingram House, Stockwell.

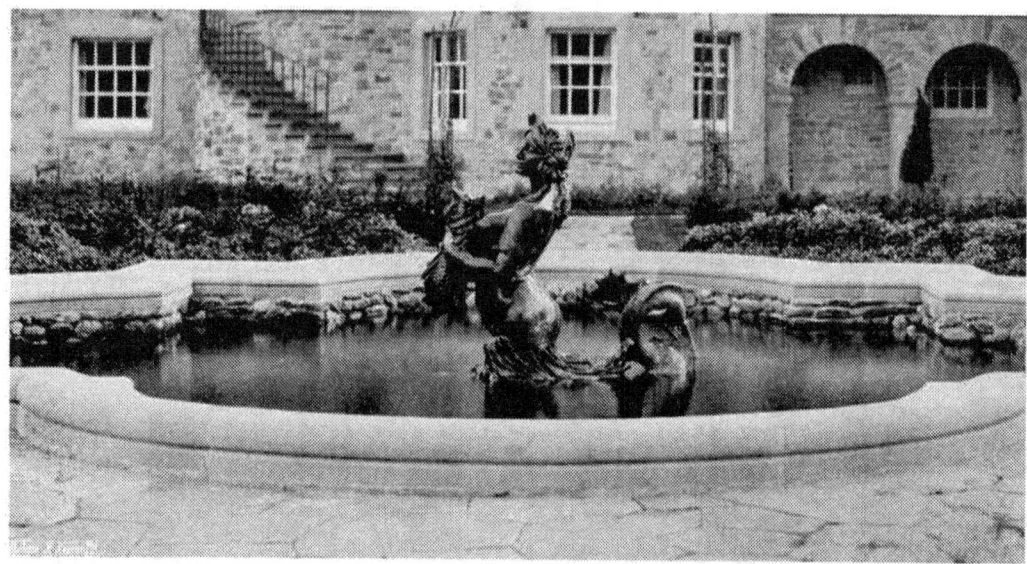

FIG. 423.—The Mermaid's Fountain.

Fagan, and cast in lead by Signor Petretti. The whole composition is successful. There is enough life in the Tritons to make them interesting, but they are sufficiently subordinated to the whole to prevent any sense of restlessness.

The figure of Apollo at Ingram House, Stockwell (Fig. 422), is another excursion into architectural leadwork by Mr Bolton. The sun-god and his attendant eagle and owl are cast in one piece, which measures about 6 feet in width, a considerable casting. It is stiffened at the back by iron bars, which are sunk partly in the lead and partly in a cement backing. The nimbus was cast separately, and its rays were ridged to secure the needed stiffness.

Fountains.

Among many charming modern garden ornaments there are none more attractive than those modelled by Lady Chance. *Neptune's Horse* (Fig. 424) spouts water from the mouth, and has been successfully used in fountain composition. The *Dolphin* (Fig. 425) also emphasises the water note in gardens. Mr Bankart made the fountain of Fig. 427, a very pleasant work, which now stands in the middle of a fine octagonal lead tank. Its design was obviously greatly influenced

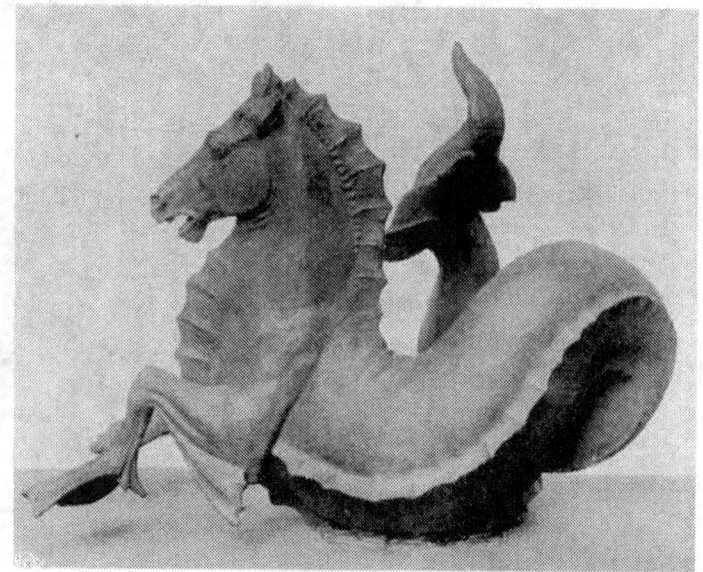

FIG. 424.—Neptune's Horse.

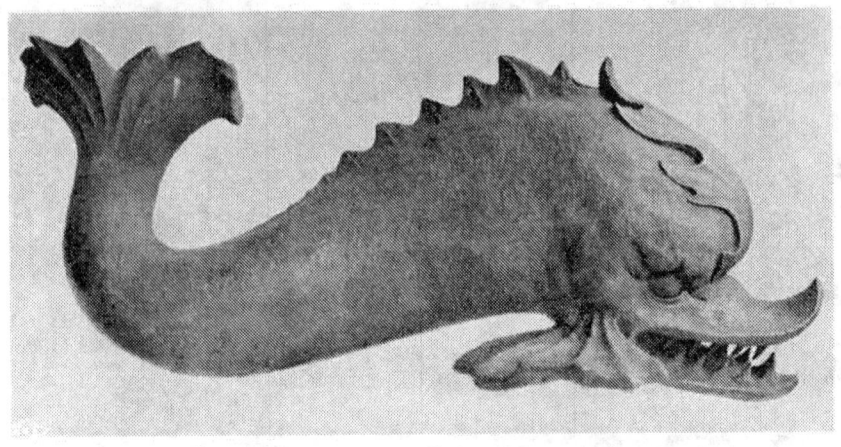

FIG. 425.—For an Italian Garden.

by the Dutch example in the South Kensington Museum.

Of quite another character is the very fine fountain modelled for Mr John Belcher, R.A., by Mr Alfred Drury, A.R.A. (Fig. 426). The strong modelling of the *putti*, and the fat, easy lines of the bowl are entirely admirable.

In all that concerns

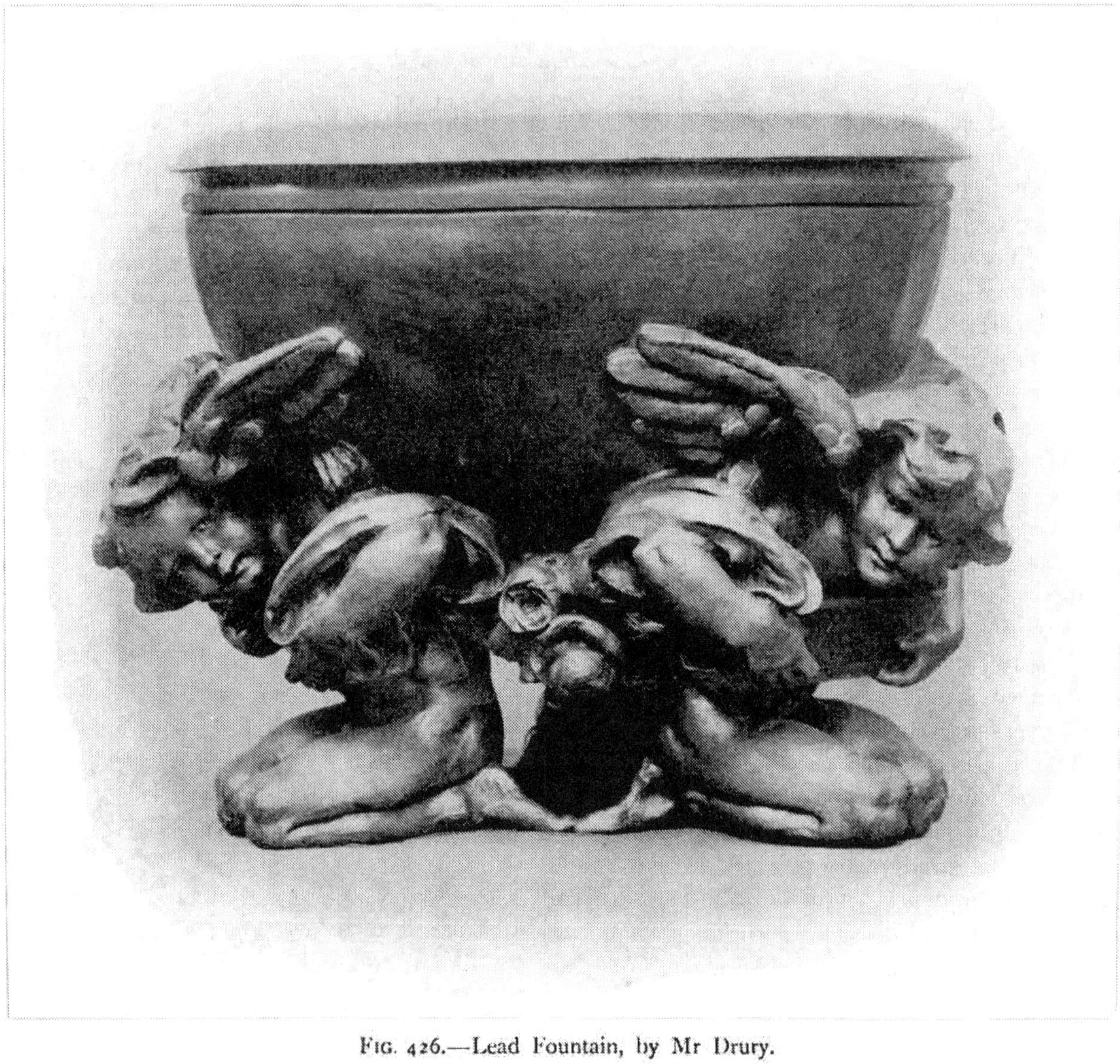

FIG. 426.—Lead Fountain, by Mr Drury.

the leadwork of the garden, the activities of the artists who compose the Bromsgrove Guild have been various and honourable, and their fountains are not the least pleasant of their output. For a garden in the West of Scotland the Guild made to Mr R. S. Lorimer's design the charming mermaid fountain of Fig. 423. This lady of the waters is grasping an unwilling fish, and the modelling is full of vigorous grace. We have the same motive of the fish in the attractive fountain of Fig. 418, p. 240, also made by the Guild. Cupid holds his dolphin, ready to spout into the vase, and his pose is lively without being unrestrained.

Vases, Sundials, &c.

For garden vases no material equals lead, for stone and terra-cotta are markedly perishable. The example of Fig. 429 was designed by Messrs Wimperis & Best; that of Fig. 428 by Mr John Belcher, R.A. Both were cast by Messrs Singer & Son. The former owes something in idea to the pair of magnificent vases at Hampton Court Palace, where nude female figures form the handles, but the design of the vase itself is quite different. The treatment errs perhaps rather on the side of sharpness,

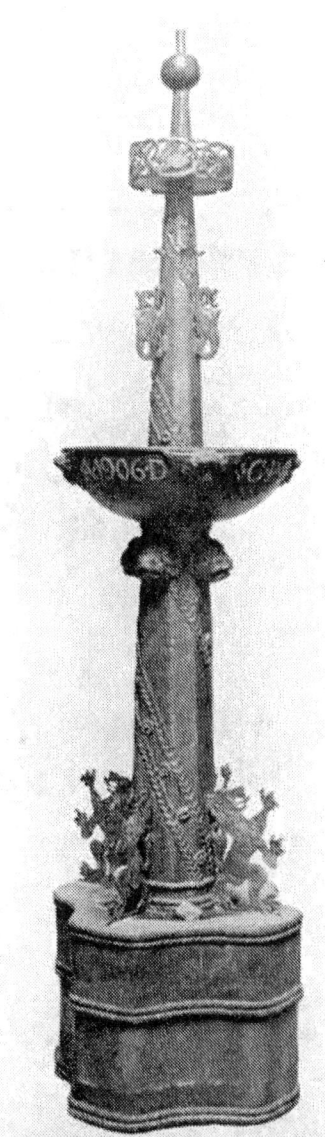

FIG. 427.—Fountain by Mr Bankart.

but it is a successful composition. The squatness of Mr Belcher's vase is peculiarly appropriate to the material, and seems to demand growing plants.

The Bromsgrove Guild has made vases of many diversified types, as is shown by Figs. 430 to 432. The

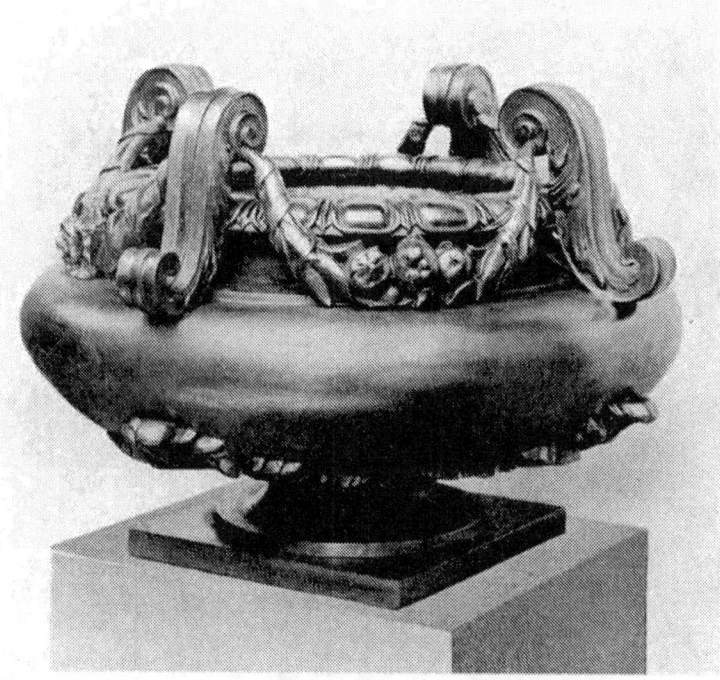

FIG. 428.—Flower Pot at Instow Park.

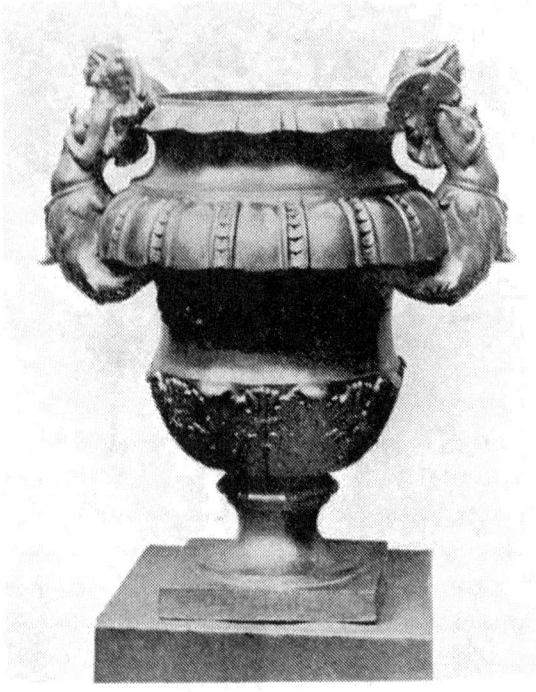

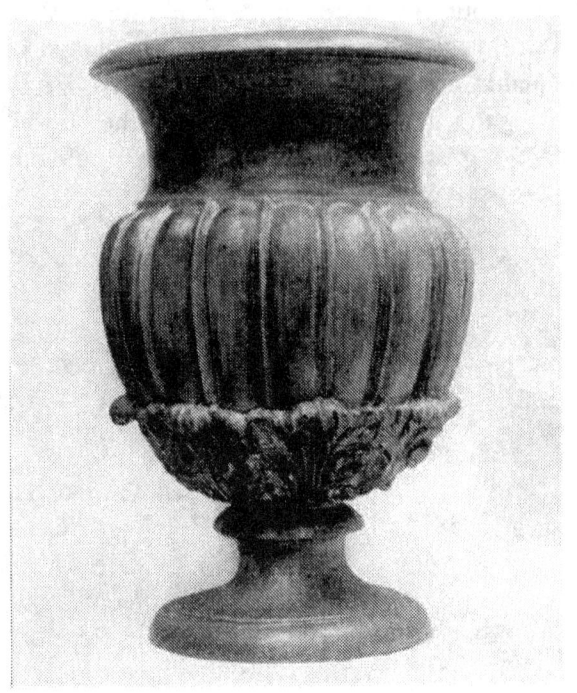

FIG. 429.—Designed by Messrs Wimperis & Best.

FIG. 430.—A Simple Design.

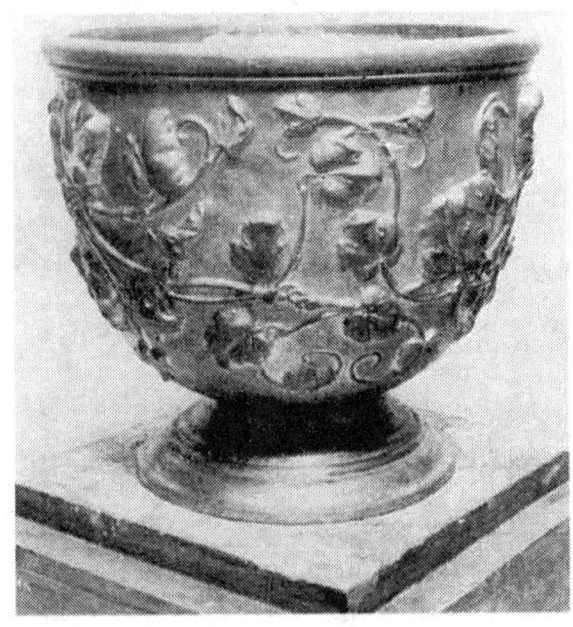

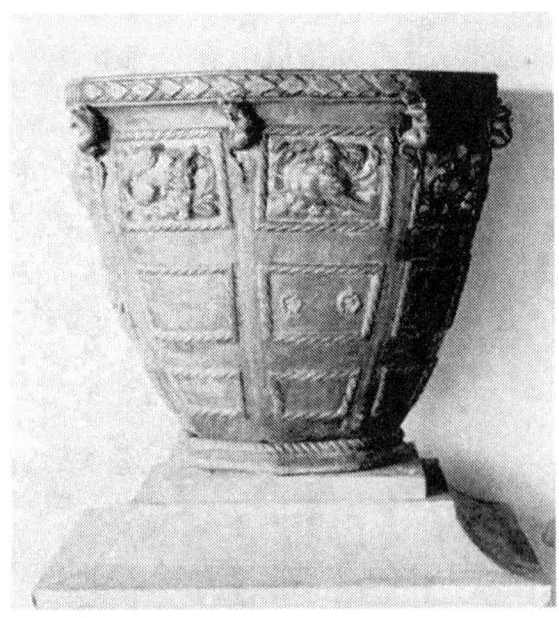

FIG. 431.   Vases by the Bromsgrove Guild.   FIG. 432.

first is very simple, with bold mouldings. The second seems to err on the side of too naturalistic a treatment of foliage, but the third (Fig. 432), with its little cable-moulded panels, is quite delightful, and is as perfect an ornament for a modern garden as the severer example of Fig. 430 would be if added to an old garden of the eighteenth century.

Professor Lethaby has been so often quoted in these pages that it is a particular pleasure to illustrate the very attractive and rightly treated pot of Fig. 436.

The flower-pot of Fig. 433 is illustrated, not for any beauty or fitness of design, but rather as a technical *tour de force*. No part of it is cast. It is entirely beaten up, and,

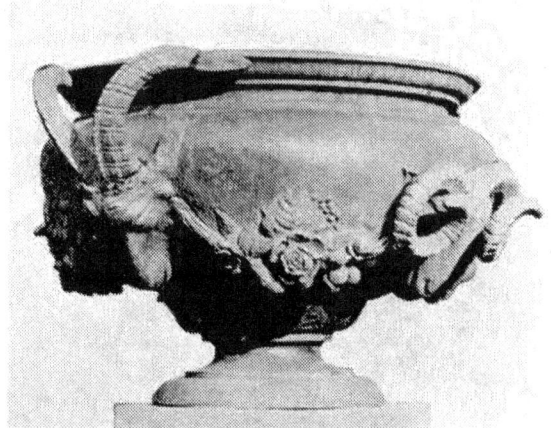

FIG. 433.—By Mr A. B. Laidler.

with the exception of the horns, out of a single sheet of 10-lb. lead, 6 feet 6 inches by 6 feet 6 inches. There are eighteenth-century vases with the same ram's horn treatment. The maker, Mr A. B. Laidler, is a capable worker in cast lead as well as wrought, but it is refreshing to find technical skill in the working of sheet lead put to some other uses than mere sanitary plumbing.

He has since done work of more artistic value, *e.g.*, the memorial tablet of Fig. 441, and the sundial of Fig. 434, designed by Mr D. W. Kennedy. It is a pleasant example of the simplest and cheapest treatment proving effective. The pillar of the dial consists merely of four lead pipes with bead and reel mouldings in the hollows between. The top is decorated with Old Time and his scythe, the hour-glass, and cherubs' heads. It is altogether a masculine bit of work.

The art of modern leadwork owes a great debt to Mr F. W. Troup, and his own designs always strike the right note. The sundial of Figs. 435 and 437 is a pleasant object, suitably decorated, and the blank clock-face of Fig. 439, is an example of an unusual but entirely suitable use of lead. Messrs Henry Hope & Sons have recently made a clock dial with cable edging, which is simple and successful.

The sundial of Fig. 440, by Mr James

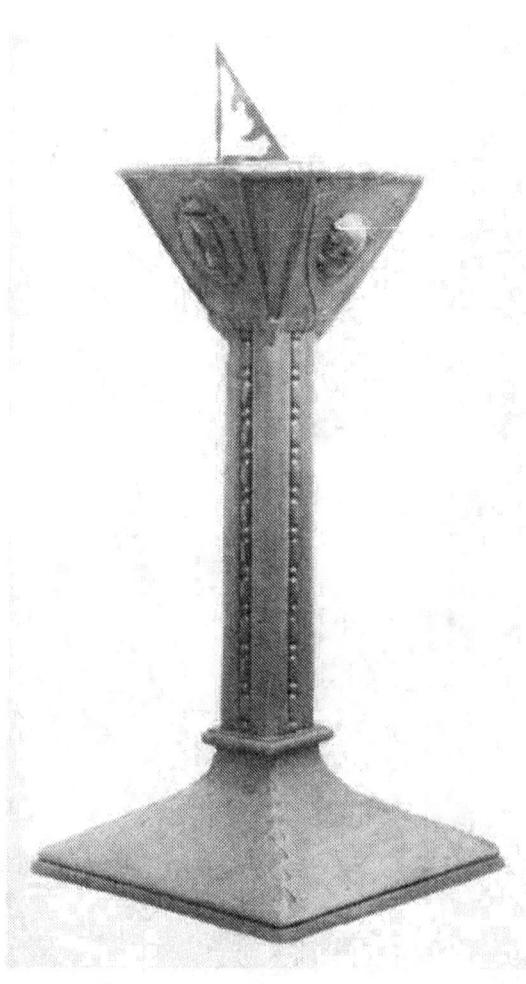

FIG. 434.—Inexpensive Sundial in Lead.

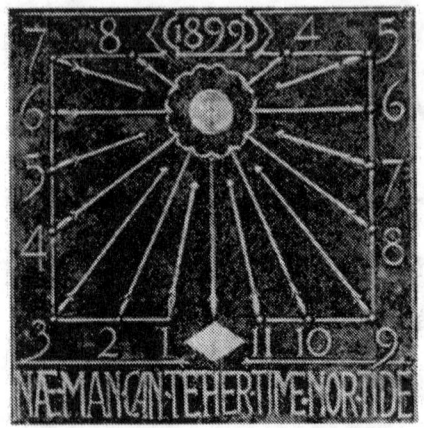

FIG. 435.—Face of Sundial
Illustrated Below.

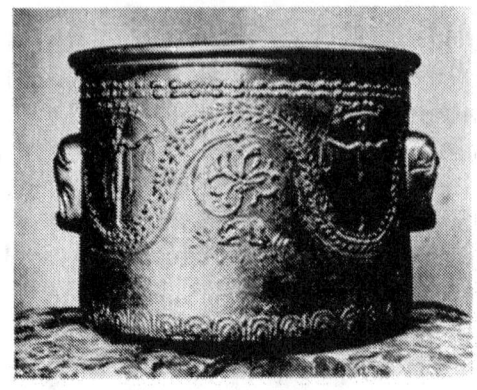

FIG. 436.— Pot designed by Professor
W. R. Lethaby.

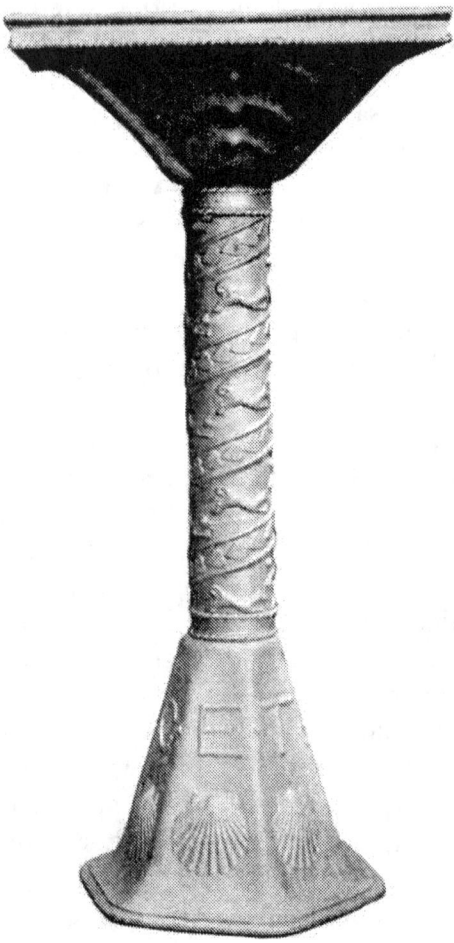

FIG. 437.—Sundial with
Tinned Face.

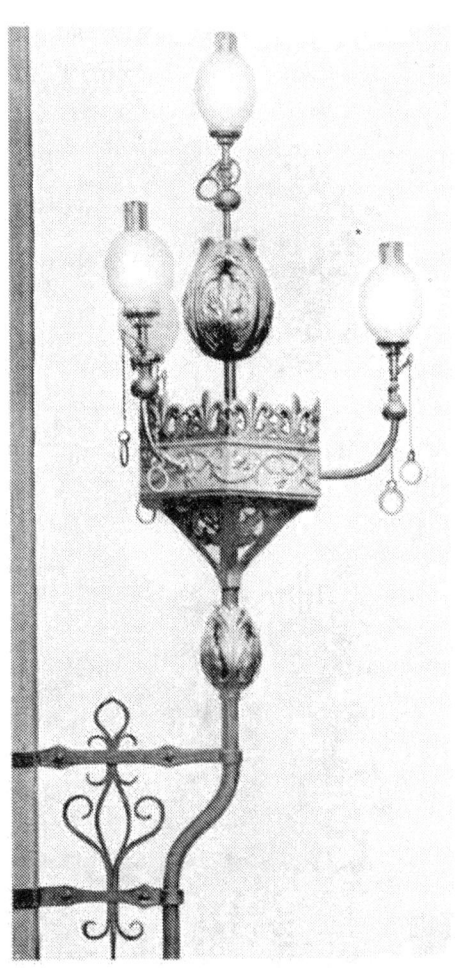

FIG. 438.—Gas Fitting, with Ornaments
of Lead Parcel Gilt.

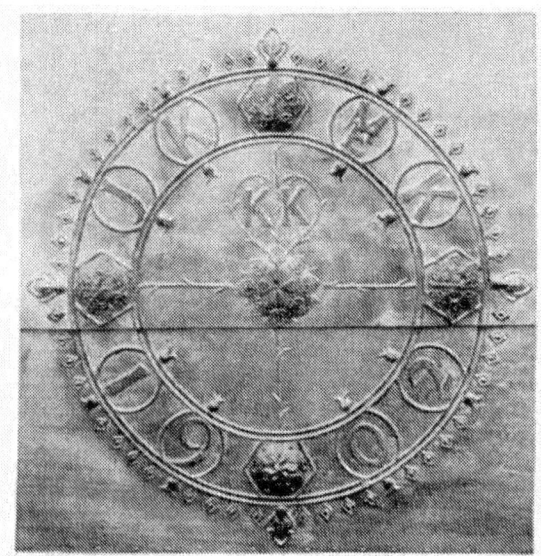

FIG. 439.—Blank Clock Face.

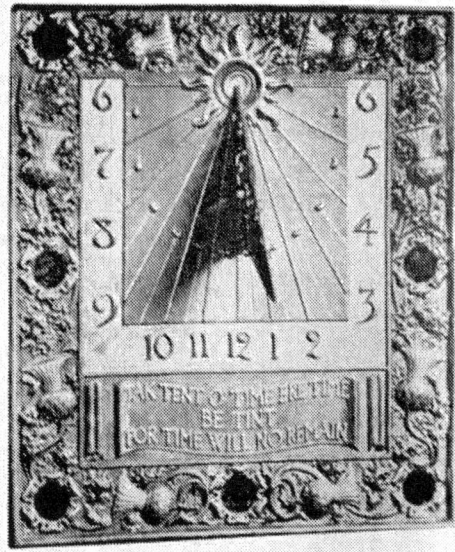

FIG. 440.—Sundial with Jasper Discs.

Cromar Watt, is like goldsmith's work in large. He has called in aid discs of jasper, dull red and greyish-green alternately, and the ornament is a good deal relieved by gilding. The whole effect is rich and interesting.

Unusual amongst ecclesiastical leadwork are the gas standards designed by Sir Charles Nicholson for the Catholic Apostolic Church, Gordon Square, W.C. Messrs Lockerbie & Wilkinson, of Tipton, made them (Fig. 438). The whole of the work, except the piping and stays, is in cast lead parcel gilt. For bowls such as that from which the burners issue, cast lead seems as reasonable a material as repoussé brass or copper (which are ordinarily used for such work), for these latter, when pierced, have a thin and papery look.

The unusual bending of the standard is a practical device to

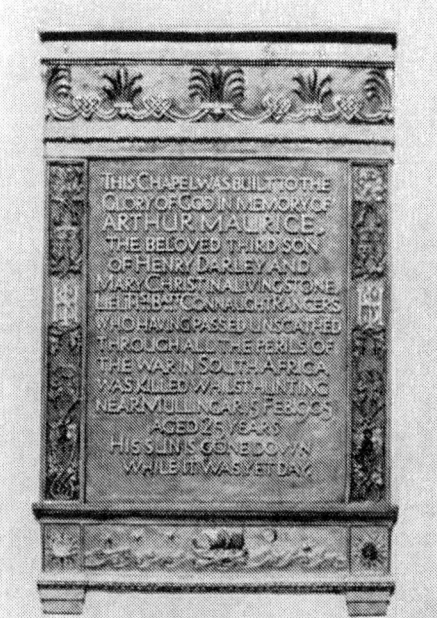

FIG. 441.—At All Saints', Belclare.

avoid a stall. In the beautiful little chapel of All Saints', Belclare, County Mayo, is the lead memorial tablet of Fig. 441. Some parts of the background are painted a strong blue, and the lettering and ornaments are gilt. The tablet has a quiet charm which has distinguished few memorials of the war. Sir Charles Nicholson was architect for chapel and tablet, and the latter was modelled and cast to his design by Mr Laidler.

A FIRST ATTEMPT AT
A BIBLIOGRAPHY OF PUBLICATIONS
RELATING TO
THE HISTORY OF ENGLISH LEADWORK

BOOKS AND ARTICLES IN TRANSACTIONS OF SOCIETIES, &c.

SOURCES OF LEAD, ROMAN PIGS, &c.

THE EARLY METALLURGY OF SILVER AND LEAD: PART I. LEAD. By William Gowland. *Archæologia*, vol. lvii.

A valuable and learned paper. Describes early processes and development of smelting. Illustrates many Roman pigs, and a few early objects, coffins, &c.

SOME ROMANO-BRITISH SOURCES OF LEAD. By Charles Perks. *Birm. and Mid. Inst.*, xiii. 1-12.

RELICS OF ENUMERATION OF BLOCKS OR PIGS OF LEAD AND TIN DISCOVERED IN GREAT BRITAIN. By Albert Way. *Arch. Jour.*, xvi. 22-40.

NOTICES OF ROMAN PIGS OF LEAD FOUND AT BRISTOL, AND OF METALLURGICAL RELICS IN CORNWALL, IN OTHER PARTS OF ENGLAND AND WALES, AND ALSO ON THE CONTINENT. By Albert Way. *Arch. Jour.*, xxiii. 277-290.

PIGS OF LEAD OF THE ROMAN PERIOD IN BRITAIN. By J. D. Leader. *Brit. Arch. Assoc. Jour.*, N.S., iv. 267-271.

ROMAN PIGS. By J. Roach Smith. *Collectanea Antiqua*, vol. iii.

ON ROMAN INSCRIBED PIGS OF LEAD FOUND IN BRITAIN. By W. de Gray Birch. *Brit. Arch. Assoc. Jour.*, N.S., iv. 272-275.

ACCOUNT OF TWO PIGS OF LEAD FOUND NEAR RIPLEY, WITH THIS INSCRIPTION ON THEM: "IMP. CAES. DOMITIANO AUG. COS." By Rev. Sam. Kirkshaw. *Phil. Trans. Roy. Soc.*, xli. 560.

REMARKS ON AN ANCIENT PIG OF LEAD LATELY DISCOVERED IN DERBYSHIRE. By Rev. Samuel Pegge. *Arch.*, v. 369-378.

DESCRIPTION OF A SECOND ROMAN PIG OF LEAD FOUND IN DERBYSHIRE; NOW IN POSSESSION OF MR ADAM WOLLEY, OF MATLOCK, IN THAT COUNTY, WITH REMARKS. By Rev. Samuel Pegge. *Arch.*, vii. 170-174.

DESCRIPTION OF ANOTHER ROMAN PIG OF LEAD FOUND IN DERBYSHIRE. By Rev. Samuel Pegge. *Arch.*, ix. 45-48.

ON THE DISCOVERY OF A ROMAN PIG OF LEAD FOUND ON MATLOCK MOOR, DERBYSHIRE. By Rev. J. C. Cox; and ON ITS INSCRIPTION, by F. J. Haverfield. *Proc. Soc. Antiq.*, 2nd S., xv. 185-189.

THE ROMAN NAME OF MATLOCK, WITH SOME NOTES ON THE ANCIENT LEAD MINES AND THEIR RELICS IN DERBYSHIRE. By W. de Gray Birch. *Brit. Arch. Assoc.*, N.S., vi. 33-46, 113-122.

ON THE EARLY HISTORY OF WIRKSWORTH AND ITS LEAD MINING. By William Webb, M.D. *Jour. Derbyshire Archæol. and N. H. Soc.*, vol. vii., p. 63.

Illustrates two pigs. Gives references to working in Romano-British and Saxon times and later. Wirksworth provided the lead coffin in A.D. 714, for the body of St Guthlac of Croyland.

Illustrates dish for measuring lead ore.

ON THE DISCOVERY OF A FOURTH INSCRIBED PIG OF ROMAN LEAD IN DERBYSHIRE. By the Rev. J. Charles Cox, LL.D.; Prof. F. Haverfield, F.S.A.; and Prof. Hubner. *The Antiquary*, vol. xxix., 218-223.

Gives illustrations of pig found and of two others.

LEAD MINING. VICTORIA COUNTY HISTORIES OF ENGLAND: VOL. II., DERBYSHIRE, pp. 323-349. By Mrs J. H. Lander and C. H. Vellacott.

A full history of the most important industry of Derbyshire in bye-gone days. It deals fully with all evidences from early documents as to the customs and regulations of mining.

THE TRAFFIC BETWEEN DEVA AND THE COAST OF NORTH WALES IN ROMAN TIMES. By George W. Shrubsole. *Chester and North Wales Arch. and Hist. Soc.*, vol. i. (N.S.).
Illustrations of three pigs.

THE ROMAN PIGS OF LEAD DISCOVERED NEAR CHESTER. By Rev. Rupert H. Morris. *Chester Arch. and Hist. Soc.*, N.S., lv. 68-79.

PIG OF LEAD IN CHESTER MUSEUM. By Egerton Phillimore, M.A. *Arch. Cambrensis.*, 5th S., viii. 137.

THE CHESTER PIGS OF LEAD. By Professor J. Rhys. *Arch. Cambrensis*, 5th S., ix. 165-166.

THE ROMAN PIGS OF LEAD DISCOVERED NEAR CHESTER. With a Letter by Professor John Rhys of Oxford. *Jour. Chester Arch. and Hist. Soc.*, N.S., iv. 68-79.

EARLY LEAD MINERS BROUGHT FROM THE HIGH PEAK TO WORK IN FLINTSHIRE. By Henry Taylor. *Chester and N. Wales Arch. and Hist. Soc.*, N.S., viii. 112-114.
Notes on an entry in the Patent Roll of 4 Richard II.

INCIDENTS IN THE BUILDING TRADES OF LONDON IN THE FOURTEENTH AND FIFTEENTH CENTURIES. By W. Culling Gaze. *Builders' Journal*, 26th June 1907.
Included are some interesting records of mediæval plumbers.

ON THE PRICE OF LEAD IN THE REIGN OF HENRY VIII. (ORIGINAL DOCUMENTS). By W. H. Black. *Jour. Arch. Assoc.*, vii. 304-306.
A fother equalled 19½ cwt. Lead cost a halfpenny per lb.

ON LEADWORK GENERALLY.

LEADWORK OLD AND ORNAMENTAL AND FOR THE MOST PART ENGLISH. By W. R. Lethaby. With 76 illustrations, 8 in. by 5 in., pp. 148. Macmillan & Co., 1893.
This altogether admirable little book, often quoted in the preceding pages, did more than anything to revive interest in the art of leadwork.

LEADWORK. By W. R. Lethaby. A paper read before the Society of Arts, and printed in their *Journal* of 9th April 1897.
A footnote to Mr Lethaby's book.

ORNAMENTAL LEADWORK. W. Burges. *The Ecclesiologist*, December 1856.
This admirable paper has been used largely by Mr Lethaby in his book. but as it deals chiefly with French work it has been little drawn upon for the purposes of this volume.

LEADWORK. By F. W. Troup, F.R.I.B.A. *Jour. Roy. Inst. Brit. Architects*, 3rd S., vol. xiii., No. 10.
Chiefly practical notes on working in lead.

ORNAMENTAL LEAD AND LEAD-CASTING. By F. W. Troup, F.R.I.B.A. *Jour. Roy. Inst. Brit. Architects*, 3rd S., vol. vii., No. 13.
A full review of methods and processes, with long quotations from Burges, Viollet-le-Duc, and Felibien.

EXTERNAL LEADWORK. By F. W. Troup. A Chapter in *The Arts connected with Building.* Published by B. T. Batsford, 1909.

LEADWORK, ANCIENT AND MODERN. By Charles Hadfield, F.R.I.B.A. A lecture before the Sheffield Art Crafts Guild. *The British Architect*, 1904.
Deals with leadwork generally, and prints extracts from building-roll of York Minster dealing with plumbing work.

ENGLISH ORNAMENTAL LEADWORK. By J. Lewis André. *Arch. Jour.*, xlv. 109-119.
This paper ranges over the whole subject.

THE REVIVAL OF THE HANDICRAFTS: LEADWORKING. By J. Starkie Gardner. *The Magazine of Art*, May 1900.
A general article with illustrations of the Melbourne leadwork, of ventilating quarries, and of a modern dragon in lead on a wrought-iron terrace screen.

LEAD ARCHITECTURE. By J. Starkie Gardner. *Jour. R.I.B.A.*, xi. 141-157.
An excellent paper followed by an interesting discussion. It deals largely with the historical evidence for the larger architectural uses of lead.

OLD LEADWORK IN EXETER AND THE NEIGHBOURHOOD. By Harbottle Reed. *Exeter Diocesan Arch. and Arch. Soc.*, 3rd S., i. 165-172.
Deals with general leadwork with special reference to pipe-heads. Some fine gutters on houses now demolished are illustrated.

ON DERBYSHIRE PLUMBERY; OR WORKINGS IN LEAD. By J. Charles Cox, LL.D. *Derbyshire Arch. and N.H. Soc.*, vol. ix.
A good general review of the county leadwork. List of fonts incorrect. Illustration of very early gutter at Derby.

OF GARDEN ORNAMENT: THE USE OF LEADWORK IN GARDENS. Anonymous. *Country Life*, 15th July 1899.
Illustrations include the "Cain and Abel," a fox with fowl in his mouth, a sportsman levelling a gun, and two of the vases at the Villa at Chiswick.

OF LEADEN GARGOYLES, MAGDALEN COLLEGE, OXFORD. By Richard Davey. *Country Life*, 27th October 1900.
If the gargoyles illustrated were not unquestionably of stone this article would be of value. As things are, however, the comparison with the Notre Dame *stone* gargoyles and the regret that Victor Hugo never "beheld the *leaden* gargoyles of Maudlin" fail to impress.

OF GARDEN ORNAMENT: LEADWORK AS GARDEN DECORATION. By Richard Davey. *Country Life*, 14th April, 28th April 1900.
In addition to several photographs of the Melbourne leadwork are "The Rape of the Sabines" at Painshill, the vases at Drayton House, a "Faun" at Peover Hall, and "Flora" at Drayton.

FORMAL GARDENS IN ENGLAND AND SCOTLAND. By Inigo Triggs. Published by B. T. Batsford.
Leadwork illustrated includes the following :—
Longford Castle: "Flora," by Sir Henry Cheere, in garden temple. Belcombe Brook: "Perseus" in garden temple (not the same as at Melbourne). Stoneleigh Abbey: vases on gate piers. Rousham: "Bacchus." Canon's Ashby: "Shepherd" playing flute. Nun Moncton: statues. Wilton House: amorini. Chiswick House: two vases. Enfield Old Park: vase. Penshurst: vase. Iford Manor: vase. Victoria and Albert Museum and Enfield: Cisterns. Drayton House, Northants: four vases.
Also other objects not noted above as they are illustrated in foregoing chapters.

THE DECORATIVE TREATMENT OF METAL IN
ARCHITECTURE. By George H. Birch.
Society of Arts, Cantor Lecture, April 1883.
Contains an eloquent plea for leadwork and a number
of references to examples. Also states that the statue of
Shakespeare on the porch of Drury Lane Theatre is of
lead.

FONTS.

OBSERVATIONS ON FONTS. By Richard Gough,
Dir.S.A., 1789. *Archæologia*, vol. x. 183-
209.
This appears to be the first reference to lead fonts.
Gough mentions four only—Brookland, Dorchester,
Wareham, and Walmsford. The last is not of lead now,
but perhaps since 1789 the font Gough refers to has been
destroyed.
The Brookland font Gough attributes to the time of
Birinus. As he died in 650 A.D. we must reject this date.
Ashover is mentioned as having lead figures on a stone
font.

LEADEN FONTS. Alfred C. Fryer, Ph.D.,
F.S.A. *Arch. Jour.*, lvii. 40-51.
An altogether admirable and exhaustive paper which
has been drawn upon freely in the foregoing chapter on
fonts.

NOTES ON FONTS. Alfred C. Fryer, Ph.D.,
F.S.A. *Arch. Jour.*, vol. lxiii., No. 250,
97-105.
On Penn, Greatham, and Burghill fonts, and the
vessels at Gloucester, Maidstone, and Lewes all described
ante.

BROOKLAND, KENT, DESCRIPTION OF CURIOUS
LEADEN FONT IN THE CHURCH OF.
Arch. Jour., vi. 159-164.

SOME OBSERVATIONS OF THE LEADEN FONT
OF BROOKLAND CHURCH, ROMNEY MARSH.
By Herbert L. Smith. *Arch. Cant.*, iv. 87-
96.

THE LEADEN FONT AT BROOKLAND. By Rev.
Grevile M. Livett. *Arch. Cantiana*, xxvii.
255-261.

LEADEN VESSEL, PROBABLY THE LINING OF
A FONT NOW AT GREATHAM. By R.
Garraway Rice. *Proc. Soc. Antiq.*, 2nd S.,
xviii. 294-303.
Dealt with in "Fonts" chapter. Mr Garraway Rice
rejects idea of the vessel being a font in favour of theory
that it is a lining.

AN ANCIENT LEAD COFFER FOUND AT WILL-
INGDON. By M. A. Lower. *Suss. Arch.
Coll.*, i. 160.
The object now in Lewes Castle, dealt with in chapter
on Fonts. It was found in a cutting in 1847. This
paper claims it as Anglo-Saxon of tenth century.

FONTS AND FONT COVERS. By Francis Bond.
1908. Henry Frowde, Oxford University
Press.
This admirable book illustrates fourteen of the lead
fonts, and the classification follows that of the present
author.

SEPULCHRAL LEADWORK.

REMARKS ON THE ORNAMENTATION OF ROMAN
COFFINS WITH ESCALLOP SHELLS. By
Henry Charles Coote. *Lond. and Middl.
Arch. Soc.*, ii. 268.
Escallops symbolise the sacrifice made to the *manes*
of the buried.
This paper also gives an account of two lead Roman
coffins found at East Ham.

ROMAN LEAD COFFIN DISCOVERED AT CANTER-
BURY. By Charles Roach Smith. *Arch.
Cant.*, xiv. 35, 36.
Roman: the coffin had two diagonal lines of cord
moulding on the top, with well-modelled rose at intersection and four simpler circular ornaments half-way
between intersection and corners.

LEADEN COFFIN, RHYDDGAER. By W. Wynn
Williams. *Arch. Camb.*, 4th S., ix. 136-140.
Remains of a Roman coffin. Has lettering CAMVLO-
RIS HOI east in relief; lettering is most unusual on
coffins, indeed this is perhaps a unique example.

NOTES ON SOME LEADEN COFFINS DISCOVERED
AT COLCHESTER. By Charles Roach Smith.
Brit. Arch. Assoc., ii. 297-303.
Roman: ornaments were bead and reel rods, escallops
and rings. C. R. S. also gives sketch of coffin found
in 1794, with attractive design of escallops and rope
moulding.

ROMAN LEADEN COFFINS DISCOVERED AT
COLCHESTER. By Henry Laver. *Essex
Arch. Soc.*, N.S., iii. 273-277.
Roman: beaded rim and beaded crosses; a queer
2-inch pipe issued from lid above where face of corpse
would be. Also child's coffin with beaded crosses.

LEAD COFFIN FOUND IN THE MINORIES, 1853.
By J. Y. Akerman. *Proc. Soc. Antiq.* First
Series, iii. 17.
Romano-British with escallops and beaded rods. Now
in British Museum.

NOTICE OF A LEADEN COFFIN, OF EARLY
FABRIC, DISCOVERED AT BOW. By Charles
Roach Smith. *Arch.*, xxxi. 308-311.
Roman; with cable moulding.

COLLECTANEA ANTIQUA. By J. Roach Smith.
For Roman Coffins and Ossuaries, see vols.
iii. and vii.
Some subjects dealt with in the *Collectanea* are re-
statements of finds that had already been described in
Archæological *Proceedings*.

ROMAN COFFINS OF LEAD FROM BEX HILL,
MILTON, NEXT SITTINGBOURNE. By George
Payne. *Arch. Cant.*, ix. 164-173.
Roman: three found. One is in Maidstone Museum,
with crosses of bead and reel rods and Medusa heads;
another had, in addition, lions, jug-like ornaments, and
a sword blade.
The lions are unique as coffin ornaments.
Note infrequency of use of escallops on Kentish
Roman coffins.

ROMAN LEADEN COFFINS AND OTHER INTER-
MENTS DISCOVERED NEAR SITTING-
BOURNE, KENT. By George Payne. *Arch.
Cant.*, xvi. 9-11.
Roman: rope moulding, rings, oxen yokes. A lead
ossuary was found near by.

ROMAN LEADEN COFFIN DISCOVERED AT PLUM-
STEAD. By George Payne. *Arch. Cant.*,
xvii. 10-11.
Roman: bead and reel ornament all round the lid
near the edge.

LEAD COFFIN FOUND AT CHATHAM. By George
Payne. *Proc. Soc. Antiq.*, vii. 415.
Romano-British: escallops and billet ornaments.

ROMAN COFFIN OF LEAD AT CHATHAM. By A.
A. Arnold. *Arch. Cant.*, xii. 430-431.
Found between Crayford and Bexley.
Roman: beaded ornament on seams and escallop
shells.

NOTICE OF A LEADEN COFFIN DISCOVERED AT HEIGHAM. By Robert Fitch. *Norfolk and Norwich Arch. Soc.*, vi. 213-216.

Unornamented; probably Roman.

THE DISCOVERY OF LEADEN COFFINS IN LEICESTER. By G. C. Bellairs. *Leicester Architect and Arch. Soc.*, iv. 246-249.

Roman: three, one with slight striated pattern, two without ornament.

DISCOVERY OF A ROMAN LEADEN COFFIN NEAR BISHOPSTOKE, HANTS. By Francis Joseph Baigent. *Proc. Soc. Antiq.*, 2nd S., ii. 327-329.

Devoid of ornament.

WEEVER'S "FUNERAL MONUMENTS." Ed. 1631, p. 30.

Reference to Roman coffin of about 239 A.D., with escallop shell ornaments—found at Stepney.

ACCOUNT OF A LEADEN COFFIN TAKEN OUT OF A ROMAN BURYING-PLACE NEAR YORK. By Ralph Thoresby. *Phil. Trans. Roy. Soc.*, xxiv. 1864-1865.

A ROMAN COFFIN FOUND AT BRAINTREE. By G. F. Beaumont. *Essex Arch. Soc.*, vii. 401-402.

SOMERSETSHIRE ROMAN LEAD COFFINS. Notes by H. St George Gray. *Somerset and Dorset Notes and Queries*, vol. ix. 8, 58, 230.

At Taunton Castle Museum is a small piece of a coffin, with plaited-work design, found near Ilchester. Lead coffins are scarce in Somersetshire.

LEAD COFFIN AND TWO OSSUARIES FOUND AT ENFIELD. By R. A. Smith. *Proc. Soc. Antiq.*, xix. 206.

Romano-British : coffin has rope mouldings in saltire and star arrangements with scallop shells. Ossuaries plain. See for notes on inhumation and urn burials.

ACCOUNT OF TWO LEADEN CHESTS, CONTAINING THE BONES, AND INSCRIBED WITH THE NAMES, OF WILLIAM DE WARREN AND HIS WIFE GUNDRAD, FOUNDERS OF LEWES PRIORY, SUSSEX, DISCOVERED IN OCTOBER 1845, WITHIN THE PRIORY PRECINCT. By W. H. Blaauw. *Arch.*, xxxi. 438-442.

Blaauw suggests that the bodies were put into the lead coffins about sixty years after Gundrada and William died (1085 and 1088 respectively), making date of coffins about 1150.

ON THE DISCOVERY OF THE REMAINS OF WILLIAM DE WARENNE AND HIS WIFE GUNDRADA, AT LEWES. By C. L. Prince. *Sussex Arch. Coll.*, xl. 170-172.

THE ANCIENT STONE AND LEADEN COFFINS, &c., IN THE TEMPLE CHURCH. By Edward Richardson. Published 1845.

Deals fully with the mediæval lead coffins and illustrates them. Richardson attributes them to the beginning of the thirteenth century.

DISCOVERY OF STONE COFFINS, LEADEN SEPULCHRAL CHEST, SKELETONS, AND INCISED SLAB OF THE THIRTEENTH CENTURY AT DRAYTON. By J. Wodderspoon. *Norfolk and Norwich Arch. Soc.*, vi. 132-141.

The "leaden chest" described was, in fact, a lead shell enwrapping the body like an Egyptian mummy case, which was placed inside a stone or wood coffin, or buried without.

EFFIGY OF KING RICHARD, CŒUR DE LION, IN THE CATHEDRAL AT ROUEN. By Albert Way. *Archæologia*, xxix. 202-216.

In addition to the effigy the lead heart casket is described. It consisted of two boxes one within the other. The lettering engraved inside the inner box has been reproduced by Mr Lethaby in *Leadwork*. The heart was found "withered to the semblance of a faded leaf." The lead casket was enclosed in a sumptuous gold and silver casket, which was sold towards the ransom of St Louis in 1250.

ST EANSWITH'S RELIQUARY IN FOLKESTONE CHURCH. By W. A. Scott Robertson. *Arch. Cant.*, xvi. 322-326.

This is illustrated and described in "Cisterns" chapter. W. A. S. R. gives details of its finding.

LEAD RELIQUARY OF ST WITA AT WHITCHURCH CANONICORUM. By C. Druitt.

With early thirteenth-century inscription, otherwise plain.

LEAD COFFINS AT WEST THURROCK CHURCH, ESSEX. *The Antiquary*, 1906, p. 326.

Thirteen were found of mummy case shape, one being dated 1607.

WOOLLEN CAP AND SHROUD DISCOVERED IN A LEAD COFFIN AT WINDSOR. By Charles H. Read. *Proc. Soc. Antiq.*, xvii. 225-228.

The "lead" interest here is that Mr Gowland notes that the first record of rolled lead in England is in 1670, when a company was formed for its manufacture. The coffin was of rolled lead and of about 1670.

LEAD COFFIN REMOVED FROM ST MILDRED'S, BREAD STREET. *The Antiquary*, 1906, p. 402.

Of Sir Nicholas Crispe, 1665. Of mummy case shape, "with the form of the body, head, and neck roughly followed—the arms crossed in half relief, the nose represented by a sharply-cut and raised triangle, the eyes, brows, and wide smiling lips by incised lines."

OBSERVATIONS ON THE MONUMENT IN CANTERBURY CATHEDRAL CALLED THE TOMB OF THEOBALD, AND AN ACCOUNT OF TWO ANCIENT INSCRIPTIONS ON LEAD DISCOVERED IN CANTERBURY CATHEDRAL. By Henry Boys. *Arch.*, xv. 291-299.

The inscription on lead sheet found in the lead coffin of Archbishop Theobald, the immediate predecessor of St Thomas à Becket, is in a good Roman lettering.

LEAD LETTERING IN GRAVE SLAB. By C. Hodgson Fowler. *Proc. Soc. Antiq.*, xii. 411.

Date about 1300.

CAMDEN'S BRITANNIA. Folio, vol. i., p. 59, edition 1789.

An illustration is given of the inscribed lead cross which was *reputed* to have been found in Arthur's (*also reputed*) grave at Glastonbury.

LEADEN BOX AND CROSSES FROM RICHMOND. By Ed. Charlton. *Arch. Æliana*, N.S., ii. 46-50.

Box contained earth and four rude crucifixes in lead. Origin, date, and purpose doubtful.

NOTES UPON THE DISCOVERY OF A NUMBER OF LEADEN GRAVE CROSSES NEAR THE GREY FRIARS' MONASTERY, NEWGATE STREET, LONDON. By F. G. Hilton Price. *Proc. Soc. Antiq.*, N.S., xxi. 12-20.

Eighty-nine were found varying in length from $6\frac{1}{2}$ to $2\frac{1}{2}$ in. long, without ornament, and very roughly cut out of sheet lead with a chisel and then roughly hammered. They are absolution crosses, and were doubtless made in a great hurry to bury on the bodies of the friars who died during the Black Death.

OBSERVATIONS ON CERTAIN SEPULCHRAL USAGES OF EARLY TIMES. By W. M. Wylie. *Archæologia*, vol. xxxv.

Deals with absolution crosses found near Dieppe. They were rudely cut out of sheet lead. Long absolutions were scratched on. The crosses were laid on the breast of the buried. There are interesting references to Abelard and Heloise and to similar crosses found at Lincoln and Chichester, and illustrations.

SÉPULTURES CHRÉTIENNES DE LA PÉRIODE ANGLO-NORMANDE, &c. By L'Abbé Cochet. *Archæologia*, vol. xxxvi. 258-266, and xxxvii. 37-38.

These two papers deal with the same subject of lead absolution crosses as Wylie's paper, but more fully.

LEAD CROSSES FOUND AT BURY ST EDMUNDS. By Samuel Tymms. *Proc. Soc. Antiq.*, iii. 165-167.

Three absolution crosses, two inscribed.

A LEADEN CROSS FOUND AT BURY ST EDMUNDS, &c. By Edmund Waterton. *Proc. Soc. Antiq.*, 2nd S., ii. 301.

An absolution cross, inscribed, and a lead matrix of a seal.

A LEAD CROSS. By J. Y. Akerman. *Proc. Soc. Antiq.*, iv. 212-213.

An absolution cross inscribed, also bearing date 1136.

EXCAVATIONS AT ST AUSTIN'S ABBEY, CANTERBURY. By W. H. St John Hope. *Arch. Cantiana*, xxv. 237.

Mr Hope here illustrates and describes a lead memorial plate and an absolution cross.

AN INSCRIBED LEADEN GRAVE CROSS FOUND AT SOUTHAMPTON. By W. Dale. *Proc. Soc. Antiq.*, 2nd S., xx. 169.

Found at a considerable depth when excavating. It commemorates one Udelina, and is thirteenth century or earlier. On the reverse side is engraved "Ave Maria . . . mulieribus." Illustrations of both sides given.

AN ACCOUNT OF HUMAN BONES FILLED WITH LEAD. By J. Worth. *Arch.*, iv. 69-72.

An odd account written in 1774 offering no intelligent explanation of a queer find.

LEAD SEALS, BULLÆ, AND TOKENS.

ON ROMAN LEADEN SEALS. By Charles Roach Smith. *Lond. and Middl. Arch. Soc.*, v. 433-435.

ON ROMAN LEADEN SEALS. By Robert Blair. *Arch. Æliana*, N.S., viii. 57-59.

Actually of pewter. Used on strings like papal bullæ.

LEADEN SLABS FOUND AT BROUGH CASTLE. By B. Williams. *Proc. Soc. Antiq.*, First Series, iii. 222.

Seals for letters or for marking clothes of Roman soldiers.

ON PAPAL BULLÆ FOUND IN SUSSEX. By Ambrose P. Boyson. *Sussex Arch. Coll.*, xlviii. 99-103.

The author is particularly indebted to Mr Boyson for kind permission to draw on this admirable and lucid paper. See *ante*.

ILLUSTRATION OF PAIR TONGS WITH DIES FOR FORGING BULLÆ OF PIUS II. *Jour. Arch. Assoc.*, vol. ii. 97.

NOTES ON PONTIFICAL BULLÆ, WITH REFERENCE TO THAT RECENTLY DISCOVERED IN CHETWODE CHURCHYARD. By E. P. Loftus Brock. *Bucks Records*, v. 71-73.

Of Innocent VI. (1352-1362).

ON A LEADEN BULLA FOUND AT WARMINSTER. By Rev. John Baron. *Wilts. Arch. Soc.*, xvii. 44-45.

On obverse: "Bonifatius P. P. VIII." (date, 1389-1404).
On reverse: SPA (St Paul), SPE (St Peter), and the two heads with beading round each.
This is common type of bulla.

DISCOVERY OF A LEADEN BULLA AT HAUGHMOND ABBEY. By Rev. W. G. D. Fletcher. *Shropshire Arch. Soc.*, 3rd S., i. 283-284.

Of Pope Urban VI. (1378-1389).
Refer also to *Brit. Mus. Catalogue of Seals*, vol. vi., plate vii., No. 21889. Also pp. 286, 287 of same volume.

NOTES ON THE LEADEN BULLÆ OF THE ROMAN PONTIFFS. By Edmund Bishop. *Proc. Soc. Antiq.*, 2nd S., xi. 260-270.

A learned review of the whole history of papal bullæ, with special reference to examples in British Museum.

ON A LEADEN SEAL OF HENRY IV., FOUND AT CATCHBURN, NEAR MORPETH. By W. Woodman. *Arch. Æliana*, x. 191-192.

The seal of the Chancery of Berwick.

PILGRIMS' SIGNS. By Cecil Brent. *Arch. Cant.*, xiii. 111-115.

Ampullæ here stated to have contained blood of Thomas à Becket mixed with water.
The religious guilds sold the tokens.
Paper includes a descriptive schedule of various signs.

BRENT'S "CANTERBURY IN THE OLDEN TIME." 2nd edition, p. 51.

Moulds for casting lead tokens.

NOTES ON A COLLECTION OF PILGRIMS' SIGNS OF THE THIRTEENTH, FOURTEENTH, AND FIFTEENTH CENTURIES. By Rev. T. Hugo. *Arch.*, xxxviii. 128-134.

Two good plates illustrating examples are given.
T. H. says ampullæ were lacrymatories (*vide* other theories).
Quotes the *Colloquy* of Erasmus, which crops up in nearly every paper on Pilgrims' Signs.

PILGRIMS' SIGNS AND LEADEN TOKENS. By Charles Roach Smith. *Brit. Arch. Assoc.*, i. 200-212.

Among the signs are described "Vernicles," or likenesses of our Lord, and the head of St John Baptist. Some such signs were used as "medals of presence" (much as modern factory hands use numbered discs) in great churches by those whose duty it was to attend choir.
Tokens were issued by tradesmen for local circulation.

NOTES ON PILGRIMS' SIGNS OF THE MIDDLE AGES, AND A STONE MOULD FOR CASTING LEADEN TOKENS, FOUND AT DUNDRENNAN ABBEY. By Dr Joseph Anderson. *Proc. Soc. Antiq. Scot.*, xi. 62-80.

The custody of the moulds for casting pilgrims' signs was often vested in the sacristan, as at the church of St Mary Magdalen at St Maximin, Provence. The plant at Walsingham greatly mystified one of Thomas Cromwell's Visitors.
The Dundrennan mould cast six signs at once, an indication of their extensive use.

REMARKS ON A LEADEN AMPULLA IN THE YORK MUSEUM. By Charles Baily. *Jour. Arch. Assoc.*, vi. 125-126.

Part of the substance of this paper is incorporated in the text, *ante*.

PILGRIMS' BADGE. By A. W. Franks. *Proc. Soc. Antiq.*, iii. 242.

Of St Thomas of Canterbury.

MOULDS FOR CASTING PILGRIMS' SIGNS FOUND AT WALSINGHAM AND LYNN. By Rev. C. R. Manning. *Norfolk Arch. Soc.*, ix. 20-24.

Made of white lias stone. The signs were stars, including representation of the Annunciation, &c., and were cast five in a row.

MATHRAVAL, MOULD FOR CASTING TOKENS FOUND AT. *Powysland Club*, vi. 217-220.

COLLECTION DE PLOMBS HISTORIÉS, TROUVÉS DANS LA SEINE. Par Arthur Forgeais. Paris, 54 Quai des Orfèvres (published in 1865).

Only the third volume of this monumental work has come into the author's hands. It deals with *Imagerie Religieuse*, and illustrates and identifies a large series of pilgrims' tokens.

PILGRIMS' BADGES. By A. W. Franks. *Proc. Soc. Antiq.*, iii. 302.

Byzantine: very similar to English badges.

FORGERIES AND COUNTERFEIT ANTIQUITIES. By T. Sheppard. *The Antiquary*, vol. xliv. 209.

Illustrates several "Billys and Charlies" of the pilgrims' sign variety.

LEADEN TOKENS. By Rev. D. H. Haigh. *Num. Chron.*, vi. 82-90.

Deals largely with the mock coinage of the Boy Bishops.

LEADEN TOKENS. By G. C. Yates, F.S.A. *Trans. Lanc. and Chesh. Antiq. Soc.*, x. 111-121.

The use of lead tokens by way of additional coinage of small value arose owing to the small supply of Royal coinage. The practice flourished despite constant laws and edicts against it. Erasmus notes, in 1499, the "plumbeos Angliæ" then in common circulation. They were used chiefly in the sixteenth and seventeenth centuries, and bore generally very rough representations. See Cleveland's *Midsummer Moon*, where he writes, "the King's image is sometimes stamped on lead, and nature's mint coynes mousters."

CATALOGUE OF LEADEN AND PEWTER TOKENS ISSUED IN IRELAND. By Aquilla Smith. *Kilkenny Arch. Soc.*, N.S., ii. 215-221.

Earliest of 1578, with beautiful cable edging. Majority of end of eighteenth century.

Tradesmen's tokens: many illustrated. One Cork example cast in brass mould.

COLLECTANEA ANTIQUA. By J. Roach Smith :—

Lead Tokens in vols. i., ii., iv., vi., vii.
„ Bullæ in vol. i.
„ Medals in vol. i.
„ Seals (Roman) in vols. iii. and vi.
„ Lawsuit in 1857, arising out of forgery of Pilgrims' Signs, in vol. v.

SUNDRY.

LEAD CELT. By C. H. Read. *Proc. Soc. Antiq.*, xvi. 329.

A mould for bronze celts: illustrated.

LEAD CELT FOUND AT ANWICK. By E. K. Clark. *Proc. Soc. Antiq.*, xx. 258.

Now in Leeds Museum. Appears to have been an experimental casting used in making of bronze celts.

LEAD COIN BROOCH FROM BOXMOOR. By R. A. Smith. *Proc. Soc. Antiq.*, xix. 211.

ON THE USE OF THE SLING AS A WARLIKE WEAPON AMONG THE ANCIENTS, ACCOMPANYING A PRESENT TO THE SOCIETY OF A LEADEN PELLET, OR SLING-BULLET, FOUND LODGED IN THE CYCLOPIAN WALLS OF SAME IN CEPHALONIA. By Walter Hawkins. *Arch.*, xxxii. 96-107.

A learned and dreary treatise on sling-bullets.

A SLINGER'S LEADEN BULLET FROM NAUPORTUS. By J. B. Pearson. *The Antiquary*, vol. xliv. 69.

LEAD LAMP, SAUCEPAN, ETC. By H. M. Scarth. *Proc. Soc. Antiq.*, vi. 190.

Romano-British objects found in Somersetshire.

ROMAN AND OTHER OBJECTS FROM VARIOUS SITES IN CHESTER. By R. Newstead. *Chester and North Wales Arch. and Hist. Soc.*, vol. viii. (N.S.).

Illustrations of Roman water pipes.

REMAINS OF LEAD QUADRANGULAR VESSEL. By A. W. Franks. *Proc. Soc. Antiq.*, iii. 93.

Decorated with scrolls, a human figure and inscription, CVNOBARRVS FECIT VIVAS.

ON A ROMAN PATELLA AND A LEADEN VESSEL FOUND IN REDESDALE. By T. Stephens. *Berwickshire Nat. Club*, xi. 128-130.

ON A LEADEN MEDALLION OF DIOCLETIAN AND MAXIMIAN. By Mdme. La Saussaye. *Num. Chron.*, N.S., iii. 107-111.

Trial piece of a medallion evidently intended to be struck in a precious metal.

NOTES ON FOUR LEADEN WEIGHTS, OF SUPPOSED ROMAN ORIGIN, IN THE GROSVENOR MUSEUM, CHESTER. By Thomas May. *Chester and N. Wales Arch. and Hist. Soc.*, N.S., ix. 129-131.

SOME CONSIDERATIONS ON TWO PIECES OF LEAD WITH ROMAN INSCRIPTIONS UPON THEM, FOUND SEVERAL YEARS SINCE IN YORKSHIRE. By John Ward. *Phil. Trans. Roy. Soc.*, xlix. 686-700.

LEAD OBJECTS FROM THE SEINE. By A. W. Franks. *Proc. Soc. Antiq.*, iv. 75.

Face of a Gaul and kneeling female figure.

METALLIC ORNAMENTS AND ATTACHMENTS TO LEATHER. By Rev. A. Hume. *Lanc. and Chesh. Hist. Soc.*, N.S., ii. 129-166.

Some lead tags or pendants attached to ends of straps are illustrated.

NOTICE OF SOME REMARKABLE INSCRIPTIONS ON LEAVES OF LEAD, PRESERVED IN THE MS. DEPARTMENT OF THE BRITISH MUSEUM. By W. de Gray Birch. *Arch.*, xliv. 123-136.

The inscriptions are in Greek and Latin, and of doubtful date from the eighth to thirteenth centuries.

LEAD AS A COVERING FOR SAXON CHURCHES. J. Park Harrison. *Arch. Oxon.*, part 4.

THE INSCRIBED LEADEN TABLET FOUND AT BATH. By W. de Gray Birch. *Jour. Arch. Assoc.*, xlii. 410-412.

Roman: attributed to between second and fifth centuries A.D. See text of book.

ON A LEADEN TABLET OR BOOK COVER, WITH AN ANGLO-SAXON INSCRIPTION. By Thomas Wright. *Arch.*, xxxiv. 438-440.

The lettering is an inscription by way of preface to the manuscript of Alfric's homilies which the cover originally encased. Date probably about A.D. 1000.

DECORATED LOZENGE OF LEAD. By Albert Way. *Proc. Soc. Antiq.*, v. 475.

Anglo-Saxon: a curious object, use conjectural: illustrated.

LEAD MATRIX FOR IMPRESSING CONSECRATED WAFER. By W. D. Bruce. *Proc. Soc. Antiq.*, First Series, i. 179.

Unfortunately merely noted, not illustrated.

COLLECTANEA ANTIQUA. By J. Roach Smith.

Lead cover of box or cup, decorated with the Visit of the Magi, &c., found in Thames in 1846, vol. i.
Lead cover of Reliquary found in the Somme, vol. ii.

LEADEN VESSEL, POSSIBLY A CHRISMATORY, FOUND AT EVESHAM. By J. A. Johnes. *Proc. Soc. Antiq.*, First Series, ii. 186.

An illustration is given: vessel much damaged. Ornament apparently represents murder of St Thomas à Becket.

LEAD WEIGHTS OF THE FOURTEENTH CENTURY. By C. V. Collier. *Proc. Soc. Antiq.*, xx. 13.

LEAD HERALDIC PLAQUE. By Archdeacon Pownall. *Proc. Soc. Antiq.*, xi. 112.

German: a fine decorative work: illustrated.

THE PARISH AND CHURCH OF GODALMING. By S. Welman. Published 1900 by Elliot Stock.

Mr Welman conjectures that in the fourteenth century the present spire was built, replacing a collar-type spire of about 1220. His examination of the evidence afforded by the existing timbers led him to believe that originally the spire was parapetted, and that the broaches were added about 1716, and are therefore comparatively modern. The fact (referred to *ante* in the text) that the lead does not "drip" the wall, gives colour to this theory, which need not, however, be too readily accepted. I do not regard it as proven.—L. W.

ANNALS OF WINDSOR. Tighe and Davis. 165-166.

Extracts from building accounts dealing with the great lead fountain that stood once at Windsor Castle in the Upper Court.

ON A FILTERING CISTERN OF THE FOURTEENTH CENTURY AT WESTMINSTER ABBEY. By J. T. Micklethwaite, F.S.A. *Archæologia*, liii. 161-170.

The cistern was of lead, but was havocked in 1544, and it does not appear that it had any decorative character.

ON ANCIENT MOULDS FOR CASTING METAL HORN BOOKS FOR CHILDREN. By Sir George Musgrave. *Arch.*, xxxiv. 449-450.

Moulds made of hone-stone for lead casting.

A LEADEN CHARM MADE UNDER THE INFLUENCE OF SATURN. By E. J. Pilcher. *Bibl. Arch. Soc.*, xxviii. 284-285.

Disc 2¾ in. in diameter, incised with symbols of Saturn. If engraved under an unlucky aspect of the planet the charm would inevitably cause the ruin of buildings.

A SIXTEENTH CENTURY LEADEN CHARM FOUND AT LINCOLN'S INN. W. Paley Baildon. *Proc. Soc. Antiq.*, 2nd S., xviii. 141-147.

See text of book.

INSCRIBED LEADEN TABLET FOUND AT DYMOCK, GLOUCESTERSHIRE. By E. S. Hartland. *Reliquary*, 1897, 140.

An imprecation on one Sarah Ellis. There is also described a similar plate from Gatherley Moor.

LEAD INKPOT FROM WILSFORD. By J. E. Nightingale. *Proc. Soc. Antiq.*, xiii. 240.

Illustrated.

ON A LEADEN TOBACCO STOPPER FOUND AT CASTLE EDEN. By R. M. Middleton, jun. *Arch. Æliana*, N.S., vol. x.

Of the seventeenth century. Shaped like a Runic cross with an included ring. Foot of cross used for pressing the tobacco into the pipe.
Other examples in Guildhall Museum.

THREE LEAD TICKETS OF THE EIGHTEENTH CENTURY. By F. Willson Yeates. *Num. Chron.*, 4th S., ii. 74-77.

Admission Tickets—

1. Of 1732 for the Glasgow Assemblies (public dances).

2. Of 1772 for the Pantheon Gardens in Spa Field, Clerkenwell.

3. Of 1773-1774 for Cox's Museum.

Note.—There have been omitted from the above bibliography the titles and details of over forty contributions to various magazines from 1905 to 1909, by the Author of this book, as all that seemed likely to be of permanent interest has been incorporated in the foregoing pages.

INDEX.